THE FACE OF GOD:
WHAT ENOCH SAW IN HEAVEN

THE FACE OF GOD: WHAT ENOCH SAW IN HEAVEN

Henry Kakembo, M.D.

Formatting and cover design by Jinxie G at jinxiesworld.com

Published in the United States of America

For more information about this book, and author Henry Kakembo, M.D., please visit: WWW.FOGENOCH.COM

The text for this book is set in Times New Roman type

ISBN: 979-8-9862542-0-3 (Paperback Edition 2023)

ISBN: 979-8-9862542-1-0 (Hardcover Edition 2023)

ISBN: 979-8-9862542-2-7 (Digital Ebook Edition 2023)

ISBN: 979-8-9862542-4-1 (Paperback Edition 2024)

ISBN: 979-8-9862542-5-8 (Hardcover Edition 2024)

ISBN: 979-8-9862542-3-4 (Audiobook Edition – coming soon)

DEDICATION

In honor of Charles Darwin
for helping to enlighten humanity to our origins
and establishing that all life belongs in the same fold.

ACKNOWLEDGEMENTS

This book took a long time to write, what with a demanding work schedule and extensive research its wide scope entailed. Some people, hoping to see it in print in a jiffy, drifted away, and a few others passed on, or their whereabouts are unknown to me today. Nevertheless, I am grateful for their input and encouragement.

In particular, I would like to thank those who read the manuscript and gave constructive feedback, including Ronald Brooks, Leitha Curtis, and Bernadine Francisco. To Willa McClean, thank you for your comments and literature recommendations that helped to enrich the book. I'm also indebted to those who, by playing devil's advocate, encouraged me to expand the citations list to back up my information, support my findings, and improve the focus, clarity, and validity of the narrative. Kudos to them.

Writing a book takes a toll on family life. I missed opportunities to spend quality time with them, but they took my absence in stride, it seems, urging me on and giving me tips. I appreciate their understanding, patience, and help. I hope their support will prove worthwhile.

Henry Kakembo, M.D.

CONTENTS

INTRODUCTION

Everything we hear is an opinion, not a fact.
Everything we see is a perspective, not the truth.
~Marcus Aurelius

Since we arrived on Earth 200,000 years ago, we have endeavored to understand the mystery of life, and the world. How did we actually get here and why, and what or who was responsible? Since we're born spiritual, we tend to attribute these things to supernatural powers, which most of us sense or feel regardless of lack of empirical evidence for them. Our convictions aside, we still would welcome answers to these mysteries and others in nature. We already have explanations for some of them, but the nature of the creator still eludes us. Even then, the construct of a higher power – widespread among most cultures – is so entrenched that it is virtually impossible to dislodge by reason. But, as Jesus said in the Bible, "… you will know the truth, and the truth will set you free." Jn 8:32. We hope that with an answer, it should be harder for those with an agenda to sway us.

Finding answers to the nature of the *Powers That Be* entails collecting information from a lot of sources. Such sources include ancient historical records, mythology, philosophy, evolution, religion, mysticism, and astrophysics. For centuries, Christians relied on the Bible, a good repository of wisdom which until relatively recently had limited availability and which very few people could read.

As a teen, I received a bible for Christmas. It contained the Apocrypha, not found in many Bibles. I was mesmerized by its gilded edges, and I would open it carefully lest I damage the pages. Having my own bible was probably the reason I performed well in scripture class. But a few years later in the early 1960s, I lost it in the turmoil following a coup that overthrew Sir Edward Mutesa. Mutesa was my guardian, and he was also the first president of Uganda and King of the Baganda people. His family gave that Bible to me. The army seized the king's palace, scattering the family, and the king escaped to England. That Bible had been great reading and I miss it to this very day.

I was born into the Anglican Church and attended services regularly in boarding school at Kings College, Budo in Uganda. My mother went to church regularly, too, and my father mostly on religious holidays. I

i

wasn't with him that long as he passed away when I was only fifteen years old.

In scripture class, questions about passages we studied arose that the teacher didn't answer. For instance, Luke 3:21-22 said that, as Jesus was being baptized, Heaven opened and the Holy Spirit descended on Him. Then God's voice came down, too, expressing love and approval for His son. The teacher couldn't explain how that scene was possible. The Apocrypha aside, some Bible stories sounded farfetched. If God were everywhere, as people said, why didn't we see Him, let alone hear from Him? The Bible doesn't describe Him, either. When Moses asked for His name at the burning bush, His answer was evasive, "I am who I am." Ex 3:13-14. In Uganda, people also worshipped other gods, mainly Katonda, vernacular for creator, the Baganda god; his name was used interchangeably with the God of the Bible. Not much was said about Katonda, either. He was a sky god, whom people sometimes envisioned sitting in the canopy of a big tree, as they sacrificed a chicken below. But, even when that close, Katonda didn't show or reveal himself; no rustling of leaves, snapping of a twig, or a big voice.

According to the Bible, we were created in God's image, so we should look like Him. Yet, we are not sure. Despite lack of proof of His existence, we keep the Ten Commandments, wary that He will punish us and throw us into hellfire if we disobey. As children, we trembled whenever lightning struck a schoolmate, believing it was God's punishment for that person's sins and we could be next. Nonetheless, there were doubts. As El Shaddai, God visited Abraham and shared a meal with Him. And as Yahweh, He spoke to Moses at the burning bush. But such encounters are only found in a two-thousand-year-old book and nowhere else. So, for evidence of God's existence, we need a reliable eyewitness, a photograph, or a recording of His voice. Then, we can be assured that our belief in Him is justified. But if the Bible is inconsistent, or allegorical as some think, including Paul the Apostle, why should we trust what it says about creation, particularly as it contains different versions of creation, none of them mentioning evolution?

Evolution was not taught in our schools, even though fossils of our early ancestors showing progression from nonhumans to humans cropped up in neighboring countries like Kenya, Tanzania, and Ethiopia throughout the twentieth century. Uganda, then a British protectorate, billed by Winston Churchill as the pearl of Africa, followed the British school system, in which evolution wasn't taught until 2011. So, Uganda's school children long believed that we came from Adam and Eve.

After I graduated in medicine from Makerere University, Uganda, in the 1970s, I went to England, then Germany for graduate studies. On completing these studies, because of political instability in Uganda, I came to the United Sates, where I have lived since 1983, going back to visit the family in Africa now and then. The travelling exposed me to various ethnicities, cultures, and religious beliefs. Of the beliefs, it was clear that none could rightly call its faith the "only one." They all had something in common. Actually, as we will find, they were based in the same reality and, therefore, they were all true, branches of the same tree. Differences in their doctrines, however, suggested that they were not grounded on hard facts, but that their tenets were shaped by cultural norms. Otherwise, they would all worship in the same way.

Belief in a supernatural power or force is shared by many religions. Such a belief seems to be rooted deeply in our subconscious and hence would be very hard to change or dislodge, even when evidence contradicts it. Some school systems in the United States, favoring biblical Creation, used to ban teaching evolution until the year 2000, when courts weighed in. For example, in Kansas, which had also banned any classroom references to the Big Bang theory – the prevailing scientific explanation for the origin of the universe – several anti-evolutionist school board members were ousted.

Even among scientists, though, the spark for the Big Bang lacks universal consensus. It looks as though, then, that to find how everything came into existence, demystifying the nature of the creator should be the first step.

To understand the nature of God, some people, such as logical positivist philosophers, long preferred a scientific approach, as they believed that the knowledge science provided was the most reliable. The advance of science and technology in the last two centuries already solved many mysteries with innovations such as telescopes, high resolution microscopes, X-rays, magnetic resonance imaging (MRI) scanners, and jet propulsion. These innovations facilitated space exploration and enabled researchers to go deep into the seas, to observe microscopic organisms such as bacteria, to detect subatomic particles and split them, and so on.

With all of this new information, scientists are now looking for a single theory, the *Theory of Everything,* uniting all the four forces of nature: gravity, electromagnetism, and the strong and weak forces. In his book, *A Brief History of Time,* Stephen Hawking suggested that, were such a theory understandable to everyone, it would allow us to take part in the discussion of the question of why the universe (and we) exist. "If we found the answer to that," he concluded, "it would be the ultimate triumph of human reason, for then we would know the mind of God."[1]

Clarifying in a television interview, Hawking said he used the word God as a metaphor for "the laws of nature governing the universe." Such laws would include those governing the four fundamental forces of nature mentioned above.

But we are far from realizing Hawking's dream. Instead of constructive reporting, the media seems to be more interested in politicians' missteps, celebrities' escapades, sexual misconduct in high places, or "faking the news" in a partisan way. Scientific breakthroughs rarely make the headlines.

Meanwhile, many people may find it nigh impossible to let go of deep-rooted beliefs. Some of these beliefs may have been drummed into them as children, setting them up to spend their lives in an alternate universe. But behaviors such as belief in supernatural powers may be based on inherited memories, which psychoanalyst Carl Jung called the collective unconscious, and so abandoning them would be a pipe dream. Indeed, Sigmund Freud and Andrew Newberg noted that belief in God, since it's based in spirituality (transcendental experiences) whose mechanism is inborn, or belief in a higher power, would also be virtually impossible to dislodge. No wonder Russia and China hit a roadblock when they attempted to abolish religion. In fact, Christianity in China is thriving underground, and the country is on track to have the world's largest Christian population by 2030.[2]

Questions about the purpose and meaning of our existence are ages old. Common sense tells us that we are here for a reason, otherwise, what's the point? Finding this reason is the answer many people give for their belief in God. Meditation gurus and mystics discover meaning through transcendental experiences, which they find beneficial. They promote mental balance, ease stress, and bring happiness and good health for them. Inexplicably, though, the gurus cannot describe the meaning. They can't find the words to do so, only saying that the experiences are ineffable, or indefinable, an oddity due to the way the brain functions. Purpose, however, though related to meaning, eludes the gurus altogether.

Meaning achieved through meditation is realized only by a few people: those with special talent to achieve profound transcendental experiences. The rest of us, some suggest, could create our own. For instance, according to English physician-writer, Henry Havelock Ellis, citing *The Endowment Thesis,* "Life has no meaning unless it is endowed by meaning."[3] American writer Henry Miller agreed, suggesting, "Life has to be given a meaning because of the obvious fact that it has no meaning."[4] Psychologist Erich Fromm affirmed that there was no meaning to life except the meaning man gave his life by the unfolding of his powers, by living productively."[5] And arguing along the same lines – that words don't make sense unless we endow them with meaning – English philosopher Daniel Hill suggested that we could achieve meaning likewise through a do-it-yourself approach.[6]

What about ultimate (real) meaning to satisfy everyone? People tend to look for it through belief in God, Vishnu, Buddha, Quetzalcoatl, or Bondye (the good god) of the Vodouisants (Vodou devotees) found in Haiti and New Orleans, and others, who are supernatural power symbols, or metaphors, for our spirituality. We worship the symbols because worship evokes the spiritual experience, which gives life meaning. Other pathways providing meaning include meditation, yoga, rituals such as concerts, love, and serotonin-like chemicals. Such experiences consist of a world, which, though it appears to be very real, cannot be described. So, we believe in the powers that the experiences reveal on faith alone, and say that they are hidden and unknowable. But if these powers really exist, wouldn't the experiences be a window to glimpse them? An answer could help us to reveal their nature. Perhaps, then, we could take on the biggest question of all. What was there before the universe, and when did time really begin?

In *The Face of God: What Enoch Saw in Heaven,* we look for answers to the mysteries in nature that we otherwise attribute to supernatural powers. The idea of such powers endures despite lack of solid proof that they exist, so there must be a need for them. The book investigates what that need might be by taking a multidisciplinary approach to search for answers. This approach includes scientific methods which allow for questioning and testing to validate possibilities.

First, though, since the past shapes the present (and the future), the book looks at our ancestors' ideas about nature's mysteries. What did the ancients and early philosophers believe created the world? We find parallels between their ideas and modern physics. For example, in the East, in one of the main ancient Buddhism scriptures, the *Avatamsaka Sutra,* the world was described as a perfect network of mutual relations where all things and events interacted with each other in an infinitely complicated way.[7] This is supported in the West by David Bohm, a main proponent of quantum theory. Bohm confirmed that the universal interconnectedness of things and events seemed to be a fundamental feature of the atomic reality. According to physicist Fritjof Capra in *The Tao of Physics*, this feature did not depend on a particular interpretation of the mathematical reality.[8] Such views are the very essence of spirituality, suggesting a mystical nature to our reality.

Since our distant ancestors generally attributed nature's mysteries to supernatural powers high above in the sky, we craft a spaceship to zoom us into space to investigate. We might even come across Enoch, Noah's grandfather and "the seventh from Adam," Jude 1:14, who was "lifted" there by Angels when he was 365 years old. Reportedly, as he ascended the seven levels of Heaven, he saw Hell and Paradise on the third level, where Paul, in a vision, had also found them. 2 Cor 12:1-4. Enoch witnessed the wicked languishing in Hell, while the righteous basked in bliss in Paradise. On the same level, he also saw the garden of Eden with all its trees in bloom, and the tree of life under which God rested after taking His walk.[9] Very likely, then, there was an afterlife. Unless we behave ourselves, preachers warn us, instead of going to Paradise, we will go to Hell and be punished.

Topping the list of mysteries is the creation of the universe and life. Was divine intervention involved, as is commonly assumed? Although many cultures believe that everything was here from the beginning as-is, this doesn't seem to be the case. To make sure, we will look to archeology for evidence of species' progression, or lack thereof, over the billions of years since life emerged. We humans occupy only a blink of an eye on this timescale.

It appears that we developed religion as soon as we emerged. I am defining "religion" here as "the organized system of beliefs and rituals centered on symbols, such as gods, used to invoke spirituality, or mystical—transcendental—experiences." Scientists find that such experiences are interwoven with human biology, thus making faith an innate behavior, or instinct. This would explain why religious behavior has passed from generation to generation, accounting for its staying power. For a behavior to be selected by evolution indicates that it confers survival benefits. What could such benefits be for religion? Although religion fosters community cohesion and social support, and promotes healthy lifestyles by frowning on vices, such as alcoholism and drug use, these actions contribute only modestly towards a healthy life. Those who don't belong to a religion shouldn't have them, yet they, too, achieve benefits similar to those of the church goers, though by other means, which we'll discuss. Besides imbuing believers' lives with meaning, we will find that spirituality also provides other significant benefits that promote longevity. What about purpose? Isn't it related to meaning? Actually, it is, and the two join forces in the preservation of life. Aptly, God said to Noah and his sons, "…be fruitful and increase in number, multiply on the earth and increase upon it." Gn 9:7.

Since the brain is responsible for storing and processing all our knowledge, it behooves us to know the way it works. After all, the Bible says that the Kingdom of God is inside us. If that is true, then the brain should be responsible for the concept of a higher power, or God, of which we're reminded by practices such as meditation and prayers. Some worshippers who incorporate chemicals like psilocybin in their religious practices report spontaneous transcendental experiences. Others report the presence of a higher power, or God, before finally dissolving into the higher power and everything else in the universe to unite and become one with it, called *Union Mystica*, during a trance. A type of epilepsy also triggers these experiences. With modern technology, then, it should be possible to locate the source of the experiences, the "God spot," in the brain, or the Kingdom of God.

Before writing this book, I had doubts that the God I grew up with was real. My intensive research suggests that we only experience Him in a trance, or mystical experience. In his book, *The God Delusion*, Richard Dawkins maintained that God was a delusion. But as studies show, biology is behind the concept of God, making faith an instinct. As such, it is universal, making all religions kin.

Lacking direction from a higher power to standardize religious doctrines, we formulate them based on our cultures, which has resulted in a multitude of creeds and deities. Even when differences among beliefs are small, though, conflicts are common. Were people to understand that a shared spirituality was the impetus behind religion, such strife could be prevented. So, I encourage you to be receptive to knowledge and concepts we could not before imagine, with the hope that as we become more tolerant of one another, more unified and whole, we will recognize that our lives and fate are the same regardless of our differences. This could help to break down barriers that separate us and hinder us from understanding each other.

To fully appreciate what this book is about, you may need to suspend your beliefs and keep an open mind to new information. By unveiling God's nature, the book is not meant to change you or your beliefs, but rather to enlighten you about those beliefs and their significant benefits and to encourage you to lead a life that will keep you in good health, and realize meaning and your evolutionary purpose. As practices such as meditation or prayers that evoke the spiritual experience improve focus, they could help you maximize your potential and live life to its fullest.

For peace and harmony, we should strive to understand one another universally. We should not hesitate to reach out, and we should assist instead of destroy, love instead of fight, embrace rather than discriminate, share and share alike.

NOTE: Unless noted otherwise, all scriptural citations will reference the New International Version (NIV) of the Bible. Other versions will be used only where the interpretation between texts varies. Shorter abbreviations for chapter names will be used and parentheses omitted.

PERIODS: Eras are denoted by the terms BCE (Before the Common Era) and CE (Common Era).

ILLUSTRATIONS: Unless otherwise credited to an outside source, all illustrations/images herein were created specifically for and incorporated into this book and are included in the copyright. Each illustration is numbered to correspond to its chapter location and appears in the *Illustrations* appendix immediately following the conclusion of the last chapter.

CHAPTER 1

THE QUEST

As small children when our bellies are full, we play, oblivious to destiny. Occasionally, though, wondering, we ask where babies come from. When I put this question to Mummy, her dismissive answer, "The stork," wasn't always satisfactory, and she continued busily knitting whenever I threw the question at her. Her answer didn't square with what my classmates were saying. But curiosity feeds humanity's quest to understand itself and the creator and to make sense of life.

Because we know next to nothing about the creator, we would like to find out what He looks like and where He lives. We endeavor to know His plan for us, to endow our lives with meaning and purpose. Even though He is elusive, we wish for His intervention in our lives, to uplift us when tragedy strikes or when we are heavy-hearted. We would like to tell Him about our problems and needs, and it gives us comfort to feel that He's nearby, protecting us.

The ancients, to account for life, attributed it to supernatural powers, creators such as gods. The gods differed by culture, but invariably behaved like us. Belief in gods made life tolerable. In some cultures, people anticipated an afterlife, where their spirits would rest in bliss among the gods as the reward for those who led exemplary lives. In anticipation of this reward – or perhaps as a way of earning it – they devoted their lives to such deities, praised them, offered them gifts and sacrifices, going to great lengths to please them.

One such supernatural power in the Western world is the God of Abraham. In the Far East, where the universe has always existed, the same idea would be expressed as the cosmic principle Brahman. People found that repetitive, rhythmic rituals produced a feeling of self-transcendence and union with a higher sense of reality, a closing of the distance between self and God.[1] When performed regularly, the rituals would turn into worship and develop into organized religion. But besides religion uniting people spiritually with a higher sense of reality, it also promoted social cohesion and good health, and made life meaningful.[2] Therefore, in addition to upholding values, philosophers considered religion a theory of reality.

Why are we preoccupied with finding purpose and the meaning of life? According to American sociologist Ruth Cavan, during our middle years, from ages twenty-three to sixty, the search for meaning intensifies and we become more reflective and active in religion. By age ninety, most people feel they have found purpose and believe in the afterlife one hundred percent.[3]

What drives our quest? Aristotle said, "All men by nature desire to know."[4] According to Einstein, "The important thing is not to stop questioning." And writing in *The Mystical Mind,* Andrew Newberg said that one part of our brain looks for reasons behind everything. It forces us to question why things are as they are, why we are here and what created the universe, all fundamental questions to religion, philosophy and science.[5] We look for patterns, explanations, and meaning in everything. If patterns don't exist, we create them in ways that make sense to us, such as the constellations.

To search for answers to the mystery of life, philosophers in the West took a rational approach. But, without facts, they became mired in debates. For instance, they didn't know life was billions of years old, or, that the universe was expanding, or that continents were drifting and causing earthquakes and volcanoes. In Plato's otherworldly "Theory of Forms," he postulated that objects were transitory, a copy of permanent forms that existed outside space and time. Aristotle's theory of species, that they were fixed and unchangeable, sounds speculative. Aristotle also believed in aether, a fluid that supposedly pervaded space and transmitted light, which later Einstein scientifically summarily dismissed. Since philosophy and science couldn't provide purpose or meaning, people still relied on God and religion for them.

A HIDDEN AND UNKNOWABLE GOD

Very few people have visited Heaven or have seen God there. Among them is Ezekiel, *see* Ez 1: 26-28; Daniel in a dream, *see* Dn 7:9; someone known to Saint Paul, *see* 2 Cor 12:1-4; and Enoch in a vision, *see* 1 Enoch 14. On reaching the seventh Heaven, Enoch, from afar, saw the Lord seated on a very high throne on the tenth Heaven, called Aravoth. When he arrived on this, the highest Heaven, Enoch fell down and bowed to the Lord, who spoke to him and asked him to stand up.[6]

People may experience God while in an altered state of consciousness or a trance, but after they return to normal, they can't describe Him. They report that He is hidden and unknowable. To believe in Him is to do so blindly. On leaving Africa some seventy thousand years ago, people's brains were hardwired to seek such transcendental experiences. While in the West some used hallucinogens to do so, others prayed. In the Far East, they meditated, but instead of interpreting their experiences as coming from encountering Jesus or God as in the West, Buddhists interpreted theirs as the Ultimate Reality or Nirvana (meaning "extinguishing the fire"), the goal and fulfillment of human life that ended all pain and suffering for them.

For some in the West, ultimate meaning and purpose are achieved not only by finding God but also by uniting with Him. The meaning of life and purpose are tied to God's existence as a whole. For instance, Karl Barth, a nineteenth century Swiss protestant, found meaning in the idea of God. At about the same time, Max Horkheimer, a German social theorist of the Frankfurt school, suggested that without the idea of God there was no absolute meaning (true or ultimate meaning of life).[7] Barth and Horkheimer, however, didn't explain the meaning of "meaning." Also, each of their gods could have been different depending on the faith of their youth. In some belief systems, such as Jainism in India, gods don't even exist. The concept of god, then, varies depending on the culture. With absolute meaning, such confusing variations would be eliminated and allow us to better understand why we're here.

In talking about the nature of whichever higher power created the world, religious explanations hardly ever make sense and questioning is discouraged. For instance, Judaism, Christianity, Islam, and Buddhism all dislike rigorous "adult logic." Rather, to understand divine truths, they stress child-like simplicity, insisting on blind faith. In the fourth century BCE, the Chinese philosopher Confucius wrote in *Doctrine of the Mean* 16, "The presence of the spirit: It cannot be surmised. How may it be ignored?"

About *Brahman,* the sacred power that sustains all existing things and the inner meaning of existence, Hindus say, "It [God] is unknown to the learned and known to the simple." And, admonishing his followers, disciple Muhammad Abduh, a nineteenth century Egyptian Muslim reformer, quoting the *hadith* (tradition) said, "Reflect upon God's creation but not upon his nature or else you will perish."[8]

In the fifteenth century, German reformer Martin Luther insisted that God strictly forbade speculating about His nature. Luther believed trying to reach God by reason alone could be risky and lead to despair. Certain that God was full of anger, harsh, and quarrelsome, Luther believed all rebellious peasants should be killed.

If the sages warn us against asking questions, do we dare not heed their warnings? Are we flirting with damnation by sleuthing, or, as experienced by Abdul Hamid al-Ghazali, courting frustration? Al-Ghazali was an early eleventh-century mystic Islamic philosopher and director of the Nizamiyyah Mosque in Baghdad, then part of the Persian Empire. The lack of proof of God troubled him. Because the reality called "God" could not be tested experimentally, how could he be sure his beliefs were not delusional? So, he set out to search for concrete proof. For three years he studied Falsafah, the Muslim philosophy that claimed to have knowledge of God, without success. He broke down and was so depressed he could not speak or lecture. Today, he might have been given a bipolar disorder diagnosis and a tranquilizer or antidepressant for relief. Eventually, after staying with the Sufis, a mystic sect of Islam, for ten years, he found the answer. He realized that proving God's existence logically or rationally was impossible and only those blessed with mystical, or "prophetic," talents could experience the reality called God. He recovered and resumed teaching.[9]

Likewise, the esteemed holy men weren't privy to the nature of God because it was hidden and unknowable. But today, since we know more about nature than the sages did, we're in a better position to investigate the mystery. As God made us in His image, wouldn't He also consist of tangible stuff like us?

Without ironclad evidence of God, we envision Him in many forms. In the Far East, Hindus worship not one, but 330 million gods, all of them likely intermediaries of Brahman, the one true reality, the individual soul also called *Atman*. The idea that the soul and Atman are one is the essence of the *Upanishads*, the last of the four ancient Hindu scriptures, or *Vedas*. In contrast, some beliefs, like Confucianism and Jainism, don't have gods. But, because they emphasize the basic unity of the universe, an awareness known as enlightenment, an experience that in its ultimate nature is religious, most Eastern philosophies are essentially religious philosophies.[10] They also believe in incarnation and souls that make life meaningful.

THINKERS INCUR CHURCH WRATH

Like religion, philosophy (from Greek *philosophos,* which means "lover of wisdom," as Pythagoras styled himself) is universal. It's said that most people have a personal philosophy –wisdom they follow for guidance in life. Thales, a Greek from Miletus on the Ionian coast of Asia Minor, is credited with initiating philosophy in the West at the turn of the sixth century BCE. He was followed by such bigwigs as Socrates, Plato, and Aristotle. The three belonged to the same school of thought, as Socrates may have taught Plato, who then taught Aristotle. Instead of myth, they looked for a rational explanation of nature, insisting on proof, with right and wrong answers, ushering in a form of inquiry that helped pave the way to today's scientific approach of searching for truth.[11]

Philosophers treaded warily to avoid falling victim to religious and state watchdogs looking out for "dangerous" ideas. Heresy accusations flew easily and heretics suffered severe penalties: exile, imprisonment, or death. In Rome, writing "God didn't exist" was a crime, as was philosophizing. Emperor Domitian exiled Stoic philosopher Epictetus, along with countless others who were lucky not to be sentenced to death.[12] So fettered, how could they advance? No wonder, then, that many Greek masters, such as Thales, Pythagoras, and Democritus of Abdera, travelled to or opted to study and work in Egypt, which placed no such penalties on knowledge. The drawback there was, for students to be fully instructed in the Egyptian system, they had to be initiated into the mysteries of the priests and take an oath not to disclose what they learned there. Such secrecy kept the priests below Egyptian law radar, launching the Mystery Religion System, or "Secret Order."[13] Notably, Pythagoras failed to keep his oath, as we see in the next chapter.

Conditions in Europe remained hostile to new ideas well into the nineteenth century. Socrates was an early victim. In 399 BCE, the Athenian state accused him of impiety and corruption of youth (a common accusation against progressives) and gave him two choices: to renounce or die. He chose the latter on principle. Turning down an opportunity to escape with the help of his supporters, he drank a cup of hemlock and died. Aristotle came next. The authorities drummed him out of Athens into retirement on his family estate, where he died in 322 BCE.

As for Plato's Academy in Athens, Byzantine emperor Justinian I closed the school in 529 CE as part of his war against "pagan" teaching. But this was small potatoes compared to what Italian philosopher, Giordano Bruno, endured. After eight years of interrogation, accused of "loose" thinking, like insisting that the Earth moved, the universe was infinite, and heaven didn't exist, among other "heresies," Bruno was burned at the stake in 1600.

More recently, in 1744, the University of Edinburgh denied philosopher David Hume a professorship, suspecting he was a religious skeptic. Skepticism may have disqualified Hume from an academic post, but in practice it enables us to separate fact from fantasy, a process essential for human survival.[14] Hume was indeed a staunch supporter of Skepticism, a school of philosophy started by the Greeks in the third century BCE. Like fourth century BCE Greek skeptic philosopher Pyrrho, Hume believed certainty was not available, so we couldn't be sure of anything. People in a particular place might believe in one thing while those in another opposed it, effectively neutralizing the argument on both sides.

In the West, in a conflict between religion and philosophy when evidence was inconclusive, religion generally prevailed because of fear of offending local authorities. Meanwhile in the Far East, philosophy, not myth or magic, has always been behind religion. Philosophy was more in tune with nature, human nature, and the nature of human society. This appeals so much to New Agers in the West today that they have embraced Eastern practices like Zen and Transcendental Meditation with zeal.

Fear of the church was not the only reason early philosophers were more often wrong than not. Science was in its infancy then. For instance, Aristotle taught that the universe consisted of five elements: fire, air, earth, water and *aether* (also known as *quintessence*). Aether was everywhere, making up the heavenly bodies "above" the moon. Before Einstein discredited the idea, scientists believed Aether transmitted light. Aristotle's elements were compounds made up of one or more of the ninety-four elements found in nature today. And when, in the early 1900s, the Andromeda Galaxy was identified as being very far away, it became clear there was more galaxies than ours in the universe.

Aristotle also estimated that fifty-five perfectly spherical planets and stars revolved around Earth. He could not have known that our small planet did not orbit only the sun. Along with the rest of the planets in our solar system, it circled the center of our galaxy, the Milky Way, home to a bevy of 200 to 400 billion stars. Our galaxy has about the same number of stars as each of the other ten trillion galaxies in the universe. So many are the galaxies, counting one per second would take more than a lifetime. Imagine the number of stars in the observable universe: over a sextillion, one followed by twenty-one zeros. To that add the unobservable universe, and the number grows to a googol stars, one followed by a hundred zeros. Aristotle didn't have a telescope anywhere near powerful enough to give him a clue about these numbing figures.

Most philosophers in ancient cultures, including the Egyptians and Greek philosophers (except Aristotle), believed the Earth was flat. The writers of the Bible went along with that theory, saying, "…. he will assemble the scattered people of Judah from the four corners of the Earth." Is 11:12. The Earth supposedly rested on four pillars at the center of the universe, as "…the world is established, firm and secure." Ps 93:1. Since the sun, moon, and planets revolved around the Earth, for them to screech to a halt for a day, as Joshua commanded, would have been a Herculean feat. Jo 10:12-13. During Jesus's crucifixion, the sun went on the blink for three hours only.

Earlier, in 250 BCE, Aristarchus, a Greek from Samos, had figured out that the Earth circled the Sun, but no one paid him any attention.[15] His "heretical" idea wasn't revived until eighteen hundred years later in 1543 when a Polish astronomer, Nicolaus Copernicus, published his work, *On the Revolutions of the Celestial Spheres*, wherein he argued that the Earth indeed moved. In one stroke, Copernicus contradicted the ancients, the Bible, Aristotle, and Claudius Ptolemy. Ptolemy was the second century CE Alexandrian philosopher, who advanced a system of celestial spheres in which various heavenly bodies revolved around the Earth.

Copernicus's impertinence infuriated religious reformers such as Martin Luther. At a dinner table a decade before the book was published, Luther lambasted, "The fool wants to turn the whole art of astronomy upside down. However, as Holy Scripture tells us, so did Joshua bid the sun to stand still and not the earth."[16] By holding off publication until his deathbed, Copernicus left it to others to defend the theory. This chancy task fell upon supporters like Galileo Galilei, a sixteenth century Italian physicist-astronomer. Galileo wasn't fazed, even though a German theologian, Andreas Osiander, had secretly added a disclaimer to the preface of Copernicus's book that the theory was not necessarily true. Galileo confirmed it with a self-built telescope. But it did not help that he lampooned the church despite warnings not to discuss the theory. The authorities now had a fall guy on whom to vent their anger.

At times, philosophy sounds like wordplay. For example, Heraclitus' sayings such as, *We step and do not step into the same rivers* (we cannot step into the same river water twice), *We are, and we are not.* He also said, *"Good and ill are one," "The Way up and the way down is one.* And according to René Descartes, *"I think, therefore I am,"* all sound like piffle. Aristotle's eleventh century scholastic philosophers puzzled over the seemingly simple question: *"What is slush?"* They seemed unable to determine whether it was a specific kind of snow versus a quality of snow. Also, they argued that: *Your mother was twenty-five years old when you were born; so, the woman now fifty is not your mother!*

Although relatively free of church fetters today, philosophers still get bogged down in semantics. They may have fun with such debates, but members of the general public would be baffled, unable to make sense of such arguments. This gap in common understanding led twentieth century linguistic philosopher Ludwig Wittgenstein to remark that the analysis of language was the sole remaining task for philosophy.[17] To unravel the nature of God, then, concrete evidence is essential.

In *Critique of Pure Reason,* eighteenth century German idealist Emmanuel Kant seemingly ruled out the possibility of gaining knowledge of God's existence, saying it could neither be rationally shown nor disproved.[18] This is generally western philosophy's stand on the subject today. Logical Positivists, however, begged to differ. They regarded Kant's argument as "meaningless" since it couldn't be confirmed according to the *verification principle* popularized by philosopher Alfred Ayers. Many early modern philosophers believed science provided a dependable source of knowledge, as it could be tested empirically.[19] Logical positivism, though, lacked staying power.

It ultimately came under such strong criticism from postpositivist thinkers like Karl Popper that it had fallen out of favor by the late 1960s.

SCIENCE AT ODDS WITH RELIGION

Science reached where it is today in the usual non-linear, two-steps forward, one-step back fashion. The Catholic Church ruled supreme in the Mediaeval Era. Hostile to new ideas, it dragged "heretics" before the Inquisition, sentencing many to death. Scientific progress slowed to a crawl.

Few topics stimulate debate or provoke confrontation like our origins do (those of other creatures' origins stimulate only experts). Until a few centuries ago, contradicting scriptures was like a death wish. In 1619, Italian philosopher Lucilio Vanini pronounced that humans originated from apes. Three years later, he was condemned for blasphemy and atheism, his tongue was cut out, then he was strangled and his body burned in Toulouse.

When Galileo corresponded with a student on Copernicus's theory, the Cardinal Robert Bellarmine passed an edict forbidding Galileo from maintaining or defending the heresy. But, the next pope, Urban III, was Galileo's friend and in 1632 he allowed Galileo to write a book, *Dialogues Concerning the Two Chief World Systems*, about the theories of the universe.

In the book, three people discussed the theory. The traits of the fictional character Simplicio closely resembled the Pope. Although in hindsight Simplicio was right, the Pope wasn't amused. In 1633, at the behest of Aristotelian philosophers, whose theories Galileo had discredited, the Inquisition put Galileo on trial. He was adjudged blasphemous and sentenced to life imprisonment. As he left the hall, he reportedly defiantly muttered under his breath, "Still, it [Earth] moves" (Italian: *si muove ancora*).[20] His sentence was promptly commuted to permanent house arrest. When he died eight years later, he was denied burial on consecrated ground. In 1992, a commission sanctioned by Pope John Paul II finally acknowledged the conviction was a mistake. To prevent future conflicts and allow religion and science to compare notes, the Vatican initiated biannual scientific conventions.

No wonder Darwin put off publication of his book on evolution for twenty years. When it finally went into print, the book shook the foundations of Christianity to the core, a storm that persists to this day. Despite Darwin, however, the majority of Americans believe that God created us as we are today. In an April 8, 2007 *Newsweek* poll, 48 percent of Americans believed that God created us in his image and created all other living things in their present form, "according to their kind." Gn 1:26-27. An additional 30 percent thought He had something to do with it.

Christian fundamentalists and Creationists, who believe literally in the Bible, have been hell-bent on yanking Darwin out of the classroom in the United States. In 1925, when a substitute Tennessee schoolteacher, John T. Scopes, disobeyed a ban prohibiting teaching of evolution, the state sued him. In the test case, dubbed the "Monkey Trial," Clarence Darrow, the defense lawyer, did not call Scopes to the stand. Instead, he ripped into the prosecutor, William Jennings Bryan, a former presidential candidate, calling him an ignoramus.[21] Although Bryan ignored evidence for evolution in taking the case, it's even more perplexing that scientists, some of whom we may trust with our lives, could—through citing their belief in God—also side with the Creationists. Even renowned surgeon Ben Carson, when running for president of the United States in 2016, said he doubted evolution took place.

Scopes lost the case, not because it lacked merit, but because he broke State law. He was fined $100, an outcome that paled in comparison to Bryan's. During the trial, Bryan was eviscerated by *Baltimore Sun* reporter H.L. Mencken—Mencken called Bryan a tin-pot pope and a buffoon—and then Bryan died within a few days after the trial. Coincidence or Stress? (*See* Chapter 11 herein for a discussion of the havoc stress can wreak on health). The Supreme Court reversed the ruling, as only a jury could levy a fine over $50.

That the law was finally repealed in 1967 did not faze fundamentalists one bit. They re-branded creation into a "science" to sneak it back into the classroom. The Supreme Court struck that down, too, ruling that since creationism could not be tested or falsified, it was not a science, just like hundreds of similar stories around the world. But Creationists are resolute. In 2006, as Pennsylvania was blocking a similar effort, the Kansas State Board of Education voted Intelligent Design into the school curriculum. And in 2008 in Louisiana, a Science

Education Act kept Intelligent Design in the schools disguised as "academic freedom."

Scientific progress accelerated in the twentieth century, as science, unlike religion, explained the universe more accurately. Unfortunately, as Hawking observed in *A Brief History of Time,* science had become too technical and mathematical to be properly understood by anyone except for a few specialists.[22] Galileo's argument that math was the language needed to understand the universe seems to have held up over time, although the development of computers has made scientific research much more accessible to interested persons from all disciplines.

In search of answers about our creator and the world, the book has adopted a multidisciplinary approach in which science takes the lion's share. Science enables us to confirm people's statements and not just to take their word for it. Despite initially being greeted with disbelief, ridicule, hostility, or censure, valid scientific discoveries are eventually recognized and accepted. Like the Logical Positivists, philosopher and writer Bryan Magee bills scientific knowledge as the most reliable and useful information human beings possess. [23]

IN A NUTSHELL

To account for our origins and cope with the anxiety caused by the uncertainty of life and death, the ancients immersed themselves in myth and magic. Myths were assumptions about creation and the behavior and temperament of gods. These assumptions answered people's need for purpose and meaning and were a large part of religion. Religion didn't provide all the answers, however. For instance, in *The Brothers Karamazov,* a novel by Fyodor Dostoyevsky, one of the characters, Ivan, thought God failed to provide ultimate meaning for the tragedy of life because the Bible talked about God and salvation, but not meaning.

For those like Karl Marx, who believed neither meaning, value, nor purpose existed outside the historical process, the idea of God couldn't help humanity. Nineteenth century German philosopher Arthur Schopenhauer went further, declaring there was no Absolute, no Reason, no God, and no Spirit at work in the world.

Unravelling God's nature, then, seems to be an uphill battle, as all we have to go on is a feeling that God exists, a feeling likely of the

11

mind's making. Meditation gurus experience the divine by looking within themselves, and the "revelation" they experience gives meaning to their lives. They attain emotional and spiritual growth as well as mental balance. As after a near death experience (NDE), their appreciation of nature, concern for others, and giving increase, and their outlook on life becomes more positive. They lead healthier and longer lives while their competitiveness and egos abate and fear of death goes away, all expressions also used by mystics.

Brushing away the obscuring cobwebs of our past may reveal breathtaking truths. Besides burying myths and hearsay, established historical "truths" may also require revision. Such results have challenged history textbooks, eliciting revisions. In his book, *African Origin of Civilization,* Senegalese historian anthropologist Cheikh Diop wrote that because of his research, Egyptologists came to recognize "the cultural unity of black Africa and ancient Egypt." Textbooks, then, would reflect it,[24] and the ancient Egyptians' contributions to civilization were reluctantly included.

As we mature, we outgrow fairy tales. We put Santa Claus in perspective and play along. Belief in witches, devils, demons, goblins, forces of darkness, ghosts, Satan, or God, however, may never fully go away. According to British psychologist Michael Argyle, in adolescence we inquire into religion, converting and reconverting. [25]

When I was fifteen years old, I wrote an essay about God for an English class. The burning question for us then was, if God was real, where was He? Why didn't we see Him? Also, who wrote the Bible? Even though the Persian Lord, Ahura Mazda, brought the "Book of Law" to a high mountain, we could not imagine God likewise handing down from on high a Bible written in the third person and with inconsistencies. A nagging feeling told me that it was written in Egypt or thereabouts, very likely in the Nile Delta, as stories like the Exodus and the hiding of baby Jesus originated or took place in that country. The Good Book seemed to stretch some things, too, such as Noah living 950 years. Noah's grandfather, Enoch, was 365 years old when God took him alive into Heaven. Gn 5:23-24; Heb 11:5. But if people (and not God) wrote the Bible, how did they know about events before we were created? God dictated only the Ten Commandments to Moses, not the full Torah.

I received no comment about the paper from my English teacher. She simply smiled. Without a second thought, I continued to play piano for Sunday church services.

That's not where I am today, though. To find the truth, as Bryan Magee says, scientific means are the best. Nevertheless, some people are cagey, believing that science is out to get religion. For example, doesn't science contradict scriptures to say the Earth is round, implying it has no edge where one can jump off? Although science may – like philosophy – contradict religion, its purpose is the same, to find verifiable answers to the mysteries of life. Besides calling on philosophy to craft its texts, and employing technology to spread its message, religion may also alert science to problems with ethical standards, or dissuade those inclined to play God from doing so. Tinkering with genes, human gene editing and cloning, or weaponizing organisms in the lab all present such ethical challenges.

What, then, were our ancestors' views on the origin of the world and the nature of God? What did they believe was our purpose? As alluded to in the introduction and as we will see later, the foundation of religion – spiritual reality – was in us from the outset. Religion and rituals were the means through which spirituality and superpowers were experienced.

In the East, religious practices have remained pretty constant over millennia, while those in the West have tended to splinter into various conflicting creeds over the past 2000 years. As mentioned in the introduction, parallels between beliefs in the West and East are evident. For example, in the East, in Buddhism the essence of the world view is the unity and inter-relation of all things and events,[26] while in the West, physicist John Bell's theorem shows that the universe is fundamentally interconnected, interdependent, and inseparable.[27] The school of "deep ecology" founded by Norwegian philosopher Arne Naess also sees the world as a network of phenomena that are fundamentally interconnected and interdependent. Humans are just one particular strand in the web of life, "…the experience of being connected to nature, of belonging to the universe."[28] This connectedness becomes clear when you consider the physicists' view that matter is a form of energy, sometimes appearing as particles, sometimes as waves, like light.[29] The Eastern intuitive view and the Western scientific view appear to be in sync.

Regardless of the world-view parallels between the past and the present in the East and West, religious practices cannot be more different. Whereas practice in the East is philosophical and meditative, that in the West tends to be based on a belief in deities who evoke supernatural powers, who created the world, and who look and have families like us. People worship the powers, sacrifice to them, and beseech them for help in times of need. These powers may die, too. Scriptures or tenets have remained the same in the East but changed a lot in the West, at least until, in Christianity, say, the New Testament was canonized. In the Old Testament, God cut human lifespan from just under 1000 years to 120 years (ostensibly because He didn't want to contend with humans forever. Gn 6:3). The Bible authors wrote about people's attitudes and way of life in their time, but since our values and basic needs remain the same as theirs were, many of their views, such as the moral codes, haven't changed. Anyway, according to the Good Book, "There is nothing new under the sun." Eccl 1:9. So, Biblical stories and beliefs, such as creation of the world out of nothing, which is also the Chinese and scientific view, give us clues in solving the mystery of God and gods.

CHAPTER 2

X-RATED ORIGINS

If a vacuum (that's to say empty space) is compressed,
particles appear where there were none before.
~Tenzin Gyatso, the fourteenth Dalai Lama.[1]

For ages, creation has served as topical fodder for mythological stories about our origins. For instance, one narrative tells us that in the beginning Nantabale created the universe, then created Earth (it's not known where Nantabale came from). Not a single soul moved on Earth then; the planet was bare and boring. So, Nantabale made a wish, and life arose, fish in the sea and plants on land. Next, animals, then people sprang from the ground in their present form. When hungry, they set on each other something terrible. The chases pleased Nantabale and he wouldn't lift a finger to help on behalf of either side. To prevent them dying out, he made them reproduce nonstop.

Because four legs were faster than two, people devised ways to protect themselves against the faster predators. Being smarter and tired of climbing trees all the time to evade the carnivores, they invented tools and weapons, controlled fire, and set traps. All the while, they lived in caves, but eventually, they built houses, settled and tilled the land, and tamed the beasts. They found religion, organized their societies, and launched civilization. The problem with this story is that I made it up. The only evidence I have to support it is that we are here. This is how many myths might have started: someone makes up a story to explain the inexplicable. If the gullible buy it and pass it on, a myth is born.

FANTASYLAND

Before the Greek thinkers' era dawned in the sixth century BCE and the effort to explain our origins took on a rational approach, people invented such stories and myths to demystify nature. According to the father of psychology, Sigmund Freud, myths were peoples' fantasies about the creation of the universe, kept alive through rituals, celebrations, and worship. A former nun, Karen Armstrong, wrote that the inventors of such stories knew they weren't true; they were merely an attempt to represent a complex and elusive reality.[2] But today, many people believe such stories about their own most high or supreme being are true, while finding the beliefs of others problematic.

Created by people who never strayed far from their villages and remained unaware that other lands and cultures existed, many myths were set close to home. Most of them were about gods and their activities and families. These gods often behaved much like humans. Other stories glorified the gods as heroes (a word derived from Heru, an Egyptian form of Horus, the young god conqueror of the dragon of darkness). In the Greek version of the same story, the hero was Heracles—Zeus's illegitimate son, while in the Italian version, he was called Hercules, Jupiter's son. Heracles was the strongest man on Earth. As a youth he killed a lion with his bare hands. To avenge his birth, Hera, his stepmother, gave him twelve impossible tasks, called the "Labors of Hercules," one for each sign of the Zodiac. Heracles found these seemingly impossible tasks a piece of cake. Later, Heracles's feats were ascribed to Dionysus, and then later to Jesus.

Other cosmic myths explain the origin of the world. Such stories generally assume that matter has always existed, but that the world before creation was disordered and uninhabitable. It was seemingly a void, a mess of elements, either a sea or a cosmic egg with all things in a disorganized, embryonic form. Cosmic forces – either Gods or an Intelligent Designer –intervened, separating the original elements out of the chaos.

THE WORLD EMERGES FROM CHAOS

Threatened by the desert and yearly floods, the Egyptians believed the will of the gods was not chaos but order, *Maat*. One of their creation stories featured a cosmic egg that burst out of the ocean. From the egg emerged the sun god Ra, or Re, who organized the world by separating light from dark, up from down, good from evil.

Other tales of how the Egyptian world came about included that of Thoth, known to the Greeks as Hermes Trismegistus, the god of writing and learning. Then Thoth created things by naming them: whatever he named came into being. For the god Ptah, though, creation didn't come so easily. He created a giant metal plate that was believed to be the floor of heaven and the roof of the sky, but he needed the help of Earth-gnomes to get his creation into place.

Ra, being a hermaphrodite, reproduced without a mate. Ra had four children: gods Geb (Earth) and Shu (Space), and goddesses Nut (Sky) and Tefnut (Atmosphere, or clouds). From Memphis, to bring order to the chaos, using Space (Shu), Ptah separated the Sky (Nut) from the Earth (Geb).

More detailed accounts of this story get steamy: at Heliopolis, Atum, the legendary mound of creation, or the conical stone *Benben* (meaning copulation), created Shu and Tefnut by masturbating with his divine member, the shaft, or phallus. Upon the Benben, or the Bennu, the Phoenix alit every 600 years to lay its "seed" the sperm, or capstone. Obelisks and pyramids represented the mound.

Humans, then, found purpose in worshipping the gods and pharaohs, the gods' representatives on Earth. People found meaning from doing the pharaohs' bidding. The myths gave them hope of a happy afterlife among the stars, and they remained loyal to the kings or gods. Such devotion sometimes went to extremes, particularly during early Christianity beginning from around the first century CE. Christianity reached Ethiopia by the fourth century CE. Skeletons found in caves around Ethiopian churches attest to the people's undying dedication and support the belief that only the soul, and not the body, may have a place in Heaven. Comparable devotion is found in monasteries today, without the remains.

HELL, NO!

In a period portrayed in a Scandinavian Icelandic poem in Norse mythology, the *Poetic Edda,* giants emerged first from chaos, then gods, followed by humans. One of the gods, Odin (or Wodan, remembered every Wednesday), was depicted with a white beard riding the skies and seas on an eight-legged white horse. He was by far the best known. Later, in Holland, he was united with the gift-giving Turkish Saint Nicholas and became Sinterklaas, who Coca-Cola renamed Santa Claus.

Odin had a son, Thor, by his wife, Frigg. Thor was the god of thunder and the source of the name Thursday. He was the strongest in the family and represented the star Sirius, the brightest in the night sky, symbolized by a pentagram. Thor also represented the sun, just as Egyptian Osiris, and biblical Baal, did. As the upside-down gavel and hammer, he became a Masonic emblem, denoting the male member, a symbol of God the creator.

Since many of the Scandinavian gods did not seem to have any duties, their importance waned as Odin's grew. Odin's blood brother Loki, a sometimes-reverse trickster god, had a daughter, Hel (Hell in English), who presided over the lower levels of the netherworld where undeserving souls languished.

LIFE SPRINGS ETERNAL

In Hinduism, the creator, Brahma, remade the world after it fell into ruin every 2,160,000,000 years, which was only a day in his life. To break the endless cycle of death and rebirth and stop the suffering, one strove to achieve Moksha and join the *Ultimate*, or *Brahmin*, the power underlying the universe from which everything came and returned. Other systems had a different take. Confucians, Jains, and Buddhists did not believe gods created the universe. For Buddha, evil and suffering were too much to believe in a creator.

For the Central American Maya people, life had always existed as a circle of rebirth and death without beginning or end. As in the Far East, the Maya believed the universe had been and would continue to be created and destroyed, only their cycle started over every five thousand years, which was about twice the length of a zodiacal age. The zodiac consists of a circle of twelve constellations that the sun traverses over during the course of the year. Each constellation has a name and a sign, such as the ram for Aries. A zodiacal age lasts 2147 years, the time it takes the vernal (spring) equinox to move through a constellation into the next one starting a new age. We're in the age of Pisces the fish today that likely began in 1 CE (255 BCE, according to Gerald Massey, since timing is imprecise). Aquarius, the water pitcher, will be the next age. Eventually, thousands of years hence, Christmas will be celebrated in the northern hemisphere during the summer. For the spring equinox to complete the cycle through all the twelve zodiacs and return to Aries takes about 25,800 years, called the Great Year. The Maya's cycle, then, is about a fifth of the Great Year. This is in line with seventeenth century Irish Bishop James Ussher's calculations in 1650 CE based on Bible chronology. Ussher pronounced that God created the universe at high noon on Sunday October 23 at 9 a.m., in 4004 BCE, a date that roughly coincides with the end of the age of

Gemini and the beginning of Taurus in 4,300 BCE. The world would have been 5,600 years old in Ussher's time. But Noah's Flood, dated to 2349 BCE, must have spared the Egyptian pyramids built between 2800 BCE and 2500 BCE, as they show no water damage.

Referring to the Mayan Calendar, doomsayers announced the end of the world would come on December 21, 2012. But we're still here, as the date actually meant the coming of the age of Aquarius. When the five-thousand-year Mayan cycle returns in 3114 CE, and as each new universe is an exact copy of the previous one, we should all return.

MOTHER EARTH BEARS FRUIT

The Hopi people, sticking to a theme common among indigenous North American groups, accounted for their origins through emergence. Creatures came from the Earth as if from a mother's womb. The first beings were the son of god, Tawa, and the Earth goddess, Spider Woman. Spider Woman then created humanity from a shadowy cave-like underworld where she lived, and Tawa's gaze animated it. There are different versions to this myth since people who told these stories knew they were only symbolic of a great mystery.

Native American creation stories describe tricksters that were generally bungling imitators, antisocial braggarts, and troublemaking buffoons. Sometimes, though, their mischief was beneficial by pure chance. When Coyote, the trickster in Navajo culture, scattered the stars and plants Holy Persons had carefully placed in the sky and on Earth, the present world formed. All was well until Coyote kidnapped the Water Monster's baby and caused a great flood, forcing human beings to the surface world. He also seduced a virgin and taught her witchcraft, regularly provoked disagreements and fights, and was partially to blame for the finality of death.

THE ENUMA ELISH EPIC (POEM OF CREATION)

Originating in the Tigris-Euphrates valley, now in Iraq, a Babylonian creation story, *Enuma Elish* (translated as "when on high"), also brought order to chaos. The earliest known occupants of that region, Sumerians, lived in the valley from around 4000 BCE, building one of the earliest civilizations. The story was probably composed in the eighteenth-century BCE and captured on seven tablets in an

Akkadian cuneiform script no later than the twelfth century BCE. It was recited during the New Year's Festival in the spring.[3]

The poem includes elements of Egyptian Ptah's creation story, especially the act of separating the sky from Earth. The story begins with the gods emerging two by two from a formless, watery waste without boundary, definition, or identity. Then fresh waters, Apsu, and salt waters, Tiamat, mixed, producing many gods. Because the young gods were noisy and unruly, Apsu planned to destroy them. The young gods got wind of this and killed Apsu. Tiamat then created Kingu to rule over all the gods and to be her own lover. But, similar to power struggles in ruling human families of the time, the highest of the gods, Marduk, slew Tiamat and cut her in half. One half became heaven and the other Earth.

There are striking parallels between this poem and the creation story in Genesis. For instance, in the poem, creation was described on seven tablets; in the Bible, God created the world in seven days. Sticking to the separation theme, water was divided into upper and lower waters in the poem as well as in the Bible. And, as man was created in the sixth tablet in the poem, man was created on the sixth day in the Bible.

ADAM AND EVE ARE BORN

Better known than the *Enuma Elish* epic is the Judeo-Christian creation story. Christian fundamentalists, Orthodox Jews, and others swear by it. As in the Mesopotamian epic, Ptah's story, and Chinese creation stories, the Judeo-Christian world came about by separation. For instance, in the Chinese version, the universe's energy, chi, existed in a chaotic void before it separated into pairs of opposites, such as bright and dark, and up and down.

After God created heaven and Earth, He made "lights in the firmament to divide the day from the night and let them be for signs, and for seasons...." The longer days would be for summer and shorter days for winter. In Hebrew cosmology, weather was stored in Heaven on the first level, where Enoch found it. Then, God created the greater light to rule the day and the lesser light, the night. Next, He set stars in heaven to illuminate the Earth and divide light from darkness. Gn 1:14-18.

Who recorded these seminal events, and where did God come from, though? John's gospel says He was a Word that was made flesh, Jesus. Wallis Budge, an early twentieth century English Egyptologist quoting from the Egyptian *Book of the Dead* (1,600 BCE), said that Osiris was "the word of what cometh into being and what is not." Anyhow, the ancients had the cosmos all wrong. They believed heaven was the seventh and last celestial sphere and stars were holes in the firmament affording us a glimpse of the blazing lights within. The sun was not a star to them, nor was the moon a satellite. It took centuries to overthrow the idea that the moon emitted light. Today we know that the universe, spanned by cosmic webs of galactic clusters, stretches ninety-three billion light years across, and that its edge is invisible, out of reach for mere mortals. Finally, regardless of the explanation, people emerged. They emerged free to pluck fruits from any tree. The Garden of Eden and the wily snake *are not* in this version of creation.

In Genesis, Chapter 2, however, God made Adam from dust. It took some practice, but after several tries, He succeed in creating man (in Hebrew, Adam, and *Adamah* stands for ground or earth, "...so, man was derived from the ground or earth."[4] Because Adam was alone, God created animals to keep Adam company. None of them would do, however, not even dogs—they couldn't talk. So, God made Eve out of Adam's rib. Gn 2:4-25. (To this day, some people swear men are one rib short). Then, a crafty snake tempted them to partake of the forbidden fruit in the Garden of Eden. Gn 3:1-6. In an older Sumerian story, a fertility god, Enki, whose phallus filled the goddess Uttu's womb with semen, ate some herbs (forbidden fruit) and became terminally ill. Then a goddess, Ninhursag, sent him a woman of the rib, Ninti, to heal him.

Bible scribes don't mention extinct animals like dinosaurs, which disappeared 65 million years ago. These animals were first recorded in the nineteenth century. But coming across giant fossil bones, biblical writers would have suspected that gigantic beasts once existed. They likely represented them in the many-headed, fire-breathing dragons like Tiamat and Leviathan. Because they could see the whales, they likely represented them as the great fish Leviathan that swallowed Jonah. Bacteria were too small to detect and pestilences (plagues) were thought to be God's retribution for human wickedness. This fallacy lasted until the seventeenth century, when the offending critters were identified.

But, why do we have two versions of creation? The story in Genesis Chapter 1 is way different from that in Chapter 2. We'll find the answer in Chapter 9. For now, who are our gods, besides Elohim (plural for *Eloah*, or God), the seven children of the ancient Mother Sophia?[5]

THE MIGHTY EGYPTIAN GODS

Although they seem largely forgotten, ancient Egyptian gods are very much still with us today. Only their names have changed. To explain, we must go back several millennia and pick up Ra's story.

Incest among gods was normal then. When Ra's two children, Geb and Nut, married, they produced two sons, Osiris and Set, and two daughters, Nephthys and Isis. Isis married her brother Osiris, who succeeded Ra as king of the Earth. Set, having been bypassed, went ballistic. He tricked Osiris into lying in a coffin, which Set then threw into the Nile. The coffin washed ashore in Phoenicia at Byblos (Greek for "papyrus" and the origin of the word *biblia,* Bible or book) and lodged inside a tree trunk. Isis found the tree, removed Osiris's body, and returned it to Egypt. Set, however, found the body, tore it up into fourteen pieces, and scattered them throughout the land. Isis dutifully gathered them, resurrected (raise-erected) Osiris, and they had a son, Horus—a miracle since a fish had swallowed Osiris' hacked-off male member (which formed the origin of the Immaculate Conception idea). Today, secret societies like the Freemasons represent Osiris' organ with a broken column over which Isis (now represented as Mary) weeps. Seeking revenge, Horus and Isis fought Set without end.

So hated was Set that Thoth called him "Stinking Face" in the *Edfu Texts.* Horus once knocked Set out cold, then smote him in the mouth with a club for good measure. And of course, Set was the mighty "roaring" serpent hiding in the ground. Set was intensely despised, likened to the Devil and called Set-*an* (the Egyptians added the suffix for emphasis). All animals identified with him, like the crocodile, ass, and pig, were shunned, prompting the Egyptians and Hebrews to ban pork. Eventually, Set was identified as the harbinger of chaos and driven out of Egypt. On the other hand, Horus was the upright one, the ruler of the sky, and champion of order.

Osiris, Isis, and Horus formed the pre-Christian trinity of Abydos. In folklore, the trinity reflected the three phases of the Sun. Sunrise was related to growth, noon to maturity, and sunset to decay or the Holy Ghost. This doctrine was adopted by early Egyptian Christians, such as Alexandrian Pope Athanasius, and applied to the Biblical Messiah.

Wars among Egyptian gods reflected conflicts between groups. The tribe of Horus, from Sudan in the south, would have clashed with that of Set, from the eastern desert. When groups merged, their gods did so, too. The god from the union of Amun and Ra's tribes became Amen-Ra, god of the gods, and later the One and Only God of the Copts. On arriving in Egypt, the Horus group, the Cushites, overthrew the Osiris group, the Anu, who inhabited the country then.

Later, Osiris was paired with the virgin goddess Isis. Isis hailed from several places, one being the *Mountains of the Moon,* or Ruwenzori, in Uganda, the highest source of the Nile. Isis was the goddess of the Ruwenzori. Sometimes identified with Hathor, Isis was mainly worshipped in the Delta as the virgin goddess. She and Osiris eventually merged with Horus, who reverted to an infant and became their son. Osiris and Horus are the same in some accounts.

THE EGYPTIAN'S ANCESTRAL LAND, PUNT

Egyptians always claimed their gods and ancestral spirits came from the land of Punt in the south like themselves. For example, Horus belonged to the Mesnitu (Metalworkers) there. In the fifteenth century BCE, Queen Hatshepsut's fleet sailed there along the East African coast and past "terraces" opposite Zanzibar. The flora and fauna – such as fragrant trees, myrrh, ebony, ivory, cosmetics, baboons, monkeys, leopard skins, and Pygmies – brought back by expeditions of the time are native to Central Africa. Other expeditions travelled there overland between the Fifth and Twentieth Dynasties.

Punt was a district of Ta Netjer, the Holy or Khui Land, Amun's personal pleasure garden south of the Sudan (Kush or Cush), through which products bound for Egypt from Nmay and Irem passed.[6] According to English poet and ancient Egypt enthusiast Gerald Massey, the Egyptians also referred to Punt as Khent, the farthest south, "…which now can be traced as far as Ganda (the U-ganda) …." and the lake country. It was "the feminine abode," their birthplace and native land in the remote past.[7] As they migrated north from East

Africa, people took with them their gods like Hathor and Pygmy Bes, Lord of Punt.

Supporting the Egyptians' Central African origins, DNA Tribes reported genetic studies showed a high frequency of King Tutankhamen's DNA in sub-Saharan Africa. Analysis of eight short tandem repeats (STR), as used in paternity testing, mostly matched those of Africans. Besides a smattering in the Mediterranean region and increased frequency from tropical West Africa to Southern Africa, most of his DNA markers matched those of the Bantu people from the Great Lakes region, supporting the view that this population was closest to the ancient Egyptians. As reported in 2012, Ramses III's Y-DNA Haplogroup (hg) E1b1a, is shared by Bantu and Niger-Congo language speakers. His STRs also are mainly distributed in the Great Lakes region.[8, 9]

Besides DNA, the language family tree could help with tracking a population's movements, for instance, Bantu speakers from Cameroon to east and southern Africa, and Indo-European speakers from the Central Asia steppes to Europe, the Americas, Australia, and throughout Asia. On leaving Eastern Africa, people would have spoken a few dialects.

A non-biblical explanation for the many different languages spoken today would be that they diverged as people spread out. God didn't have to muddle their language for them to speak in "tongues." Gn 11: 6-9. For instance, Egypt, was settled from Central Africa by Negrillos, pygmies known as the Anu, 20,000 years ago. They spoke several languages, but for unity, they developed a *lingua franca,* which the Cushites adopted when they arrived there from the south in about 4,000 BCE. In 2007, linguist professor emeritus Théophile Obenga, then at the San Francisco State University African Study Center, confirmed that Egyptian originated from a Black African group of languages from Central Africa, particularly the highland Great Lakes region. Its source was not Afro-Asiatic as was originally believed.[10]

After Osiris was resurrected, he became king of the dead. Together with the god of mummification, jackal-headed Anubis, he weighed the hearts of the dead against the feather of Maat, goddess of truth, balance, and order, at the gates of the Underworld. The heart, then, was believed to contain the soul, and had the deceased been charitable to the poor and free of sin, it would be lighter than the feather and the soul would be allowed to go to the next life, hence the word, "lighthearted." Were the heart heavy with sin, the soul was fed to a ravenous monster, Ammit, who was waiting hungrily nearby. Otherwise, Osiris would send the person back into the world as a pig or crocodile. If the two balanced, the heart was hauled before fourteen jurors, and if found guilty, the sinner to the monster went. Likewise, Jesus will judge us at the end of days.

Other countries adopted Horus, or Osiris, transforming him into a Mithra, Adonis, Dionysus, Krishna, or Tammuz. In Ezekiel, God pointed out women sitting at the gates of the Temple mourning Tammuz. Ez 8:14. Even the Diaspora Jewish god, according to G.R. Mead in *Fragments of a Faith Forgotten*, was identified with *Iao*, Dionysus' ancient name, by many ancient authors, including Plutarch, Diodorus, and Tacitus. According to Gerald Massey in *Ancient Egypt – The Light of the World*, Jehovah was derived from Horus, or Iu, as the son of Ptah. [11]

Similarities among regional gods were such that on visiting Egypt in the fifth century BCE, Herodotus, reputedly the first historian, observed that Greek gods were originally African gods. Brian Flemming of the DVD "The God who wasn't there," also noted, "Contemporary Christians are largely ignorant of the origins of their religion. [12] In *The Jesus Mysteries*, spiritualist philosopher Timothy Freke and pagan mysteries expert Peter Gandy traced it to Africa.

After merging and turning into the manifestation of the Sun god Ra during the Fifth Dynasty, Amun became Amen-Ra, "the One of One," "without a second," and the father of Ra's four children. Amun's temples included one in Upper Egypt at Thebes, now Luxor, and Gebel Barkal (the mountain of his Birth) in Nubia. At Thebes, one of his shrines, the Great Temple, stood in the center of a 200-acre Karnak temple complex inside a 61-acre sacred enclosure. King Solomon's Temple, as described in the Bible, was an Egyptian replica complete with a courtyard for the public, an outer court for priests and sacrifices and an inner room, God's sanctuary. As in Egyptian tradition, only special priests entered the sanctuary.

In time, Amun came to have a ram's head with horns, symbolizing Aries, the age in which his worship peaked. As the creator and god of virility, he gave rise to the term "horny." Amun was androgynous, both male and female, which meant that we were, too. As the Dogon in west

Africa do, to bring out our sex identities, the feminizing foreskin in the male and masculinizing clitoris in the female were circumcised. Jews, Muslims, and other cultures later adopted this practice without the explanation. Women are still circumcised in parts of Africa.

Also, as the affirmation *Amen* began and ended prayers of the ancient Egyptians, so it does today those of the Jews, Christians, and Muslims. The latter end theirs with *Amin*. So, we still venerate the sun as Amen-Ra. According to Dan Brown's *The da Vinci Code,* we admire Amun in a most celebrated painting, Leonardo da Vinci's Mona Lisa. Brown noted that her name was an anagram of Amon and Isis, "whose ancient pictogram was once called L'ISA," denoting the divine union of male and female.[13] And the knowing smile? Perhaps, the lady knows whom we really worship.

In the fourteenth century BCE during the Eighteenth Dynasty, King Akhenaton, the "Heretic" pharaoh, identified Amun with the Sun disk and called him Aten. He declared Aten the only god of Egypt, earning the king credit for founding monotheism. But as elsewhere, different gods often stood for the One or Supreme God, the sun. When Akhenaton died, people returned to their old ways with a vengeance. Osiris, who was never associated with the unpopular king, then eclipsed Amen-Ra and was worshipped throughout Egypt.

The Bible is full of references that show Ra is all-encompassing in Western religions. For instance, according to Charles Finch, a physician and an enthusiast of ancient Egypt, Abraham's name, (Ibrahima in Arabic), broken down in Egyptian means "the desire or wisdom of Ra's light or fire." His son's name, Isaac, (Ysak in Hebrew), means "offering by fire or burnt offering," linking it to Ra by his relation to fire. Israel, originally Yaqub (Jacob), means "the place of Ra's Creation." Based on this analysis, Finch suggests the god of Abraham must have been Ra, likely "...in the form of Aiu [Iao], the Golden Ass...," also a form of Set.[14]

Osiris and Isis were worshipped well into the Roman Empire period; even London had an Isis Temple at Southwark. In Greece, Osiris became Dionysus or Zeus, and in Rome, Bacchus or Jupiter. Isis was Demeter, the Greek goddess of motherhood. These cults lasted until Christianity absorbed them, the final takeover coming in the fourth century, when Emperor Theodosius outlawed all cults. Not to worry, we still adore Isis in the Black Madonna and Child, Horus, and Osiris in Mary's broken column, his hacked phallus.

Thus far, the origin of the world and us has been mythological with the sun as the most adored object, directly or indirectly. Since we have not yet found the creator, how about we adopt a methodical approach and shelve the myths?

MATTER UNCREATED

Unlike their mythology-oriented predecessors and contemporaries, philosophers weren't obsessed with creators, fertility, or progeny. They took a rational approach for answers, adding a heavy dose of speculation, as scientific methods had not yet been established. Around the sixth century BCE, working from his school on the Ionian coast, one such deep thinker, a Greek, Thales of Miletus, took the initial steps to find a scientific explanation of nature and was credited with introducing philosophy to the west during the flowering of Greek civilization between 600 BCE and 200 BCE. As he left no writings, we are obliged to rely on Aristotle's version of his story. And Herodotus, who called Thales a Phoenician by remote ancestry, reported Thales correctly predicted the solar eclipse of May 28, 585 BCE. Thales's work resulted in a systematic enquiry into creation 300 years before Genesis was written in the third century BCE. (See Chapter 9 for more discussion.)

Greek philosophers were indebted to Egypt. Besides Thales, Pythagoras, Plato, and Archimedes, a good two-thirds of early Greek philosophers studied there. Others with Greek-sounding names, such as astronomer Eratosthenes and mathematician Euclid, were born or lived in Egypt all their lives. Astronomer mathematician Ptolemy (Claudius Ptolemaeus) was practically Egyptian. No wonder Herodotus observed that, in addition to its gods, Greece borrowed the basics of its civilization from Egypt.

Like philosophers in the Far East, Thales did not overly concern himself with creation. He preferred to view the world in physical terms, promoting the idea that it, and we, were made of a single element: water. To his credit, the idea was insightful, correct in principle if not substance. His pupil, Anaximander, agreed but believed the element was an eternal substance. With a stroke of genius, the pupil suggested that nothing at all supported the Earth. The planet simply hung in space, suspended by its equidistance from everything else. Anaximander's student, Anaximenes, likely considered this to be nonsense, pouring cold water on the idea. Something had to hold it, perhaps air. Wrong though he was, Anaximenes's view prevailed while further inquiry proceeded at the snail's pace of science. People were as skeptical and suspicious of new ideas then as they are today, especially if such ideas challenged deep-seated beliefs.

Dissatisfied with the Ionian explanation, a Heraclitan School proposed fire as the primal source of matter, and argued that the soul was a mixture of fire and water. The school's founder, Heraclitus, known as "the weeping philosopher" because of his aloof and gloomy nature, believed stability was an illusion. Only change and the law of change were real and since the entire world was in a continuous state of change or flux, nothing was permanent. Such a view echoed beliefs from the Far East, where philosophers discovered the phenomenon through meditation.

PYTHAGORAS

Pythagoras was born around 570 BCE in Sidon to a Phoenician father, Mnesarchus, and became a well-traveled trendsetter whose story became legendary over time. Because of his Phoenician background, he was closely related to the Egyptians, which helped him gain acceptance and initiation into the Egyptian religion mysteries and to also become a priest. After twenty-two years in Egypt, Pythagoras returned home revered as a god, as mortals could be deified in those days. The Egyptian oracles would simply declare a person son of Amun, or Ra. Alexander the Great became pharaoh this way and later was buried in Alexandria. Not so long ago in 1865, President George Washington was also deified by Constantino Brumidi in a painting, *The Apotheosis of Washington*, in the dome of the rotunda in the US Capitol Building. Brumidi depicted Washington surrounded by ancient gods giving him knowledge. His bare-chested statue as the Greek god, Zeus, once in the rotunda, now stands in the Capitol's east lawn. Pythagoras' own god, the Monad, the power within all things, was omnipresent.

In 530 BCE, Pythagoras started a school at Croton in southern Italy. The school was like a religious cult. It stressed nonmaterial forms or ideas, fusing the ancient mythological view of the world with the developing interest in science. Like the Egyptians and Hindus,

Pythagoras regarded the body as the soul's tomb. When released at death, the soul would be reborn in a higher or lower form of life according to one's deeds. Pythagoras remembered his own reincarnations, including one as the Trojan hero, Euphaobus. He also recognized a friend's voice in a yelping dog someone was beating and told them to stop. Like the Hindus, he believed the universe had no beginning or end. And by holding that all things consisted of numbers and geometrical figures, his school contributed to mathematics, music theory, and astronomy.

Although Pythagoras's theorem may have surprised western philosophers, it appears Pythagoras might have violated his non-disclosure oath to the Egyptian priests.[15] The Egyptians applied the theory in problem six of the *Moscow Papyrus* at least 1,500 years before Pythagoras was born. They also knew about irrational numbers, like the square root of two, and pi, that were random with decimal places going on infinitely without falling into a repeating pattern.[16] Many philosophers since Pythagoras have also been mathematicians, most likely because a working knowledge of mathematics helps to understand the universe.

In the fifth century BCE, Parmenides founded the Eleatic School, which held that the appearance of movement and the existence of separate objects in the world were mere illusions. Anticipating neuroscience's view that the brain created the images we saw, Parmenides said movement and objects only seemed to exist. Our senses did not know Reality and True Being but only found them in reason. Also, contradicting a Chinese origins story (see Chapter 3), Being couldn't arise from Nonbeing. Being neither arose nor passed away. As with Ultimate Reality in Buddhism, nothing could be truly asserted except that "being is."

Unimpressed by talk about illusions, Empedocles, a disciple of Parmenides, proposed a materialistic explanation. He said all things were composed of four elements: air, water, earth, and fire. Like Yin and Yang in Chinese thought, or love and strife, opposing forces alternately combined and separated the elements, making the world evolve from chaos to form and back to chaos in an eternal cycle.

Anaxagoras, born in 510 BCE, introduced philosophy to Athens. He improved on Empedocles's idea, saying that all things were composed of very small particles or "seeds" (atoms), which were uncreated and had existed from eternity. To explain how the particles combined to form objects, he invoked a cosmic evolution theory, namely, that the active principle of the evolutionary process was the eternal intelligence (*nous* or mind). It separated and combined the particles, creating order. A bright idea, it set the stage for the atomic theory of matter by Leucippus and Democritus. But he also boldly announced that the Sun was a hard stone and the moon was made of earth. The claim landed him in prison for heresy!

In the fifth century BCE, Leucippus proposed the atomic theory of matter. He said that all matter was composed of tiny, indivisible particles that differed only in simple physical properties such as size, shape and weight. Leucippus' associate, Democritus, elaborated that all things were composed of minute, invisible, indestructible particles of pure matter or *atoma* (Greek for indivisibles) moving about eternally in the infinite empty space. As a natural result of their ceaseless whirling motion, they collided, forming larger clumps of matter. Hence, their quantitative arrangement was responsible for the qualitative differences in the things we saw, as well as their birth, decay, and disappearance. Leucippus argued that everything in the world came about in this way. Since the particles were uncreated and indestructible, no divine influence was involved in their creation. In all, up to this point philosophers looked only for a reason to change and not for design, purpose, or meaning.

Leucippus took his atomic theory from a very ancient idea, however. Newton traced it through Greece to a Phoenician physiologist named Moschus, whom Strabo said lived before the Trojan War, about 1200 BCE.[17] Egyptian and Phoenician mystics, Pythagoras for one, sometimes referred to atoms as monads.

Renowned philosopher Socrates belonged to a different school of thought than Democritus and worked along the metaphysical lines of Pythagoras. To him, the soul was a combination of an individual's intelligence and character. Socrates's student, Plato, felt that objects of the real world like trees, stones, and human bodies, were mere approximations or shadows of ideas of the real eternal Forms. Since Forms existed beyond the known world, we could not know them. Mystics surely agreed. They couldn't define reality either.

ARISTOTLE

In 367 BCE, Aristotle entered Plato's Academy at seventeen years old. The Academy was probably the world's second university; the Eighteenth Dynasty Karnak complex at Thebes, now Luxor, Egypt, taking first place. Some twelve hundred years earlier, Thebes had over 80,000 students.[18] Aristotle was Plato's best student. During his lifetime, Aristotle defined logic, as well as the basic concepts and principles of many theoretical sciences, including biology, physics, and psychology. He continued Plato's idealism, but disagreed with the theory of separation of form from matter. For the material universe, he thought nature was an organic system whose common forms remained the same, or immutable. Such forms existed in a hierarchy from simple to complex natural kinds, or "species," with humans perched at the top. But he allowed for some very low forms such as worms and flies to arise from rotting fruit or manure by "spontaneous generation." This process seemed so obvious it took a millennium or so to debunk. Some people still swear maggots arise from spoiled meat.

In his work, *The Physics,* Aristotle argued a divine being, a Prime Mover, was responsible for unity and purpose in nature and identified the mover with God. Although God had not created the world, as in other creation accounts, the cosmos had emanated from him as a necessary effect of his existence. The idea of emanation was not new; the Babylonians, Kabbalists, and Gnostics all mentioned it when describing creation.

Aristotle sought answers to the questions "what is?" and "why?" To fully explain anything, he said, one had to know its purpose and meaning, and its value as well as its makeup (material) and sources of its changing (causes). One needed an answer to each of them to understand and to have wisdom about something. This was particularly true of purpose and value. As designing a test for meaning and purpose was not easy, science overlooked them, focusing on makeup and, for cause, focusing on mechanical forces instead of the meaning *(logos)* in the change. Had science taken them on, we might have resolved the mystery of God long ago, obviating the trials and tribulations of subjective religious dogma.

EPICURUS

In 306 BCE, Epicurus founded a school of philosophy called the "Garden," since classes were held among flowers at his home in Athens. Epicurus admitted women and slaves as students, and thus his school became more popular than those of Plato and Aristotle. It came as no surprise that goings-on among the students there kept rumor mills busy. Had tabloids and paparazzi existed then, they would've had a field day, driving celebrities to self-destruction.

Epicurus taught his students that we should pursue happiness by minimizing harm to others and ourselves. He also taught that gods did not care about people one way or the other. The gods did not watch over them, ready to punish them if they did something wrong. So, no need to worry. Building on Democritus's theory of atoms, Epicurus said gods had not created the world but it had instead come from the accidental collision of atoms. Atoms filling space had simply been falling straight through it. As they fell, some with sideways movements ran into others, bouncing them around. In time, the jostling created the world and us. When we died, the atoms broke up and that was all. Therefore, instead of worrying, people should relax and enjoy life. It was useless agonizing over death, as this wouldn't prevent it.

At the same time as the Garden school, Zeno of Citium founded his own school for Stoics in Athens. Here, he also addressed worry, or stress, but with a twist. To the stoics, nothing mattered. Everything was fated and nothing could change it. Stoics took pain or a call to duty without complaint. Once, when a cruel Roman master reportedly twisted his slave's leg, the slave – Epictetus – calmly cautioned that if the torture continued, the leg would break. The torture continued, and the leg did break. "What did I tell you...?" Epictetus said, not upset in the least. He accepted the result as his destiny, even though he spent the rest of his life crippled. For this reason, Stoics were given jobs such as carrying out Roman law, jobs that relied on performing duties selflessly. While a slave, Epictetus studied Stoic philosophy, and on being freed, went to Greece where he became a renowned teacher in exile.

If you repress emotions, ignore pain, and are indifferent to loss or love, you are a Stoic. But if you pursue pleasure, happiness, comfort, and good food, you are an Epicurean, unlike a glutton. A glutton is at risk for obesity and a shorter life. Gluttony is one of the seven deadly sins, too, according to the Catholic church.

Although the church condemned Epicurus's theory of uncreated atoms, which left no room for God, questions arose centuries later when nothing could explain why Halley's Comet flew the wrong way. Despite Aristotle's disdain, Epicurus's free-floating atoms theory now made sense. The idea that matter consisted of atoms like those described by Epicurus and other ancient Greek philosophers was resurrected in the sixteenth century. In 1915, when explaining the reason tiny particles of dust suspended in water jiggled, called Brownian motion, Einstein put it on atoms.

During the Middle Ages, the great Muslim libraries in Egypt, Córdoba, and other places, preserved Greek philosophy and works. Many of the works were translated into Arabic, including Euclid's *Elements* and Ptolemy's *Almagest* (Arabic for *Great Work*). Ibn Yunus, whose formula was used by renowned astronomer Tycho Brahe, improved on Ptolemy's work.

Muslims in Cairo also contributed significantly to mathematics. Scientists like Copernicus, Galileo, and Newton referred to work from Africa. In 1202, Leonardo Fibonacci of Pisa, an Algeria-raised Italian mathematician, introduced Arabic-Hindu numeral systems to Europe, contributing to the simpler and more efficient modern arithmetic. He copied extensively the work of "the Egyptian calculator," Abu-Kamil (850-930), who wrote a more advanced algebra and influenced mathematicians for centuries. But the Fibonacci Sequence he's famous for had been integrated in Egyptian pyramids and temples by the third millennium BCE. Egypt remained a center of higher education for the sciences, mathematics, astronomy, and optics well into the fifteenth century.[19]

The biblical story of creation out of nothing, *ex nihilo*, was now well known in the Middle East. But Islam regarded it as a parable, a sign or a symbol, like poetry in the Quran. Muslim theology teaches that Allah created humanity, but it does not say how Allah did this. In contrast, Kabbalists related to the idea of creation more symbolically. They found the Greek or Gnostic idea of emanation preferable to the biblical doctrine of creation out of nothing. They were not overly concerned with the physical origins of the Earth.

Moses Maimonides, a Jewish rabbi and physician, was given to defending God's existence with scientific arguments such as Aristotle's. But faced with beliefs lacking proof, e.g., creation *ex nihilo* or emanation, he trusted prophets more than philosophers and erred on the side of the Bible.[20] In the thirteenth century, St. Thomas Aquinas, a Dominican monk and philosopher, combined Aristotle's science with religion. He wrote extensively about philosophy and science, saying reasoning from facts or experience uncovered the truths of natural science and philosophy. He also wrote, however, that the principles of religion were beyond rational understanding and had to be accepted on faith. Aquinas's writing had a long-lasting influence on the philosophy of the Roman Catholic Church.

During the Renaissance (1300 CE-1600 CE), views based on a mechanical and material interpretation of the universe increased. Belief in human thought was the only ultimate reality and became the basis of modern philosophy. As mechanization took hold in Europe, religious beliefs were challenged. Scrutiny of the world and rationalization intensified. Experience and reason became the sole standards of truth, as people explained things more in scientific terms.

About this time, Galileo affirmed Copernicus's heliocentric theory, that the sun was at the center of our solar system and not Earth. Scientists following him, such as Newton, also believed mechanical laws controlled the universe. Baruch Spinoza, a Dutch Jewish philosopher from Amsterdam, went even further, saying that God, matter, and nature were identical, in fact God was nature. He was not alone; Buddhists thought so, too, about the Universal Force, and, as we see later, Einstein did as well. For Einstein, God was the universe, of which we were ourselves only a small fleeting part. According to Arne Naess's philosophical school of deep ecology, this was the essence of spirituality (a euphemism for mysticism).[21] Spinoza's Rabbis were not amused. In 1656, when he was only twenty-four years old, they accused him of heresy, excommunicated him, and expelled him from the city. Christians, too, condemned him for pointing out numerous inconsistencies about Moses in the Torah, such as saying some place names did not exist during Moses's time.

Spinoza was a keen observer of nature, according to neurologist Antonio Damasio. And although single with modest means and ill-health from silicosis – a slow-progressing lung disease related to grinding glass that ultimately led to his death – Spinoza lived a contented life. In addition to Bible knowledge, he created his own meaning by acknowledging the inevitability of natural events such as death, and accepting them, and by reflecting on life in the perspective of nature.[22]

LEIBNIZ AND NEWTON FACE OFF

German philosopher Gottfried Leibniz vies for the title of greatest mathematician of all time. Born in Leipzig in 1646, he independently invented the calculus in use today. But he was forever at odds with its other inventor, Isaac Newton, who came up with his own form of calculus. Newton was not pleasant. He was vindictive, and he trashed Leibniz because Leibniz ridiculed Newton's idea that God caused gravity. That aside, Leibniz wondered, "Why is there something rather than nothing?"[23]

For the answer, he suggested the world was an infinite number of infinitely smaller conscious centers of force, or energy, called monads. Anticipating virtual particles that pervade space, such monads resembled those theorized by Egyptian and Phoenician mystics like Pythagoras. By defining monads as forces, Leibniz was looking for common ground between Eastern and Western religious ideas. His monads then were spiritual entities. God was also a monad, who created all other monads and determined their development.[24]

MORE RECENTLY

In a 1710 treatise, *Concerning the Principles of Human Knowledge,* and a 1713 treatise, *Three Dialogues between Hylas and Philonous,* George Berkeley, an Irish philosopher and clergyman, stated the only things one could observe were one's own sensations, which were in the mind as ideas. Inexplicably, Berkeley believed that this proved God existed. Enter British Philosopher David Hume. No way, José, I see him opining; no observable evidence was available for the existence of a mind, substance, spirit, or God. Small wonder, then, that Berkeley was regarded as the founder of modern idealism and Hume, modern skepticism.

Philosopher Johann Fichte at the University of Jena, Germany, held a different view about matter and our origins. He believed an absolute ego created the world. He also argued that the human will is a partial manifestation of this ego, which tended toward God as an unrealized ideal. Fichte was accused of supporting atheism and was divested of his academic chair of philosophy in 1799. Apparently, he was difficult to get along with and couldn't find work at other universities.

Building on Egyptian priests' knowledge, ancient Greek philosophers formalized the basic concepts and principles of modern science, mathematics, and the atomic theory. For instance, in addition to philosophy, Thales was credited with the introduction of geometry from Egypt into Greece. During the medieval era, however, philosophical ideas were still greeted with hostility, sometimes with death sentences, hampering progress. Work only resumed during the Renaissance as disciplines increased and discoveries multiplied.

Clearly, philosophers do not see eye to eye either on our origins or on the nature of God. Their answers are all over the place. The gods of Pythagoras (himself and Monad), Leibniz (Monad) and Aristotle (Prime Mover) were as disparate as the earlier mythologized origins were. The discerning thinkers, like Epicurus, believed life arose by itself. Hume agreed, denying divine intervention, strongly implying that the world and its occupants were a fluke. No wonder Eastern philosophers such as Buddha cautioned that questions about these matters were unanswerable and we needed to leave them alone. God was indefinable. Rather, for enlightenment, Buddha looked for answers within himself, and not for answers from the skies as the ancients in the West did.

For our ancestors, as sky events changed with time, so did their meaning and the gods associated with them. For instance, the importance of the Celestial Sphinx, also known as the constellation Eridanus, or the god HU (Egyptian for authoritative uttering or Word, described in more detail in Chapter 7) waned as it sank below the horizon and into the vaults of esoteric knowledge. Today, HU is known to only a few people. Dialing back the skies to the times of our ancestors with the astronomy computer program Starry Night Pro helps us to recover the lost meaning and to garner what the ancients believed created the world. As Emperor Marcus Aurelius found his gods in the sky – Mercury, Venus, Mars, Jupiter, the moon – let us turn our gaze upwards to the *Greatest Free Show* in the universe. We might get lucky and catch a glimpse of our God there, also.

CHAPTER 3

HEAVENLY EXTRAVAGANZA

*Man must understand his universe in order to
understand his destiny.*

~Neil Armstrong, Astronaut

John's Gospel relates the origin of the world as follows:
In the beginning was the Word, and the Word was with God,
and the Word was God. He was with God in the beginning.
Through him all things were made; without him nothing was
made that has been made. In him was life, and that life was the
light of all mankind.

~Jn 1:1-4.

Is the above true? Was the world and life created by a Word? This
story bears similarities to that of the Egyptians and HU, as we see later.
To verify it, we need to look to the universe for evidence. Darwin
cautioned that we should speculate about the origins of matter, and not
speculate about the origins of life. But he didn't say we shouldn't
explore.

Having stayed close to home all their lives, ancient people located
the site of creation near their own societies. For instance, Navajo Native
Americans say that they came up from the ground, birthed like a child
migrating from its mother to the earth's surface, one being to another.
Their DNA, though, shows that they are related to Central Asians,
where their ancestors came from, and to Europeans.[1] In Australia,
spirits sprang from the ground and created the Aborigines, then,
exhausted by the hard work, retreated. From there, people spread
throughout the world. Yet, where did matter come from?

In Genesis, God simply said, "Let there be" x, and all that there is
came about. Then, after consulting Elohim, the seven children of the
ancient Mother Sophia, He (God) created us in His image, Gn 1:26-
27.[2] Creation was always instantaneous in such stories. But as science
indicates, life developed over billions of years to reach today's stage
and couldn't have started in a flash on-command. When examined
scientifically, none of the numerous creation stories can be true.

When Einstein looked through astronomer Edwin Hubble's
telescope at Mt. Wilson Observatory in California, he was surprised.
He saw an expanding universe. Light from distant galaxies was red-

37

shifted, indicating that galaxies were flying away from us at great speed. In 1927, Georges Lemaitre, a Belgian priest, posited that the universe was once small – a singularity – a point many times smaller than the proton of an atom with infinite density without space and time, as at the center of a black hole. It was a singularity. The universe came into existence when the singularity expanded into a gigantic fireball. This understanding astounded even Einstein.

But Einstein shouldn't have been surprised, as his own theory of general relativity had predicted it, and he had even introduced a cosmological constant to rein it in. In Einstein's support and to promote his own steady state model, astronomer Fred Hoyle ridiculed the idea of an expanding universe, derisively calling it the Big Bang. To his dismay, the name stuck for lack of a better term. (Hoyle probably kicked himself whenever he heard it). The term was a misnomer anyway. Even if somebody had been around, without air to transmit the sound, there would be no bang to hear. The event was noiseless.

HOW DID IT HAPPEN?

As in our ancestors' stories, in the beginning, before the creation of the universe and matter, there was a void. But instead of a complete vacuum, the void seethed with foams of energy bubbles, particles called virtual pairs. At the time, the particles, matter, and the four forces of nature (the weak and strong forces, electromagnetism, and gravity) were one. Then, as the singularity expanded, the forces separated, giving off lots of heat, and energy froze into matter, causing temperatures to fall.

What triggered the blast? Many say it was no fluke. They see the hand of the Almighty in it. But, without proof, thoughtful people don't accept that argument. In fact, the evidence points to equally head-scratching quantum theory. In 1900, German scientist Max Planck suggested that light traveled in small packets of energy, or quanta (from which quantum derives), instead of waves. Einstein said the packets, later called photons, traveled as waves, too. In 1927, Werner Heisenberg, another German scientist and a major contributor to quantum theory, ran with the idea. He said that one could not precisely measure the position and velocity of a small particle at the same time, as doing so changed its position or speed.

The theory, called the Uncertainty Principle, predicted different possible outcomes for an observation and the likelihood of each, offering a way for the universe to come into being. As there was nothing having a precise value of zero energy, it meant that a small non-zero energy fluctuation briefly existed in the universe. When the fluctuation lasted longer, it allowed virtual pairs to borrow the energy and pop into space. Being matter and antimatter, they annihilated in a split second and gave the energy back before being detected. Traces left behind, such as a tiny shift in atom energy levels, indicate virtual pairs are real.

So, when a random energy bubble fluctuation failed to annihilate under certain positive gravity conditions, it inflated and never looked back. Space expanded so fast particles couldn't recombine and disappear and so became real. According to the Vacuum Genesis theory, symmetry was broken. "Out of nothingness could have come the spark of genesis," said Timothy Ferris.[3] From a void, then, a quantum fluctuation created the universe, what some cosmologists called, "the ultimate free lunch." According to a story shared in his autobiography, *My World Line*, Russian-born American scientist George Gamow said this theory astounded Einstein. Gamow explained that one day while crossing a street in Princeton, Einstein learned physicist Pascual Jordan had theorized that a star could be made from nothing this way. Dumbfounded by the news of Jordan's theory, Einstein stood in the middle of the street, blocking traffic.[4]

Many cosmologists see no reason why Vacuum Genesis should stop at one universe. Quantum fluctuations likely spawn baby universes all the time, each one with its own unique properties. Because conditions in our universe allow us to exist, we acknowledge it and deliberate on its origins, our purpose, and the meaning of life. To see the other babes, however, we might need new senses or a functioning wormhole opening in one of them. A crop of other worlds on the subatomic scale could account for the beginning of life. More discussion on that later.

As above, as the universe expanded and cooled, energy froze into matter like crystals forming on a pond. Form arose from formlessness and all kinds of particles were created. The beginning of the universe would then not have been grounded in a first cause, itself uncaused, as St. Thomas Aquinas argued. Unlike in scripture, intention (and therefore design) was redundant in a universe full of playful random fluctuations and uncreated energy.[5] Rather, as the ancient Chinese said in one of their many origin stories: Being arose from Nonbeing. Nothingness produced the universe, space, and time.[6] Similarly, according to the Buddhist *sutras,* a void, emptiness, or *sunyata* produced form. To Nagarjuna, one of the most intellectual Buddhist philosophers, since reality could not be grasped by concepts or ideas, he called it sunyata, "the void," or "emptiness." Such terms were equivalent to *tathata* or "suchness," used by Ashvaghosha, the first expounder of Mahayana doctrine in the first century CE.[7]

In 1981, at the end of a Vatican conference organized by the Jesuits to advise the church on cosmology, Stephen Hawking, together with several other experts, was granted an audience with Pope John Paul II. While he conceded that the theory of evolution was more than a hypothesis, the Pope told the group that there should be no enquiry into the Big Bang itself, as it was the moment of creation and therefore the work of God. Hawking was relieved that the Pope wasn't aware of the subject of his talk, "the possibility that space-time was finite but had no boundary, which means that it had no beginning, no moment of Creation."[8]

Steven Weinberg, basing creation on the standard model in his book, *The First Three Minutes,* described the beginning thus:

> At about one-hundredth of a second, the earliest time about which we can speak with any confidence, the temperature of the universe was about 100,000 million (10^{11}) degrees Centigrade. This is much hotter than in the center of even the hottest star, so hot, in fact, that none of the components of ordinary matter, molecules, or atoms, or even the nuclei of atoms, could have held together. Instead, the matter rushing apart in this explosion consisted of various types of the so-called elementary particles, which are the subject of modern high-energy nuclear physics.

To explain shortcomings in the Big Bang theory, such as why the universe had about the same temperature everywhere and appeared flat, or static, Alan Guth proposed an inflationary model. He said that during the first minute the universe expanded 10^{50} (one followed by 50 zeros) times its size in a split second. It blew up from a fraction of a proton to the size of a grape, which is comparable to a tennis ball blowing up to the size of the present-day universe. Inflation prevented the universe from collapsing. The tremendous energy for the expansion came from the four forces separating, similar to splitting an atom or fission. So, from nowhere a quantum fluctuation created the universe, leaving God no role to play.

Given his distrust of quantum mechanics, Einstein likely turned over in his grave at this. Or, maybe not. His family opted for cremation and scattered his ashes along the Delaware River, possibly to prevent his grave from becoming a shrine. But a pathologist at Princeton Hospital, Thomas Harvey, had previously spirited Einstein's brain to Wichita, Kansas, where he studied it (albeit unmethodically) for forty-three years before he returned it to the hospital. The curiosity didn't end there, though. More studies were undertaken, and what remains of the brain is now kept at the National Museum of Health and Medicine in Silver Spring, Maryland.

Other theories explaining the Big Bang and inflation include *Superstring theory* from the 1970s. This theory describes matter as being made of strings and branes. One model, *M-theory,* or *ekpyrotic* (Greek for out of fire), agrees in principle with the Hindu and Central American Mayan cyclical views of creation, in which the universe came, shrunk, and bounced back once every trillion years. Only time will tell which if any of these origin scenarios is correct. Strings are too small to detect, and do not allow for testable predictions.

The first law of thermodynamics states energy is conserved and cannot disappear from the universe. Instead, it is transformed into other forms such as light, work, life, or matter. Indeed, in his equation for all time, Einstein found matter was laden with energy:

$$E = mc^2$$

With the equation, many natural events in the universe could be explained, including the Big Bang and energy from the Sun. *E* stands for energy, *m* mass, and *c* the speed of light.

Unleashing the energy in matter may wreak havoc never imagined by Bible scribes and Apocalypse doomsayers. Think of Hiroshima or Nagasaki. In 1945, two baseball sized atomic bombs, called Fat Man and Little Boy, wasted these cities, effectively bringing about the end of World War II. On the positive side, nuclear energy generates electricity, allows submarines to run for long stretches without surfacing, heats Earth from its core, and powers the Sun.

WHY THE BIG BANG AND NOT PROVIDENCE?

Currently, the Big Bang, not design, is the best theory describing the origin of our universe. For example, take today's temperature of the cosmic microwave background radiation, the Big Bang afterglow. This radiation was accidentally discovered in 1964 by two Bell Laboratories engineers, Arno Penzias and Robert Wilson, in New Jersey. They were working on a radio telescope, investigating radio hiss, and found it coming from our universe when it was 380,000 years old. This hiss is visible as "snow" on the screen on an empty TV channel. Although Penzias and Wilson didn't think they deserved it, they were awarded the 1978 Nobel Prize for physics.

In the late 1940s and early 1950s, George Gamow predicted the presence and current temperature of the cosmic microwave background radiation, the afterglow of the birth of the universe. A Cosmic Background Explorer satellite, COBE, measured the temperature in 1989, and found it hovering just above absolute zero at 2.7-Kelvin, exactly as Gamow predicted. This drove a nail into the coffin of Hoyle's "steady state" model. In 1992, COBE detected ripples in the afterglow that astrophysicist George Smoot called "fingerprints from the maker." An unexpected onslaught of journalists and the public alike, eager for a glimpse of God, caught Smoot off guard. The discovery confirmed matter in the early universe was unevenly distributed. This allowed some parts to be denser than others, resulting in unequal gravity. The denser areas became "seeds," drawing in matter that formed the stars and galaxies. During the first 380,000 years, the ceaseless particle frenzy absorbed all photons, and light could not break through. It produced a "thermal" or "blackbody" spectrum (a blackbody theoretically absorbs all radiation striking it). Following very accurate measurements by an infrared radiation detector aboard COBE, the spectrum was as the Big Bang theory predicted. No other theory explained it.

Also, there appears to be far more helium gas in the universe than stars can make. Back when the universe was three minutes old, the ratio of the nuclei of helium to hydrogen was about 1:4. That ratio has not changed.

Then, according to Hoyle's "steady state" model, the entire sky should be bright both day and night, ablaze with stars every which way.[9] But this isn't so. Only an expanding universe explains that discrepancy. Since stars turn on at some point in time, light from distant receding galaxies has yet to reach us. This phenomenon, known as Olber's Paradox, named for the German astronomer who first publicized it in 1823, further supports the Big Bang idea. Throwing quasars into the mix strengthens the theory even more. Quasars formed in the early universe and are almost exclusively confined there today. No quasar shines close to us; they are all more than 2.4 billion light years away, evidence pointing squarely at a universe that was once miniscule before it expanded into a gigantic inferno, the Big Bang.

For clues to the maker, what about examining His handiwork? What does all that we see consist of? Scientists peel the basic unit of matter, the atom, into its constituent parts to find out. But be warned, science fiction does is no match for the weird world of the small. Should you get lost, don't despair. The best scientists sometimes get lost, too.

WHAT ARE WE MADE OF?

To avoid being misled by misinformation, or by those with an agenda, or by illusions, we should know our world well, even though Plato, Parmenides, and others doubted we could do so. Plato taught that we can only see shadows of the real world. What's the real world then?

According to Fritjof Capra in *The Tao of Physics,* in the West, objects appear separate from one another, and science describes the basic units of matter as points, individual waves, strings, or branes. In the Far East, however, nothing exists independently. For instance, Buddhists see that the smallest part of a blade of grass is continuous with everything, including us. Making the blade is a transient collection of processes. So, the concept dividing one event from another, or the idea of a separate individual, is illusory, imposed by our categorizing and pattern-seeking minds.

Which of the two main views, Western or Eastern, appeals to you? First, though, what does science say the universe is made of?

"Atoms," answers classical Newtonian physics. Ancient Greeks like Democritus were among the first to formalize the atomic theory. Lacking the aid of modern technology, however, they deemed atoms to be the smallest units of matter. Today, scientists not only split atoms into electrons, protons and neutrons, but the last two into quarks as well. Many more particles existed during the Big Bang, but disappeared as the universe cooled. They can only be recreated using incredible speeds in multimillion-dollar accelerators.

The quest for the elusive so-called "God's particle," the Higgs boson, a particle that gave matter mass, or weight, fueled the 2008 ten-billion-dollar upgrade of a seventeen-mile ring accelerator, the Large Hadron Collider (LHC), located at CERN on the French-Swiss border.[10] The collider probed further into the first second of the Big Bang than the once most powerful particle accelerator in the world, Tevetron, at Fermilab, Illinois. The Tevetron searched for the particle for ten years in vain. CERN announced its discovery in 2013.

Particle life was short during the Big Bang. Regardless, if you trust Einstein's theory, as a particle's speed approaches that of light, its life span increases and it ages more slowly, an effect called *time dilation*. The particle also gains weight, and the closer it gets to the speed of light, the heavier it becomes, until it gains so much weight that there isn't enough energy in the world to carry it farther. That's bad news for us. It means travelling at such lofty speeds is out of the question for humans and we shouldn't even think about it. Those seeking immortality in suspended time should look elsewhere. But it also means, were an identical twin to embark on intergalactic wanderings at a fraction of a second shy of the speed of light for a year, on return he would find his twin brother grown much older than himself. This effect is known as the twin paradox. So, are we sure Einstein had all his marbles?

In the 1800s, British physicist John Dalton identified the atom as the basic unit of matter. Come the twentieth century, and a zoo of hundreds of subatomic particles opened. In the standard model, scientists pared them down to two elementary particles, fermions and bosons. Fermions made up the matter we see, including electrons and protons. Bosons were associated with light and the weak force.

Subatomic particles are in constant flux. They jiggle 40,000 times a second in a nucleus and don't break down easily. Otherwise, one would crumble into radioactive dust on shaking hands or stealing a kiss. For a proton to decay, for instance, we would have to wait many times the age of the universe, over a decillion years. A decillion is represented as 1 followed by 33 zeros.

With the dawn of quantum physics, the Newtonian notion that matter consisted of separate basic building blocks became untenable. According to Capra, when Einstein discovered that matter was a form of energy, the scientific concept of a material substance died, as did the concept of basic structures. At the atomic level, nature appeared to be a network of unified relations with no separate parts. And as matter and energy were equivalent and energy was associated with activity or processes, subatomic particles were intrinsically dynamic, continually changing into one another.[11]

Capra also noted that the ability of matter to intertwine paralleled the intuitive deduction of its properties by Eastern mystics. In modern physics, the image of the universe "…has been replaced by that of an interconnected, dynamic whole whose parts are essentially interdependent and have to be understood as a cosmic process."[12] Indeed, in 1964, British physicist John Bell derived a theory from an experiment Einstein had helped devise. Einstein wanted to show that physical reality consisted of independent separate objects joined by local connections. Instead, John Bell showed the universe was fundamentally interconnected, interdependent, and inseparable. Because what happens here and now might depend on something far away in the universe or from a different time, called quantum entanglement, Bell's theorem showed non-locality (instantaneous action at a distance), or teleportation, was a valid quantum event. Indeed, in 2003, a Geneva scientific group teleported particles of light 1.2 miles.

Bell had shattered Einstein's view of reality, which contradicted that of his nemesis, quantum theory. Einstein dubbed the phenomenon "Spooky action at a distance." It sounded like telepathy. Now you understand why Heisenberg's version of reality, including the inability to precisely discern the state of the universe, appalled Einstein. But it made mystics giddy, particularly Buddhists in the Far East. Heisenberg's theory showed that things in nature were interconnected and related as mentioned earlier, so, according to Capra, individuality, or a separate self, was an illusion.[13]

Teleportation could also offer a way of rapidly travelling through space. As in *Star Trek*, with teleportation we could instantly beam people to their destination. After obtaining information from them with a measuring apparatus, a "transporter beam" would teleport them to a receiving chamber on a distant planet. For transmission to work, however, particles would first have to be entangled together at each end.[14] Many challenges remain.

Back to the present. According to two-slit experiments, in the quantum world, an electron or photon can take two paths at once. En route, it's said to be in *quantum superposition* (in many states). A detector at the slit, even when not in its path, could throw the electron out of the states into adopting one position and make it real. The detector is said to have measured it and thrown it out of the quantum state. Were the phenomenon to apply to humans, we would have an alternative alibi if caught on camera robbing a bank. Someone would have seen us elsewhere at the same time.

While in the bank, like a proton tunneling its way into impossible places, we would materialize from the walls, slither across the ceiling to evade laser beams, and slink into the vault. "Bunkum," say you. But as quantum mechanics shows, as we do not understand nature fully, we should suspend judgment and simply go along with what works. Otherwise, we could spend lifetimes disputing valid observations like Einstein did. Some people say that we will never know the real world, that we can know only how things are related to one another, if they exist. Others view properties of an object as the only fundamental category, called tropes. We realize an object when its properties bundle themselves together in a certain way.[15]

Demonstrating great foresight, Buddha taught that the reality question was likely unanswerable. To keep us rooted to the ground, the brain constructs the world we know, which differs from person to person.

The reality dilemma hinders our ability to make sense of life or find an answer to the source of the laws and order in the universe. Aristotle called the cause the *Unmoved Mover*; the Hindu, *Rta*; the Egyptians, *Ma'at*; the Pythagoreans, *Number*; the Chinese, *Tao*; the Bible, *Wisdom;* and Heraclitus logos backed up by necessity the *Enforcer*.[16]

THE FOUR FORCES REUNITED

What a relief it would be if scientists could reunite the four forces of nature. It might help to explain ourselves and everything else in the universe. The quest for such a theory of everything began with the ancient Greeks. So far, Einstein's theory of general relativity, dealing with the large structures in the universe, cannot be united with quantum mechanics, dealing with subatomic scale events. General relativity also breaks down at singularity before telling us about the first instance of the Big Bang. A unified theory would not only clarify it, but also throw light on the trigger and allow us to solve all our scientific problems, bringing about the end of the need for additional science to optimists.

Muslims by the twelfth century thought they knew all there was to know and sat back, only to see the Europeans pass them. Eight centuries later, the Europeans smugly believed they had discovered everything about science and felt darn sure they could foretell the future. At the beginning of the nineteenth century, a French mathematician, Marquis de Laplace, argued that by knowing the complete state of the universe at one time, we could predict its outcome and that of human behavior as well. Some ridiculed the idea because a deterministic universe interfered with God's freedom to run it as He pleased. The theory sidelined God's envoys, including people like St. Augustine. Anyhow, quantum theory torpedoed the argument. According to Heisenberg's uncertainty principle, it was impossible to know the state of the universe precisely as we could not measure anything accurately. The universe, then, could not be deterministic or predictable. So, nothing was destined to be, not your sweetheart or a calling to the clergy. Even nature's constants changed and, in a black hole, they disappeared.

In the last thirty years of his life, Einstein worked on a unified field theory, but the attempt was in vain because it was premature. The weak nuclear force had not yet been discovered. Even though he helped discover quantum theory, and had won a Nobel Prize for identifying the photoelectric effect (the mechanism behind automatic doors), he distrusted the phenomenon. Like the uncertainty principle, to Einstein, it smacked of chance in nature. "I can't believe that God plays dice with the universe," he said, or words to that effect, in one of his famous debates with Niels Bohr, the Dutch atomic physicist, who helped craft the Copenhagen interpretation of quantum mechanics.

"Albert, stop telling God what to do!" Bohr admonished. No one fully understands quantum mechanics. It is one of the dilemmas we face when trying to demystify nature.

Check this out: according to the Copenhagen interpretation, matter does not exist until we observe or measure it. At the subatomic scale, it normally hovers in the twilight zone of quantum superposition. Measuring collapses it into our classical world, making it real. So, it seems you don't exist until somebody notices you. Luckily, the phenomenon applies in the subatomic world where particles don't interfere with each other and, like laser beams, are uniform or *coherent*. When observed, it means ordinary light, or the environment, has interfered with them, causing them to lose coherence and become real. We humans are simply too large to prance about invisible. That's why under an intense stare we may shrink but not vanish.

In looking for a single universal force, we stand warned. Since we do not know true reality, how can we have the answer to everything? Scottish philosopher David Hume wanted to know. Newton had shown Hume that color was a brain construct representing the various electromagnetic wavelengths, thus showing that our image of the world was but a shadow of the real one. Hume also believed that manmade statements could never be known with certainty to be true. "We should therefore not give house-room in our heads to Theories of Everything..."[17] Even Leibniz, reputedly the first computer scientist, hinted that many problems, such as irrational numbers, were so complex that no theory explained them. Small wonder Einstein's endeavor to find a unified theory stalled.

Although fuzzy and inexplicable like Zen or a meditation experience, quantum theory lends itself well to experimentation. We have successfully applied it in much of modern science and technology with results such as the electric lightbulb, the CD player, PET (positron emission topography) scans, and MRI (magnetic resonance imagery). It's the basis of transistors and integrated circuits, the latter forming essential components of electronic devices like computers and televisions. Without quantum theory, the electron microscope, modern chemistry, and biology wouldn't be possible.

At the first instant of the Big Bang, when the four forces split, there might have been a fifth one, called quintessence, a form of dark energy. From standard Newtonian physics, the matter we see makes up less than 5 percent of the universe. The other 27 percent is invisible, called dark matter. This leaves 68 percent of matter, called dark energy, partially consisting of quintessence, unaccounted for. Quintessence appears to help keep the universe flat instead of flying apart or collapsing into a crunch. And were it not for the four forces, the universe would be a uniform, almost zero-density expanse with no structures or life.

HOW A STAR IS MADE

In the Bible, God set stars in the sky to give us light. Gn 1:14-18. But according to natural physical laws, stars formed over millions of years and only one, the sun, illuminates Earth. To find out about the stars' true origins, scientists study nebulae, or interstellar dust clouds.

So, for the mechanism, as the universe cooled, clouds of hydrogen banded together and collapsed under their own gravity. As they contracted, some fragmented into many pieces, a process that took about twenty million years. Were a fragment large, it would collapse further, and if it had a slight rotation, it would start spinning and became a protostar with the core temperature reaching one million Kelvin (about two million degrees Fahrenheit). Over the next 10 million years, more collapse would increase its core temperature and its density; it would enter adulthood smaller and turn into a main sequence star. Like 95 percent of all stars, it would fuse hydrogen at the core into helium for energy. The collapse of a gigantic cloud could result in a cluster of many stars forming a galaxy.

Were a star a supergiant, more than eight times the size of the Sun, its life would be short. Because of its huge size, it would "guzzle" its hydrogen, fusing it into helium, helium into carbon, carbon into oxygen, oxygen into neon, neon into magnesium, magnesium into silicon, silicon into iron, and stop there. Iron, a heavy element, requires energy to fuse it.

With iron at the core, the star becomes too heavy and collapses. But the collapse stops abruptly sending shock waves to the outer layers, blowing them off in a humongous explosion – a supernova. The intense heat cooks up small quantities of heavy elements: iron, silver, gold and so on. To their credit, Hoyle and a co-worker, W.A. Fowler, established this process. But Hoyle must have wondered what he had done wrong again when Fowler shared the 1983 Nobel Prize in physics with a third worker, astrophysicist S. Chandrasekhar, for the theory. Hoyle complained about it, but he was such a controversial figure. In the

2010s, it was proposed that the heavier elements like gold, platinum, lead, and uranium came from neutron star collisions.

Supernovae make life's building blocks: carbon, nitrogen, and oxygen, which collect in nebulae, out of which new stars and solar systems like ours come. So, like the Dogon, we are all from the stars.

SUNNY SIDE UP

Like life, stars come and go, too. The sun, now 4.6 billion years old, ambles in midlife. The sun, planets, and other masses around it spin and orbit in the same direction, indicating that they formed or fragmented from a huge rotating cloud of gas and dust. As the cloud collapsed, it ignited, forming the sun, a disk like a fried egg, and unleashed solar wind that blew the remaining clouds further afield, leaving only solid particles behind. The particles condensed, attracted matter, and – as more and more of them joined (accreted) – formed the eight planets.

Because the inner four planets, Mercury, Venus, Earth, and Mars – called the terrestrials – formed close to the Sun, their original interstellar dust vaporized, leaving behind metallic grains, making them small and rocky with dense cores. Further out, where it was colder, water ice and ammonia gas grains survived, forming the four Jovians: Jupiter, Saturn, Neptune, and Uranus. The Jovians accreted large amounts of gases, trumping the terrestrials in size by far. Fragments escaping capture became asteroids.

Our sun is a yellow third-generation star of medium size. Its core churns at 27 million degrees Fahrenheit while the surface maintains a relatively cool 10,000 degrees Fahrenheit. In the not- too-distant future – six billion years – it will throw in the towel. By then, life here will be long gone, and with it, Einstein's genius and work. Imagine.

SPACE EXPLORATION

Acting on Neil Armstrong's advice above, away from city glare, we await darkness. The moon is AWOL. And the sun is sinking into the netherworld, painting the sky a bright golden yellow color reflecting its awesome fading power as "it dies to be reborn the next day," the ancient Egyptians would say. As darkness falls, we cannot be more excited, for in pursuit of answers about our creator, we hunger for knowledge hidden in the heavens. A boon would be coming across life there, too, and a relief that we are not all alone.

THE SOLAR SYSTEM

Even without a telescope, we see a great number of gods in the sky, most of them unnamed. In the late afternoon or at twilight during the right time of year, a star, Venus—the Roman goddess of love – shines in the west, and at sunrise in the east. Venus is our closest neighbor, one of eight wandering stars (Greek planets). It's the brightest object in the sky next to the Sun and moon. In ancient cults, Venus represented the goddess of female sexual love and beauty. Its symbol, a five-pointed pentagram, denoted life and sexuality, and was linked to five-petal flowers and their multiples, particularly the rose, an anagram for Eros, the Greek god of sexual love. As it resembles female genitalia from which humanity came into the world, the rose today, like Venus, stands for womanhood.

Such fondness for Venus belies the planet's hellish nature. With an atmosphere of 95 percent carbon dioxide, greenhouse gas temperatures stay in the 800-degree Fahrenheit range. Its 92 bars of atmospheric pressure would crush you into a pulp if you survived its sulfuric acid (like car battery fluid) rain during descent. Also, its landscape is littered with volcanoes and crisscrossed by rivers of scorching lava. If Venus were truly a goddess, she would make a good match for Enoch's God. Both are searing hot and terrifying.

Next comes Mercury, a hazy planet, the Roman messenger god, hugging the horizon close to the sun in the East at dusk and in the West at dawn. Also, easy to see as it courses across the night sky is the "Red Planet," Mars, the Roman god of war. Mars has so much iron it shines bright orange-red when closest to us. To the ancients, this meant the gods were angry. But they were petrified when the planet appeared to reverse course and move backwards (as our planet moving faster passed it). Mars appears whitish at its furthest point, 233 million miles away.

Far beyond Mars orbit the Jovian planets. The closest, Jupiter, the Roman supreme god 400 million miles away, is second only to Venus in brightness. Jupiter is so big it could easily hold 1321 Earths. Next in line comes Saturn, the Roman god of agriculture, famous for its rings. The rest of the planets are too far away to see with the naked eye.

Uranus, the earlier supreme god of the Greeks, may require binoculars or a telescope to find it. Beyond it, and not even visible with a large telescope, is Neptune, the Roman god of the sea. Winds there blow at 1,400 mph—faster than a bullet. So, there is no way this planet can be a benevolent god. Farthest away from the sun courses icy Pluto, the Roman god of the dead and ruler of the underworld. The debate regarding whether Pluto was a planet began with its discovery in 1930. Since Pluto had moons, it was argued, it must be a planet. But some asteroids have moons, too. Pluto also had other ice bodies in its orbit, including one even bigger than Pluto itself. Swayed by this reasoning, the International Astronomical Society demoted it to a dwarf planet in 2006. But Pluto still has its supporters, such as NASA Administrator Jim Bridenstine.[18]

Our sun is but one star among trillions. Awestruck by its sunspots (transitory dark patches), Galileo challenged the belief that it was smooth, saying in addition to spinning, it bubbled and roiled. Occasionally, it erupts, ejecting 10 billion tons of material into space, called solar flares. A perpetual god, for instance as Helios in Greece, the sun has been revered by umpteen cultures since time immemorial. We find symbols of it everywhere. The circle on priests' shaven heads identifies the priesthood's God, the sun. In nunneries, nuns are obligated to guard the sacred, or holy fires, also a symbol of the sun-god. And in Egypt, the pyramids (*pyr* means fire) were built to revere the fire-god.[19]

The sun is praised every *Sun-day* and its "rebirth" celebrated on Christmas Day, when it embarks on its yearly return journey from a "cave" to the northern hemisphere. It's frustrating that no matter how hard we pray, it won't answer. That's not its style. Regardless, we dutifully pay homage to it—and other heavenly bodies as well—when we clasp and point our hands or lift our eyes towards Heaven, or face East when praying. By saying, "…thou art in heaven," it shows where we believe God resides. Because the sun is extremely bright, we dare not look at it directly, instead casting our gaze away as Enoch and the angels did in the second house of God.[20]

The sun, planets, comets, Kuiper Belt rocks, and asteroids make up our solar system family. More ice bodies and comets tumble behind Pluto in the Kuiper belt. Still other ice bodies and billions of comets circle further out in the Oort cloud close to the solar system's edge, the heliopause, an average of eight billion miles away. Now that we have had a good look at the sun, planets, and other objects in the solar system, we can confidently say that they are not gods. The ancients who worshipped them didn't know. Still, this was not so bad or useless. As we will see later, they would have benefited merely from the belief in them.

THE MILKY WAY

When we look at the night sky with the unaided eye, all the six thousand or so twinkling "fixed" stars we see belong to our galaxy, the Milky Way. Like other galaxies, the Milky Way is home to hundreds of billions of stars circling a common center. Galaxies date back to a time when the universe was about a billion years old. Some, such as ours, grew by seizing their smaller neighbors. Next in line for capture are the two Magellanic Clouds nearby. In a twist, from 2.2 million light years away, Andromeda likewise tugs at us, inch by inch, and in four billion years will engulf us all together.

The ancients believed most stars were fixed in the outermost celestial sphere, one of eight spheres Claudius Ptolemy proposed to explain the motions of the planets. Star movements helped the ancients create calendars and predict weather events. For example, Egyptians timed Nile flooding by tracking Sirius. By 4,125 BCE, they had devised a calendar, the predecessor of the Julian calendar, replaced by the Gregorian calendar in 1582. Stars such as Polaris, aka the North Star, guided seafarers in the Northern Hemisphere.

People recognized patterns in the stars and grouped them into constellations. Constellations served as astronomical navigation points to locate and identify stars. Many were also believed to be gods, such as Eridanus as god Hu, Orion as god Osiris, and Canis Major as the goddess Isis with her son, Sirius, as god Horus. Other constellations interacted with the gods. For example, Aquarius carried water of the gods, and Leo was slain by Heracles (Hercules). Some had families, too, such as Zeus and his daughter, Virgo.

Believing star positions had a bearing on a person's character and destiny, Babylonians practiced astrology. Based on the zodiac, the constellation the sun occupies when you're born becomes your sign, on which your horoscope, or predictions, will be based. For instance, I was born on March 21, and my sign is Aries.

In the olden days, gods were sacrificed yearly during spring, and, like the crops, reborn, or resurrected. According to various traditions, Jesus was either born during this time on March 21, on January 6, the Epiphany (Dionysus and Osiris' festival days), or on December 25 (like Horus and Mithra before Him). The ancients celebrated whenever the vernal (spring) equinox entered a new age (also known as the Great Month), which occurred every 2,155 years. As mentioned earlier, 2012 in the Mayan calendar was about the dawning of the age of Aquarius, not the end of the world. The Bible alludes to this age as the "Second Coming," by mentioning the man carrying a pitcher of water. Mk 14:13; Lk 22:10. When Aquarius dawns, Jesus's sign should change from that of a fish to a water pitcher.

In the Northern Hemisphere on a dark, clear, summer's night, you may see a white band stretching across the sky resembling a trail of spilled milk, the origin of our galaxy's name, the Milky Way. In Greek mythology, in a sight to remember, Hera, the goddess of the sky, squirted the milk there when she married Zeus. The band is located toward a ten-light-year thick bulge at the galactic center. A big black hole, four million solar masses, devours stars whole there. As eighteenth-century astronomer Thomas Wright declared, before the makeup of the galactic center was known, the consensus was that God lived there.

To find God's home in our galaxy without someone leading the way – the angels who "lifted" Enoch there, for instance – is a tall order. The Milky Way is huge, holding up to 400 billion stars. Its visible part stretches 120,000 light years across, or about 705,000 trillion miles. Adding a halo of dark matter expands the size to two million light years across.[21] Our solar system is nestled two-thirds of the way in one of the galaxy's spiral arms. Unbeknownst to the ancients, there were other galaxies, any or all of which could also have gods and therefore heavens. The only other galaxy that was visible to the naked eye – Andromeda – whirled 2.5 million light years away, as a nondescript faint oval patch in the constellation by the same name. It appears to us today as it appeared when our genus, Homo, first walked the Earth. The rest of the galaxies, too many to count, may be seen only with a powerful telescope.

The Milky Way belongs to a local group of about eighty-five galaxies spread over 10 million light years. The group belongs to Virgo, a Supercluster Galaxy spanning 100 million light years, which is embedded in an even larger web, Laniakea (Hawaiian for immense heaven), a galaxy group 500 million light years across. But this is only a small part of the universe. So where can one even begin to look for God?

When a massive star – eight or more times greater than the mass of the Sun – dies, it goes out in style with a bang – a supernova –that outshines galaxies. In 1054, the Chinese recorded such an explosion. It shone in broad daylight for a week. The remnant of this mother of explosions is known as Crab Nebula. Galaxies display a supernova about once every fifty years. This translates into about one star per ten seconds going supernova somewhere in the universe. Due to masking by dust, no supernova has been seen in our galaxy since 1604. Based on other observations, scientists estimate we have about three explosions in the Milky Way per century. Supernovae are particularly abundant in Leo and Virgo, appearing as bright stars when viewed with binoculars.

The origin of the stars in the Bible was like placing bright dots on a canvas. But space has young and old stars, so they couldn't have all appeared at the same time. They formed slowly from nebulae as described above. With a good eye, a small telescope, or binoculars, you can easily locate the Orion nebula, a hazy patch snuggled below Orion's three-star belt in constellation Orion 1600 light years away. The nebula is a nursery of star birth, glowing with the activity.

Our ancestors believed stars were gods. For example, Sirius was considered to be the god Horus and Thor. As the brightest star in the sky, Sirius was also a harbinger of good news. It was Sirius – also known as the Star in the East or the Star of Bethlehem – that led the Magi to Jesus's birthplace. Astronomically, on 25 December, 1 BCE at midnight, the three stars in Orion's belt, still known today by the same name they were given in ancient times –the Three Kings or Magi – aligned with the star pointing to Virgo (Latin for Virgin), occupying the place of the sun's birth, or sunrise, in the east. The kings then followed the star to locate the birthplace of Christ, the sun, at dawn. Besides wheat-harvesting time, Virgo also signified the *House of Bread*, Bethlehem in Hebrew. For that reason, Virgo's ancient symbol, M, denoting Mary, begins the names of Myrra, Adonis' mother; Maya,

Buddha's mother; and Maritala, one of Krishna's mothers.[22] This indicates that Jesus's story is astronomical.

WHO GOES THERE?

Travelling back in time takes us across the sky and heaven to the beginning of the universe. It could help us find the creator of the universe and the spark for the Big Bang. But by the beginning being so far away and long ago, we'll need to rev our engines to the speed of light or better to get there. Cautioning us that we could not travel faster than light, Einstein also predicted we would shrink to almost nothing and gain unbelievable weight in the process. Einstein's equations, though, allowed time travel through wormholes at the subatomic level. But a wormhole would only take us to its opening. Better still, though slower, would be a warp drive, like the one used in *Star Trek* by the USS Enterprise. Then we could go whenever and wherever we pleased, pronto, and right on into the past. Never mind Hawking pouring cold water on such ideas. He argued against the plausibility of going back in time, as then you could kill your grandmother before your parents were born.

Because we're incurably restless and ridiculously nosey, after perfecting the light-sonic machine, we embark on a fact-finding space tour blasting through the sky. We zip through space and time, gaping in awe at all the rainbow colors flying past, colors usually hidden from view by vast distances and dust. Close to constellation Centaurus in the southern hemisphere we come across the smallest and brightest constellation in the sky, Crux, the Southern Cross reflected in the Latin cross. Before it moved too far south, it was visible from the Northern Hemisphere, accounting for the mythology surrounding it.

At Winter Solstice on December 21, the Sun stops in Crux's vicinity, appearing to rest in a cave, or grave, according to ancient cultures, to be "dead on the Cross" (Crux). On December 25, Christmas Day, it resurrects, or is reborn, and ascends northward to revive lifeless crops in the spring and ward off starvation, the origin of the concept of "Savior." Horus symbolized this role in Egypt (as Sirius he signaled the beginning of Nile flooding that started the crop growing season). The role of the savior fell to Krishna in India, Mithra in Rome, Ptolemy I Soter in Egypt, and Jesus. Indeed, early portrayals of Jesus showed

Him with His head on the celestial cross with sunrays, the aura, about his head.[23]

Unable to contain ourselves, we tear through open star clusters, like Pleiades in Taurus, furiously birthing stars. The beauty of a time machine is that it fast-forwards the entire ten-million-year birth process for an average star into minutes. Most stars have a companion or two circling them. Here at home, Jupiter would be the Sun's partner were it 50 times larger. Farther on, we see what a dangerous place space can be. Black hole cannibalism is everywhere. Black holes are not really black, only invisible. Stray into one, and you and your soul will disappear.

BLACK HOLES

Could God's home be a black hole, as Thomas Wright conjectured? Black holes are like quicksand in space. Their gravity is such that they haul in stars and nebulae whole. Light is not spared either, which is the reason black holes are invisible. Only jets of energy streaming from the black holes' poles as stars are consumed may give them away. The existence of black holes (dark stars) had been speculated upon since the eighteenth-century. Even though Einstein's equations predicted them, he poo-pooed the idea and was aghast when he heard they could be real. This was typical of Einstein and he spent much of his life disputing his equations' strange predictions, usually losing the arguments. Such misgivings included gravitational waves, which were confirmed in 2016. The first black hole was discovered in 1962 when a group of American scientists studying the sun accidentally detected a bright X-ray source in Scorpius. Black hole discovery accelerated with the launch of the Hubble Space Telescope in 1994. Although incredibly powerful, the monster grumbling at the center of our galaxy is definitely not God. There's too much demolition going on there, and it's irreparable.

Black holes shaped the universe. Most galaxies, like our Milky Way, have a black hole at the center. Black holes form when supergiant stars thirty solar masses or larger die, collapsing under their own weight to almost nothing, making their gravity very strong. Were Earth to collapse similarly, it would measure only two-thirds of an inch across. In six billion years, our sun is fated to die. It'll swell into a red giant and engulf Mercury, Venus, and Earth, then shrink into a burned-out hot white dwarf thirty thousand miles across, before eventually cooling further, and contract to spend the rest of its days as a black dwarf ten miles wide. Long before this happens, though, temperatures will rise up into the 140s Fahrenheit, and in about 400 million years, snuff out life on Earth. Lowering our carbon footprint will have no effect.

Inside the Southern constellation Volan, about 300 million light years from Earth, we come across a galaxy known as the Lindsay-Shapely Ring. The ring, 150,000 light years wide, is a region of furious star formation churning out massive, hot blue stars. Along the ring are regions of pink rarefied clouds of glowing hydrogen gas, bombarded by ultraviolet light from the blue stars. The Cartwheel galaxy 500 million light years away matches this unchecked star birth. *See Illustration 3.1*

Because cutting through the universe takes us back in time, we blow through billions upon billions of old galaxies as our warp drive contraption rips past them like a speed train through a landscape of tall trees. Although we see these galaxies from Earth today, because they are billions of miles away, their light also takes billions of years to reach us. By the time we see it, many of them may no longer exist.

After its launch in 1990 and a billion-dollar repair of a flaw in its mirror, the Hubble telescope beamed pictures back from its orbit four hundred miles above the earth. The pictures' stunning detail shows the visible edge of the universe, 13.8 billion light years away. Since light from the edge is just embarking on its journey here, though, we will never know what the boundary looks like. Anyhow, Hawking said a boundary doesn't exist, absolving us of our ignorance. Will the universe expand forever, end in equilibrium, or implode into a big crunch? Most scientists believe the first scenario – eternal expansion – is the most likely.

THE HOUSES OF GOD

In the document, *The Book of the Watchers,* Enoch arrived in Heaven riding on winds, clouds, lightning, and shooting stars. There, God made Enoch an angel, a Watcher called Metatron, and invited him to sit on His left side. He would act as an intermediary between God and the Watchers (the fallen angels) on Earth. Enoch returned briefly to Earth to instruct his offspring in the knowledge he acquired in heaven and to deliver a message to the Watchers. The Watchers had lusted after earthly women, disclosed to them secrets from heaven, and sired Nephilim, violent giants, by them. God condemned the angels to stay on Earth forever. They pleaded with Enoch to petition God for forgiveness. Instead, God wiped them out, together with the Nephilim, with Noah's flood.

On arriving in Heaven, Enoch swept past a wall made of hailstones surrounded by tongues of fire, and entered a house with walls and a floor made of slabs of snow. On the ceiling there were shooting stars, lightning, and fiery Cherubim with water above them. Fire encircled the walls and the doors blazed with fire. Enoch couldn't stop shaking and trembling with fear.

The winds then carried him into a second, bigger house, whose doors and floor were made of fire. There was lightning above the path of the stars, and the ceiling was on fire. God sat on a throne like crystal with wheels shining like the sun and rivers of fire pouring from underneath. God, or "the Great in Glory," shone brighter than the sun and his clothes were whiter than the snow. He was surrounded by a sea of fire and attended by ten thousand times ten thousand angels. Neither Enoch nor anyone else could enter or see God because of, "His magnificence and glory."[24]

Ezekiel and Daniel described Heaven similarly. We wanted badly to see His magnificence for ourselves. But, no such luck. Despite our advanced technology that enabled us to see up to the edge of the universe, we couldn't find God, let alone Heaven. Of Enoch, there was no sign. The reason could be that the universe is incomprehensibly vast. But a place as bright as the sun and apparently within Enoch's human reach—since he even went there twice—should not be so hard for us to find with our warp drive.

As the laws of physics break down at the first instant of "creation," the Big Bang, we could not go any farther back in time to look for the cause or what existed before our universe. Star-dazed and beat, we miss home. The beauty of a time machine is we can return instantly, with no one the wiser that we had been away, far away.

BELIEVE IT OR ELSE

To look deep into the heavens, scientists use powerful aids like the Hubble Space Telescope. In 1929, using the 100-inch telescope at Mount Wilson Observatory, astronomer Hubble observed that the universe was expanding. Despite his equations suggesting space was dynamic, Einstein had rejected the idea. The notion of a universe that had a mind of its own cast doubt on the role of a Designer and contradicted prevailing thinking. To correct this inconsistency, Einstein introduced a *cosmological constant,* a fudge that he later regretted and withdrew in 1932 following Hubble's discovery, calling it his worst blunder. It now seems that he withdrew it too hastily. The constant could explain why the universe was not collapsing under its own gravity or flying apart, and could account for the invisible 95 percent of the universe, dark matter, and dark energy.

Einstein's doubts show how devout scientists unwittingly may put God's role in the universe into question. When he withdrew the cosmological constant, Einstein must have disappointed those who took for granted that the universe was static and stationary. Even though Copernicus didn't publish his findings challenging the long-held belief that the Earth was stationary, Martin Luther still tore into him simply for having the audacity to think the thought in the first place. Nothing provoked the ire of the faithful like such sacrilege. In our quest for the truth, though, this is but one of the heresies we'll encounter.

Although we've been to the farthest nooks and crannies of heaven, marveling at our past as recorded in the planets, stars, and galaxies, we have not come across the Architect. Planets, stars, and galaxies are not gods. We also couldn't go beyond the instant of creation; the laws of nature don't allow it. So, now, since Enoch was promoted to angel status, he wouldn't be bothered to show us all that he saw in Heaven.

Understanding the universe is key to finding the nature of our creator and making sense of life. Science looks for the *how* but not the *why*. Physical laws, as first defined by Isaac Newton, throw light on the *how*. Even then, people like physicist-philosopher Ernst Mach, he of the speed of sound, argue such laws are only estimates made from many observations to explain the behavior of the universe, and they aren't absolute; they are expressions of human observations, which are illusory.[25] Nevertheless, they help us to navigate an unpredictable world, or to explore space. To those who believe that creation and the existence of the universe was a fluke, the question of *why* – the purpose for its creation – is moot.

Physicists describe the basic building blocks of matter as strings or branes. Capra likened such particles to the Hindu dance of life by Vishnu (the second member of the Hindu god triad). Both were in flux with energy. Further, Bell's theorem showed that all things in the universe were connected, dependent, and inseparable. As mystics in the Far East maintained, life was united in a cosmic web and was inseparable from the universe.

Elegant and breathtaking as the universe is, it's also a head scratcher, as it's difficult – perhaps impossible – to explain why and how life arose on one lone planet only, as far as we know. Clearly, for life to emerge, certain conditions such as water and the right temperature, had to be present. Could the spark of life be divine? We lack evidence for this scenario. Stories invoking the influence of various invented creators, sometimes called "Gods of the gaps," don't help.

After the terrestrial planets formed, they were hot and dry. Then on one of them, carbonaceous meteorites containing water plummeted, oceans formed, and life arose. Before we investigate how life emerged, however, we need to define it, so we know what to look for. Yes, Darwin, we listened and have an idea of the nature of matter. But not life. What is life?

HENRY KAKEMBO, M.D.

CHAPTER 4

LIFE

It is not the strongest of the species that survive, nor the most intelligent, but the one most responsive to change.

~Author uncertain[1]

After Earth formed 4.6 billion years ago, life followed about a billion years later. Where did life come from? Was it divinely created? What is life? Most people are confident they can tell the living from the nonliving. Suppose you're on life support, the brain silent, but the heart beating, the eyelids moving, and a muscle here and there twitching; are you dead or alive? What about the soul in this twilight zone? Is it hovering in the rafters, preparing to bolt when you succumb? Then the car battery goes *kaput.* Gives up the ghost. The Bible says life is blood. To the Greeks, life was breath, or pneuma. And according to the pop group 'N SYNC, "No music, no life." Clearly, the answer is not so simple. But the question is worth exploring, as it could lead us to the maker.

WHAT IS LIFE?

Since understanding the nature of life is key to making sense of our existence, a working definition is crucial. Life conjures vitality in our minds. The ancients believed it began when a vital divine force, such as a soul, animated matter. Aristotle, seeing that plants breathed, afforded them souls also. Buddhists and Plato agreed. In Plato's "Timaeus," a dialog about the nature of the world and us, the universe teemed with souls. Christians ruled such thinking blasphemous and backward, drawing the line at only people harboring souls. Otherwise, there would be no shortage of things with souls, even bricks, then, walls could divulge inconvenient truths.

With time, more forms of life came to light. In 1665, using a crude light microscope with a magnification of 30 times, British physicist Robert Hooke described the structure of cork, which resembled cells in early Egyptian monasteries, thus the name for the unit of life. In 1683, a secretive Dutch linen draper, Antoni van Leeuwenhoek, surprised everyone. Because of his hobby collecting and examining small objects, he became adept at making magnifying glasses. He developed powerful glasses achieving a magnification of up to 275 times, but he wouldn't reveal how he made them. With the increased magnification, he could see bacteria in saliva, one-celled creatures in pond water, and sperm swimming in semen, showing that far more life existed than people knew.

Although van Leeuwenhoek magnified objects 275 times, he could not clearly see structures inside a cell. Today's advanced microscopes do. The arsenal includes microscopes that use ultraviolet light rays, quantum scanning tunneling, X-rays, and electrons. Electron microscopes achieve magnifications of up to 100,000 times, resolving objects up to a billionth of an inch, showing details of structures inside cells and viruses.

Armed with the new microscope capabilities, biologists identified certain qualities life ought to have. Many of these, however, are shared by things that are clearly inanimate. For instance, to be alive, it seems, an entity must be able to metabolize, grow, and multiply. But what about fire? Even the ancient Egyptians thought fire had life. Back to the drawing board. A widely cited definition was proposed by a member serving on a NASA Exobiology panel, Gerald Joyce. Joyce's definition stated:

"Life is a self-sustained chemical system capable of undergoing Darwinian evolution."[2]

In his book *Origins,* astrophysicist Neil deGrasse Tyson, wary of committing himself to definitive criteria, proposed that life should consist of sets of objects that can both reproduce and evolve. The objects must also evolve into new forms as time passes. Tyson had an eye on discovering extraterrestrial life, so the definition couldn't be too restrictive.[3]

Otherwise, physicists consider life a quantum event, rejecting the theory that a soul must be present for something to be considered alive. By conforming to the second law of thermodynamics, life is an anomaly: the universe in losing heat decays inexorably, increasing in disorder, called entropy. Life, then, is matter temporarily organized by energy, a byproduct of a blazing sun, and – according to Carl Sagan – a localized region that increases in order through cycles driven by energy flow.[4] Unable to come up with a common definitive list, scientists generally agree that every living thing is made of cells and that all cells arise from the division of previous cells.

These definitions leave viruses in limbo. Although smaller than cells, viruses are made of similar building blocks to ours. Viruses are a strange bunch, though. They hijack cells—even dead ones—to reproduce, repair dead cells, and reassemble when destroyed. So, are viruses living organisms? If not, why do we speak of trying to kill them? Their status hinges on our definition of life. But whatever they are, for the sake of our health, we dare not ignore them.

Prions, culprits of Mad Cow Disease and Creutzfeldt–Jakob disease in humans, throw yet another wrench into the definition of life. A hundred times smaller than viruses, prions are protein particles without the genetic material essential for life. Yet they seem to replicate and cause disease. We inactivate them by exposing them to protein degrading treatments.

Most people believe the way living things appear today is the same way they have always been. But Aristotle, who shared this belief, also taught that low life could form spontaneously, such that small fish in muddy ponds could arise from the mud itself. This theory stood for nearly 2,000 years before another bigwig, French microbiology founder Louis Pasteur (he of sterilized milk), overturned it. In a series of remarkable experiments, he showed that maggots did not arise from rotting meat. Aristotelian diehards didn't trust him then, and many still don't today, preferring their milk raw.

Pasteur also suggested that every living cell came from a previous cell. Going back into the deep past, then, there should be the grandmother of all cells, our *Last Universal Common Ancestor* (LUCA). This came as no surprise to observant people. Since ancient times, people had noticed similarities between humans and other animals suggesting a common origin; some even defiantly hinted at evolution.

So, evidence for evolution had been out there for centuries, but many people who chose to see only what they wanted to see simply overlooked it. For instance, in the 500s BCE, Greek philosopher Anaximander, regarded as the first evolutionist, proposed that the Earth initially existed in a liquid state. Then humans evolved from fishlike beings in the sea and stayed there until they developed sufficiently to survive on land. In the 400s BCE, Anaximander's countryman, Empedocles, suggested humans and animals came from randomly joined body parts. Structures that successfully reproduced multiplied, and those that couldn't reproduce shriveled and died out. For some two thousand years, however, religion quashed all such "loose" talk, impeding progress. Then came the nineteenth century and one Charles Darwin discreetly provided a framework on which such information could be meaningfully organized, opening new ways of understanding life, and, in the meantime, shaking the foundations of Christian belief to the core.

DARWIN'S SACRILEGE

By formulating the theory of evolution, Charles Darwin joined the ranks of the most eminent scientists arguably ever. Like similar luminaries, such as Galileo, Newton, and Einstein, Darwin transformed our view of the universe. By launching evolutionary science, he clarified much about biology, but by contradicting the biblical creation story, he kicked up a storm that has yet to abate. Suggesting that humans and apes share a common ancestor prompted accusations that Darwin was bundling us with monkeys. "This was not how Adam and Eve came about," people protested. Or, "The Bible is the word of God and all true. Contradicting it is blasphemy." For others, however, monkeys and apes were a type of human; Adam's degenerate children, they said. Not so in Tibet, however. Tibetans believed humans came from six monkeys born of rock. Indeed, apes evolved from monkeys, as we see later.

Because of the controversy, Darwin shunned debates. Many challengers didn't do their homework and were ignorant about the topic. Debating them, then, was futile and exasperating. Out of spite, some, like anatomist Sir Richard Owen, claimed to know better. Owen even prepared eloquent Bishop Wilberforce of Oxford for a debate, "Evolution versus Creationism," in 1860. Despite the help, "Soapy Sam," as his eminence was fondly known, made an ass of himself. Addressing Thomas Huxley, dubbed "Darwin's bulldog," the bishop wondered whether it was through Huxley's grandmother or grandfather that he descended from a monkey. Huxley, smelling blood, stiffened, then pounced. He advised the bishop to mind his own business and to stop meddling in scientific matters in which his competence was questionable. But the bishop confirmed the adage that the least informed often make the loudest noises. Unfortunately, the louder the noise, the more people are likely to believe whatever is being shouted.

Darwin was born in 1809 in Shrewsbury, Shropshire, England, to a wealthy family. His father, Robert Darwin, was a prosperous country physician and his grandfather, Erasmus Darwin, was a well-known physician, biologist and poet. Erasmus believed in biological change and likely contributed to his grandson's acceptance and formulation of the concepts of natural selection. Rounding off the impressive family list was his maternal grandfather, Josiah Wedgwood, a famous wealthy potter and royal supplier of dinnerware to fine stores like Harrods. Small wonder Darwin didn't have to work for a living. Wedgwood played a key role later when Darwin's father refused to finance his "good for nothing" son's historic expedition.

From childhood, Darwin showed a keen interest in natural history, collecting rocks, insects, and bird's eggs instead of doing his schoolwork. His father was disgusted. So, he bundled him off to medical school in Edinburgh, Scotland. Young Darwin showed little interest there, too, preferring to dissect marine specimens collected from dredging expeditions. Attending an operation performed without anesthesia reportedly sickened him and turned him off medicine for good. He went back to collecting biological specimens, befriending a black taxidermist, who taught him skinning and stuffing birds, a skill that later came in handy during his global wanderings. *See Illustration 4.1*

Darwin's medical career in shambles, he moved to Cambridge to study for the cloth and become a country pastor. His father was outraged and called him a disgrace to the family, as all Charles did daylong was collect insects and catch rats. Unfazed, Darwin continued hunting small animals and hauling in plants. He befriended a botany tutor, John Henslow, who helped him land a job as a naturalist by recommending him to the captain of HMS Beagle, Robert FitzRoy, who was going on a scientific expedition around the world. The position was unpaid. When Darwin's father refused to sponsor him, his grandfather, Wedgwood, stepped in and saved the day.

A naturalist, Robert McCormick, also the ship's surgeon, was already on board the ship. But such journeys were long and lonesome and FitzRoy yearned for gentlemanly company since captains did not mingle with subordinates in those days. The last ship's captain had shot himself out of boredom, which did not bode well for FitzRoy later when he learned what Darwin's tireless mind had been up to. After getting over the shape of Darwin's nose, which for some reason he thought betrayed indecision and laziness, FitzRoy reluctantly hired him. They set sail in 1831 on a five-year expedition. Darwin was twenty-two years old and FitzRoy twenty-six.

En route to South America, the ship docked in the Cape Verde Islands off the coast of Africa. Darwin observed the birds there, saving the information for later. Once in South America, he went on long collecting trips inland, having a literal field day among the many species biblical Adam never got a chance to see, let alone name. The naturalist became frustrated, as only Darwin had the captain's ear. He quit and returned to England. Darwin may have despised his predecessor, but he was more troubled by FitzRoy's political and religious views. FitzRoy was a Tory (conservative), an avowed Creationist, and believed slavery was natural. In contrast, Darwin was a liberal and flexible. Darwin was so open-minded that FitzRoy, in one of his temper tantrums, once refused to dine with him.

As Darwin sorted his collections on the journey home, a pattern emerged. He noticed a strong resemblance between fossils of extinct armadillos and skeletons of living species in South America, suggesting the extinct animals were ancestors of those living. He also observed that other species differed from their seemingly related geographical neighbors. For example, ostriches that had been long separated by geographical barriers had diverged and could no longer interbreed. This resulted in a new species, the Rhea.

The most remarkable findings came from the Galapagos Islands off the coast of South America. Birds in Cape Verde were similar to those on the mainland. Not so in the Galapagos. Thanks to long periods of physical separation, they differed from their South American neighbors. And then, despite similar climate, tortoises, lizards, and birds differed from one island to the next. The differing species of finches displayed such variations to a T.

As the 700,000-to-four million years old volcanic Galapagos Islands surfaced in the Pacific Ocean, they were settled by finches from the much older mainland. Water expanse prevented the birds from interbreeding, and after a million years or so of isolation, a close relationship between them was no longer recognizable. They had evolved into fourteen species with different shapes, diets and beak sizes, ceasing to be of "the same kind." For instance, beaks were adapted to eating seeds, insects, leaves, ticks, lice or, like that of the "vampire" finch, sucking blood from other birds. These variations effectively counter the argument from intelligent design proponents that God created all species fixed in their present forms.

Once back in England, Darwin wrote about similarities in animal behavior in his notebooks. For instance, people yawned, just like horses and dogs did. Also, some features of fossils from different species were strikingly similar, suggesting origins from a common ancestor. What Darwin lacked – the missing puzzle piece – was the mechanism that had led species to multiply and diversify into the millions today.

Coming across British economist Reverend Thomas Malthus's 1789 *Essay on Principles of Population* opened his eyes. Malthus touted a relationship between food supply and human population, whose growth, if left unchecked, could lead to food shortages and misery.

Darwin realized this was applicable in the wild, too, as species competed for limited food supplies to survive. Evolution then had no other purpose than survival of the individual. According to Richard Dawkins in the *Selfish Gene,* we strove only to increase representation of our genes into future generations, nothing more, giving support to the theory of evolution by natural selection.[5] But if we got here through competition, how are we created equal?

Referring to Pasteur, way back when, we all came from the same cell, LUCA.

Darwin presented his theory of natural selection with tons of evidence. For instance, farmers cultivated strawberries and breeders raised pigeons with desirable traits, increasing their yields and numbers, while those rejected petered out.

Darwin avoided using the term evolution, because in his time it connoted the process by which tiny humans in sperm, "homunculi," developed into embryos. Following van Leeuwenhoek's findings, interest in microscopy spiked. A Dutch enthusiast, Nicolaas Hartsoeker, announced that he saw the tiny humans in sperm, theorizing that God created all homunculi needed in the testes of Adam. People believed him.

Evolution also implied progress from simple to complex. In 1862, Victorian sage Herbert Spencer coined the phrase, *Survival of the Fittest* to describe Darwin's theory. But Darwin did not see it that way, as modification (the term he preferred) without improvement could also occur. So, at times, a simpler species was better suited to prevailing conditions and survived. For instance, our shrew-sized ancestors shrugged off a catastrophe that wiped out the dinosaurs and lived, when an asteroid ploughed into Earth sixty-five million years ago.

Darwin published his theory in the book, *On the Origin of Species,* in 1859, twenty years after he formulated it. He did it reluctantly, perhaps afraid of upsetting his devout wife and friends. Also, anxiety attacks may have immobilized him, delaying writing and publication. Rather than take the risk, he made provisions for publication after his death. His wishes might have prevailed had another naturalist not turned up. Half a world away, Alfred Wallace independently arrived at the same theory. A Scottish gardener, Patrick Matthew, however, had hinted at the theory in the appendix of an obscure book on naval timber earlier in 1831 and it was ignored.

Wallace collected specimens for museums and zoos in England, amassing an impressive array of 125,000 mammals, insects, reptiles, and shells. Although the species came from widely differing regions, many resembled those separated by geographical barriers, such as rivers. The similarities suggested that differing local environmental pressures had forced the animals to change and adapt. Wallace hit on natural selection for the driving force behind the change. He may have seen similar divergent changes on an earlier expedition to the Amazon, but that collection was lost when his boat caught fire and sank. Wallace survived only because a passing ship rescued him before the boat went down.

Wallace wrote Darwin, asking his opinion about a paper. Darwin was stunned. Someone had scooped him. To ensure Darwin would get credit for his own work, scientist friends presented his extracts, along with Wallace's paper, to the Linnaean Society in London in 1858. The society was not impressed. A year later, however, all hell broke loose when the book was published. It sold out in one day.

Although Wallace went on to publish lectures, head organizations, and earn medals of merit, he is not remembered like Darwin is. Wallace believed staunchly in evolution, but thought intelligence was supernatural. Eventually, he got religion. Remember Captain FitzRoy? He was so enraged by Darwin, he marched up and down waving the Bible during the Oxford evolution debate. In part blaming himself for

Darwin's blasphemy, he slit his own throat with a razor and was no more.

Evolution has been a hot topic ever since the publication of Darwin's research. For many, the theory not only contradicted God's word, but also implied an amoral world in which some species ate their children just to survive and others made short work of their mating partners to assure offspring sustenance, such as the female redback spider, and the praying mantis.

As nature impels us to fulfill our evolutionary duty by ensuring that our genes survive, Darwin sired ten children. This was in stark contrast to his devout critics who considered sex a sin. To St. Augustine, "Woman's only function was the childbearing which passed the contagion of *Original Sin* to the next generation, like a venereal disease."[6] But a sin it should not be, due to extenuating circumstances: the snake. The deception with the apple was a trick of the snake to grab immortality from humans. Since then, the snake has renewed itself by shedding its skin, and humans have died.

Still, some Christians ruled sex and God incompatible. Martin Luther was convulsed by a loathing and horror of sexuality and hated women. He regarded a woman as a temptress, a destroyer of men. Luther, as well as John Calvin, the Swiss reformer, did not think anyone could contribute to their own salvation anyway, not even by abstaining from sex. Nonetheless, some people swore off it, leading celibate lives in monasteries. Ironically, if it hadn't been for their alleged sin, Adam and Eve would have been the only people on Earth.

Criticism on his theory of how hereditary information passed from parent to offspring troubled Darwin. Natural selection could not be achieved by simply mixing characteristics, as species would change in the first generation. Although fossil records showed a new species could appear suddenly after very long periods of stability, Darwin wrote, "it is far more probable that each form remains for long periods unaltered, and then again undergoes modification." Keep in mind, geological findings can be difficult to interpret. Species migrate constantly and may die far from their point of origin. But rarely is new life an exact copy of the old. Were variations advantageous, nature would select them over harmful ones. As Bill Bryson put it in *A Short History of Nearly Everything*, life evolved to suit Earth's conditions. If the conditions changed and we didn't, for example by failing to adapt to global warming, we could be out of here fast.

A question troubling Darwin was the mechanism for evolution. Little did he know that the answer lay right under his nose. In 1868, a monk, Gregor Mendel, published it in an obscure journal, and a German botanist, Wilhelm Focke, reported Mendel's work in a book, a copy of which found its way to Darwin's library. The relevant page in Darwin's copy, however, was uncut – books produced in Darwin's time were generally published with pages uncut, and a reader would have to cut them open first.[7] Mendel reportedly read Darwin's book but overlooked sending him a copy of his work with peas, for which Darwin likely would have given an arm and leg. For thirty more years, Mendel's results languished in obscurity until scientists working independently came to the same conclusion in 1900. After long experiments with peas, Mendel had found the "atoms" of inheritance, or genes. Genes were the mechanism that passed on information from one generation to the next.

GREGOR MENDEL'S IDEA

In the same way that archeologists digging for evidence in support of the Bible story ended up contradicting it instead, Mendel was a man of the cloth like Copernicus, Darwin and Lemaître (he of the Big Bang theory) before him, men whose discoveries inconveniently poked holes in their faith. Other luminaries like astronomer Johannes Kepler and Einstein scrambled to carry out damage control. Too late. The genie was out of the bottle. For instance, Kepler reversed himself after finding that the supposedly perfect circular planetary orbits around the Sun (previously ascribed to God) were elliptical. Einstein also thought he could mathematically keep the universe from expanding, which no one could do. He also died denying an unpredictable world controlled by fuzzy muzzy quantum mechanics.

GENES

Gregor Mendel was born in 1822 into a family of gardeners in Czechoslovakia. His mother and her close relatives were gardeners, and his father was a farmer. Not having much money for education, he was fortunate to study in the Augustinian Monastery at Brno that encouraged experimental gardening. For over eight years he doted on pea plants, performing more than 29,000 experiments. when he cross-pollinated tall pea plants, he obtained tall plants, and when he cross-pollinated short pea plants, they produced short offspring. Then when he cross-pollinated the tall plants with the short plants, he obtained only tall plants. Befuddled, he cross-pollinated these tall plants and, to his surprise, grew three times as many tall plants as short plants.

From these results, Mendel concluded there were atoms of inheritance, which we call genes today. During cross-pollination, an offspring received a gene from each parent. So, a tall plant with the dominant **T** gene was always expressed and produced tall plants. A short plant carrying the recessive **s** gene was expressed only when paired with another carrying an **s** gene.

Even before they were aware of genes, farmers had always used such genetic engineering methods to improve yields, grow sweeter corn, and breed better sheepdogs. We, too, could theoretically likewise choose what our offspring should look like, the so-called designer babies. For instance, we could select brown eyes, dominant and always expressed. Blue eyes result from a mutated recessive gene, OCA2, that causes lack of the brown pigment, melanin, as in Chapter 7. Rarely, blue eyes occur in black populations, too, without a skin color change. Height is another matter. Unlike peas, height in humans is controlled by many genes.

Mendel presented his work at meetings in Brno, and published it in 1866 in an obscure journal where it went largely unnoticed. Likely bored, tired, or both, Mendel turned to official duties as a monk and passed away quietly in 1884 without knowing how the cell stored and passed on the genetic information.

DNA

With the dawning of high-powered microscopes that revealed cell structure details, scientists could observe reproduction. For instance, when a grasshopper cell nucleus divided, chromosomes paired up in a row in the center and then split down the middle. Each half migrated to the opposite pole of the cell where it was walled off, making two cells. In sexual reproduction, the chromosomes split again to make four cells, each with half the usual number of chromosomes. The cells became sperms in males and eggs in females.

A gene is a unit of heredity information occupying a fixed position on a chromosome. In 1944, a Canadian, Oswald Avery, isolated genetically active material from bacteria, showing genes were made of pure DNA, or deoxyribonucleic acid.

The 1950s saw the DNA structure unveiled. It was crystallized and X-rayed by Rosalind Franklin of Kings College, London. But as she was reluctant to share her research, James Watson and Francis Crick working at Cambridge University sneaked a peek. They knew a good thing when they saw one and a year later in 1956, published the structure of the elusive DNA molecule, a double helix strand resembling a long-twisted ladder, or spiral staircase. Stretched out, a strand could easily reach six feet, showing how vast amounts of information could fit in such a tiny space. The rungs of the ladder consisted of bonded pairs of four types of amino acids or bases. The arrangement of the bases (sequence) along a strand of DNA determined the genetic code that ultimately produced a product such as a protein or enzyme. The older cousin of DNA, RNA, or ribonucleic acid, found in some viruses, also carried genetic information, but being single-stranded, it was rather unstable and easily damaged. Watson and Crick, along with Maurice Wilson, a colleague of Franklin's, shared the 1962 Nobel Prize in chemistry. Franklin had died four years earlier of ovarian cancer and could not be named.

Cracking "the secret of life" confirmed that all life was made of the same building blocks. Unravelling the secret, however, is one thing. Solving the puzzle of its origins, another. Since life's origin has stumped scientists and non-scientists alike, maybe weird, spooky quantum mechanics can help. The driving force behind modern biology, chemistry, and physics, for instance, quantum mechanics has helped to resolve many knotty problems and we don't have to invoke divine intervention to account for things yet unexplained.

CHAPTER 5

QUANTUM GENESIS

Nothing in biology makes sense, except in the light of evolution.
~Theodosius Dobzhansky (1973)

As a child, whenever a new robot was announced, I felt that scientists were close to finding out all they needed to know about the human body so they could build one. It never occurred to me that making the robot self-sustaining would be a challenge. The smart teams would only have to throw a switch to animate it, I assumed.

The ancients didn't have the technology to investigate the mystery of life as we do today. Instead, they expressed the mystery in myths they knew were not "factual accounts of reality that not even the gods could explain adequately," according to Karen Armstrong.[1] Such creation myths abound, for instance, the story of the whale hunters of Kukulik Island, where a creator raven fished pebbles out of the Bering Sea and started life. Or that of the Yoruba in Nigeria and diaspora in Cuba and Brazil, where the supreme god, Olodumare, created itself and breathed life into people after Olorun, its first manifestation, and Olofi, its second manifestation, molded them from clay. And, from ancient Egypt, the god, Thoth (also known as "the word that cometh"), brought into being whatever he named. None of these stories, however, stand up to scientific scrutiny today. To debunk them convincingly, though, it's essential to do so rationally and with evidence showing that they are just stories not based on fact.

Life's emergence couldn't have been as simple as mythology would have us believe. Life would have begun as chemical evolution, which then proceeded to biological evolution. The process of evolving from inanimate matter to living organisms must have taken zillions of steps. In this chapter, we figure out how organic molecules could have formed and became self-sustaining, leading to life.

LIFE FROM A FORMLESS MESS

It is oddly coincidental that several ancient societies intuitively believed the world was created from a void or a formless mess of elements. When it came to life, though, their speculations knew no bounds. Today, we know that for life to emerge, it required carbon, hydrogen, oxygen, and nitrogen, some of the most abundant elements in the universe, plus small quantities of phosphorous and sulphur. These six elements formed the bulk of carbohydrates, lipids, nucleic acids, amino acids, and other organic substances. Such ingredients laced giant molecular clouds, dusted Jovian planetary moons, and hitchhiked on comets and meteorites, as detected by radio spectrometers. In 1969, about ninety-two amino acids were found in a 4.6-billion-year-old 100 Kg meteorite that fell at Murchison in Australia. All the evidence available thus far suggests that when Earth formed, everything life needed to start was here from the get-go. And as soon as conditions became favorable, life was up and running in no time.

In its early days, 4.6 billion years ago, Earth was bombarded incessantly by comets, meteorites, and planets. After a body the size of Mars (dubbed Theia after the Greek goddess who gave birth to the moon) most likely plowed into it, debris ejected into space joined to form the "lesser light," or moon within a year. At the time, Earth's surface temperature was a sterilizing 1800 degrees centigrade. No way could life start then. The temperature had to drop, the heavy meteoric bombardment had to stop, volcanic activity had to abate, and meteoric chondrites had to plummet from space, bringing water. Then, about 3.5 billion years ago, life began to stir. Supporting this timeline are chemical traces found in the earliest known ancient rocks from Greenland.[2] Bacteria began to fossilize about 300 million years later. Clearly, under the right conditions, the right distance from the sun, or the so-called *habitable zone*, the emergence of life was a given.

Today, imitating steps originally credited to the maker, scientists make building blocks of life in the laboratory. Acting on Russian biochemist Alexander Oparin's suggestion that life arose from chemicals and water, they have produced amino acids from scratch. In 1928, a German chemist, Friedrich Wöhler, made urea and excitedly congratulated himself for "creating" urine. In 1953, Stanley Miller and Harold Urey from the University of Chicago stunned the world. Mimicking what they believed were early Earth conditions, they subjected methane (natural gas), ammonia, and hydrogen to electric sparks (simulating lightning) in a water flask. After a day or so, the solution turned yellow with amino acids, thirteen of them essential for life and some similar to those found in the Murchison meteorite. Every element that makes up amino acids used by living organisms, except

Sulphur, was present. It turns out, however, that the early atmosphere contained mainly carbon dioxide and nitrogen, which could not react together. Also, oxygen was excluded as it would have prevented the experiment from running.[3]

A flurry of experiments produced more building blocks of life. Agitating any mixture seemed to produce complex biological compounds. American biologist Craig Venter "created" a virus that multiplied inside a bacterium, synthesized DNA, and produced a new artificial strain. Plant and animal cloning are now old news. It would seem scientists are closing in on creating humans, bypassing God.

But, since life started so close to Earth's beginning, we cannot rule out an outside source. Building blocks forming in the frigid vacuum of interstellar space could have hitchhiked to Earth on comets and meteorites. Once Earth was seeded, LUCA would have emerged, branching out into all the species today. Regardless of where life started, though, the steps involved should be the same. So, we might as well assume it began here and figure out how that happened. The process is not as straightforward as stories about our origins are, and may require donning your thinking cap.

According to John Gribbin, a British science writer and an astrophysicist, life is an obligatory property of matter.[4] As such, it would have emerged on its own, organized and capable of growing, multiplying, and progressing Darwinian style. As a chemical process, then, life would not be special or divine.

In the beginning, there was LUCA, our progenitor, probably bacterium-like with 265 to 350 genes, the minimum required for life. Maybe to assuage the religious faithful, at the end of his second book, *The Descent of Man*, Darwin suggested that the creator breathed life into such an ancestor. Privately though, in a letter to a colleague, Joseph Dalton Hooker, a botanist, Darwin imagined protein compounds forming in a warm little pond in the presence of all sorts of ammonia, phosphorous salts, light, heat, electricity, and so on. He theorized that the compounds underwent further complex change leading to life. Miller and Urey tested the idea and found it viable. Today, however, it seems that although Earth's early atmosphere was flush with carbon dioxide and nitrogen, it had little methane, resulting in mixtures that would have failed to produce amino acids. Darwin's warm little pond, then, was likely more of a deep-sea hydrothermal event.

Even after the unrelenting meteorite pummeling slowed, our planet was still uninhabitable and frozen, the sun then being half as hot as it is today. The moon was also closer to Earth, so tides rose 1000 feet. The shorter distance between the moon and us also made the planet spin faster, and a day lasted only six hours. Oxygen, then scarce, was poisonous to bacteria (there's a reason it's a good idea to let fresh air into that musty bedroom). Oxygen is still harmful to organisms today; it damages our cells and accelerates aging.

There was no ozone layer shielding the planet from deadly ultraviolet and cosmic rays either. Life couldn't have survived such harsh conditions out in the open. Most likely, it formed elsewhere, in a protected environment.

One possible sanctuary would have been around hot volcanic vents, called "black smokers," in the deep sea. Two miles below the sea surface from the ocean floor billows smoke rich in chemical nutrients iron sulfide and hydrogen sulfide gas. The two chemicals react to produce a mineral called pyrite'(fool's gold), together with hydrogen and energy.

Despite extreme deep-sea conditions - 660 degrees Fahrenheit temperatures, crushing 2000 atmospheric water pressures, and total darkness - life couldn't care less. In 1977, a submersible, Alvin, found the most primitive and ancient microbes, *Archaea,* in the surrounding tepid water of the Galapagos Islands. On the evolution tree, Archaea are genetically closest to LUCA. Organic molecules forming at the sea bottom would have been safe from destructive surface radiation, but high temperatures and pressures there would quickly destroy amino acids and proteins. How could they survive?

ROCKS TO THE RESCUE

In 1998, Robert Hazen and Jay Brandes, working at Washington's Geophysical Laboratory Carnegie Institution, showed amino acids could stay intact for days in the presence of an iron-sulfur mineral, pyrrhotite (Greek *pyrrhos*, "flame-colored") commonly found with pyrite around deep-sea hot springs.[5] These minerals could have protected the amino acids as we see below.

Although primitive organisms abound around the chemical billowing, nutritious deep-sea vents, they also inhabit rocks in the Earth's crust. Those rock inhabitants, aptly called thermophiles, thrive in piping hot geysers such as those found in Yellowstone National Park. Life could have started anywhere there was water.

Whether they originated in outer space, in a little warm pond, or in hot deep-sea vents, how could molecules assemble, grow and self-replicate without divine intervention? In an *RNA World first* scenario, proposed in 1962 by American Biologist Alexander Rich, a gene coding for a molecule that survived as RNA, or had enzyme-like properties, and was capable of facilitating self-copying reactions, developed from a primordial soup of elements abundant in the early seas.[6] In a competing *metabolism first* scenario championed by German Günter Wächtershäuser, self-replicating molecules developed on the ocean floor and multiplied; RNA came later.[7]

Early RNA probably carried the most effective blueprint for reproduction. But in the April 2000 issue of *Proceedings of the National Academy of Science,* Doron Lancet suggested that lipids might also have played a manufacturing role. Minerals in rocks such as pyrite could facilitate such reactions. Within its protective maze of tiny pores, pyrite could have helped the formation of proteins and oil droplets (lipids) while at the same time protecting them. Like soap bubbles, oil droplets assemble their own membranes forming *micelles* (a sphere of such droplets or molecules). And although the odds of forming by chance are considerable, soap bubbles are common. Likewise, oil droplets could have been common and could have trapped RNA molecules or proteins to form the cells.[8]

Hold your horses! According to Johnjoe McFadden, Professor of Molecular Genetics at the University of Surrey, England, the odds of a self-replicating molecule evolving from a possible primordial soup, or "gunk," were about as likely as those of the soup produced in the laboratory, practically nil. As Wächtershäuser pointed out, such a soup would contain too many compounds.

Worse, the soup would have as many (left) L-handed as (right) D-handed amino acids. (Nature uses mostly one form of the two mirror images of amino acids, L over D, called chirality, supporting a common ancestor). McFadden suggested instead that the directed motion of particles drove the emergence of living cells according to quantum mechanics laws. McFadden believed a few amino acids trapped inside tiny structures, such as the pyrite rock pores discussed above, oil droplets, or protein droplets, joined, forming short chains called peptides and thus creating a proto-cell (one or more peptides link together into long chains to form polypeptides or the proteins we are familiar with). One of these peptides or a chain could have emerged as a self-replicator.

David Lee and colleagues designed such a short self-replicating peptide, thirty-two amino acids long, that, acting as an enzyme, joined two bits of itself together.[9] To McFadden, however, it was hard to imagine how a replicator could have emerged randomly and given rise to a proto-cell, the ancestor of life, on the early Earth. He found that the chances of a peptide coming out from the cell as a self-replicator were next to none. One way out of this conundrum was if – by pure chance – a self-replicator's properties were compatible with the emergence of life, called the *anthropic principle*. It's not a hypothetical occurrence; as a rule, belief, or method, the principal was used by astronomer Fred Hoyle to predict the structure of the carbon atom.[10] But McFadden was disappointed with this scenario, as it would lead to only our kind of life and intelligence in the universe, leaving us all alone.

Because of the huge improbability of the first self-replicator, McFadden turned to quantum mechanical laws and invoked the multiverse or *Many Worlds* concept. The idea of such *multiverses* was first proposed by wizard mathematician Hugh Everett in the mid-1950s. Everett was so ridiculed for it that he quit theoretical physics to work for the department of defense. Nevertheless, proponents of the *string theory* jumped for joy at his suggestion, putting the number of such universes at 10^{500}, or infinite.

Now, say each peptide occupied its own universe, one of the *Many Worlds* or *anthropic multiverses*. Then, our universe alone happened to contain the self-replicator and life materialized.[11] Darwin probably envisaged this step when he suggested that the creator breathed life into our common ancestor. But although alone in our universe, I don't see why life shouldn't form in the other universes the same way, perhaps surely looking different from us. So, McFadden needn't lose heart. Going forward to form organisms, "the self-replicators must have at some stage captured lipid membranes, peptides or nucleic acids," McFadden wrote, "These would have protected the self-replicators and helped in their replication. Eventually, the first living cell would have emerged."[12]

The jury is still out on the origin of life. For a definitive answer, scientists might have to find the *Theory of Everything* and explain dark matter and dark energy first. But definitely no evidence here supports a supernatural origin of life. Going forward, DNA, being more stable and better at resisting heat and radiation, would have come easily from RNA. Because it was double stranded, DNA was better at preserving the reproduction blueprint and seized the information role. Other cells likely emerged with LUCA and swapped genes. But LUCA turned on its unsuspecting siblings and had them for lunch. Then from there, it multiplied by leaps and bounds.

By eating and incorporating other organisms, the descendants of LUCA grew bigger and more complex. They developed specialized structures as evolution shifted into high gear and shaped them. The sculpting, though, was not as straightforward as you might imagine. It probably involved a process similar to the directed emergence of living cells, the very controversial—or speculative—proposal of adaptive evolution through biofeedback.[13]

As we have seen, life's building blocks: amino acids and nucleic acids, for instance, are easy to make in a lab. Are scientists, then, any closer to making life or engineering a human? Some of the mechanics are worked out, but the prebiotic soups made in the lab still lack self-sustaining replicators. We are not even sure what to make exactly. We know life by a working definition only. Adding a soul that connotes an afterlife only muddies the waters for making a human.

Laboratory experiments aside, computer models have failed to simulate life. Instead of growing larger and complex, the self-replicating molecules they start with tend to evolve in the opposite direction. Perhaps, as McFadden wrote, computers do not take the quantum nature of subatomic particles into account. Quantum computing with its lightning speeds holds promise.[14] Then we could reenact the emergence of life according to the multiverses scenario. Since the odds against making life in a test tube appear to be too great, it's unlikely to happen. Abiogenesis scientists, however, risking sanctions, damnation, or excommunication, would be tickled silly if they succeeded and made life.

Although quantum mechanics seems bizarre, it's real. The idea of multiverses is solid and consistent with the facts, particularly at the subatomic level. The implication is mindboggling. As products of virtual particles, multiverses are common, indicating that, under the right conditions, life teems everywhere. Many scientists envisage multiverses on a large scale and hope to visit one someday. They might want to be patient, though, as we are still struggling to understand this one.

Life might have begun in fits and starts, but once it took hold, it struck out, Darwinian style, and became unstoppable.

OUR FAMILY TREE

Only a few people can trace their bloodlines back centuries. The royal family of Great Britain traces theirs to a time before the Norman Conquest of 1066. Going further back, the number of ancestors diminishes as lineages converge to a bare minimum. To go from such a small number of ancestors to billions of descendants today means we share grandparents, royals included. Going much further back, before people and terrestrial life existed, we run into our ancestors in the sea, where they frolicked for eons before some forsook the water for life on land.

For a family tree of all species, then, scientists trace lineages by analyzing DNA. This allows them to avoid the dilemma faced by Swedish naturalist Carolus Linnaeus. Linnaeus arranged species based on the binomial classification system, unreliable because it used differences in appearance. DNA analysis, on the other hand, places those with the most identical genes next to one another on the family tree, for instance, chimps and us.

In 1977, Carl Woese, a microbiologist at the University of Illinois, constructed the current family tree, consisting of three domains and four kingdoms. Of the domains, bacteria appear to be the most primitive. Although bacteria superficially resemble archaea, the next domain, in terms of size and cellular organization (members of both groups lack nuclei and are called *prokaryote*), they are surprisingly different to eukaryotes, our domain with nuclei. Their genes are as vastly different from those of archaea as they are from our domain, Eukarya.

The Eukarya domain consists of the four familiar kingdoms, classified according to their feeding style. For instance, fungi absorb food directly from their surroundings, decomposing organic matter and rocks into fertilizer and soil. We eat them as mushrooms, and we use them as yeast to ferment our beer, and as penicillin from penicillium to treat bacterial infections. Penicillin was accidentally discovered by Alexander Fleming at St. Mary's College in London, in 1928, saving many a soldier's life during World War II. Years later, I trained in the building where he found it.[15] *See Illustration 5.1*

So far, of the 8.7 million species on Earth (could be as many as a trillion or more when all microbes are included), we have classified only about 1.3 million. Biblical Adam could never have named them all, let alone been aware that they existed. After evolving into larger cells, single-celled bacteria with no nucleus dominated life for two billion years. Then bacteria with a nucleus showed up. To grow, bacteria split into two. Along the way, some bumped into others, sharing genes. Small, agile cells (nascent sperms) easily caught up with the larger, slower cells (nascent eggs), launching the first sexual reproduction. Mixing genes led to diversity, increasing the chances of offspring surviving erratic environmental conditions, and promoted sex. According to biologist Lynn Margulis, sex sealed our fate. It became linked to programmed death, aging, and mortality.[16] At first, her theory of the mergers generated controversy among her colleagues, but eventually they came around to see it her way.

Mounds of fossilized stromatolite dotting the shallow waters of Australia and South Africa show that early colonies of cyanobacteria (formerly blue-green algae) lived there over three billion years ago. Sunbathing atop the mounds, the bacteria trapped light for energy. They synthesized carbon dioxide from water, releasing oxygen that combined with the abundant rock iron, turning the landscape a rusty reddish color. By 2.4 billion years ago, all the iron had rusted. The oxygen escaped into the atmosphere, creating the ozone layer about 600 million years ago, a godsend, as it shielded Earth from deadly ultraviolet rays. Life had been simple until that time. Then, multi-cellular life such as jellyfish evolved. Bacterial cells teamed up, forming cooperative bodies, and cells specialized. Eventually, bigger and more complex organisms formed. The growing number of species led up to the Cambrian explosion 530 million years ago. Plants and animal organisms came ashore. The latter developed into amphibians and lizards that eventually evolved into dinosaurs 230 million years ago. Dinosaurs died out 65 million years ago from an asteroid wallop.

Before disaster struck, though, a small dinosaur evolved into a bird that survived the Apocalypse.

EXTINCTION'S BEAT

The last 350 million years have seen about twenty mass extinctions, five of them worldwide. It's sobering that 99.9 percent of all species that have ever lived on Earth have gone extinct. Imagine, at the end of the Permian period 250 million years ago, the mother of all extinctions, the Great Dying, snuffed out 90 percent of life in the sea and 70 percent on land. Such mass extinctions contribute to evolution, as the surviving species multiply and overrun the planet again.

Extinction mechanisms vary. Most mass extinctions are caused by global warming from volcanic activity that releases carbon dioxide and methane gas (the latter coming from cows today, the thawing of permafrost, and rice fields) into the atmosphere.[17] Other mechanisms include glaciation, which kills off vulnerable life forms. Ice ages began fifty million years ago and they return at the rate of one every 100,000 years. As ice advances, a three to four kilometers deep (average of two miles) northern glacier creeps to the southern parts of the United States, England, and deep France. As it advances, it scours landscapes and litters mountains with boulders, feats so stupendous that Christian fundamentalists credit them to Noah's flood, but rocks wouldn't float to mountaintops.

Icebergs tie up a lot of water of the Earth's supply, up to 1.7 percent in "normal" periods and up to 5.5 percent at a glacial maximum. At the maximum, with so much water locked up, shorelines recede, exposing 40 percent more land and creating new bridges. Animals and people may then cross them, as Native Americans did the Bering land bridge between Siberia and Alaska about 18,000 years ago. Because rainfall decreases when ice locks up the water, land dries and forests recede. Such tree scarcity about six million years ago may have led our ancestors to abandon a carefree canopy-swinging lifestyle in the forests in favor of the riskier open savanna. Bedeviled by climate fluctuations between wet and dry, only creatures that could adapt to the changes, creatures such as our remote ancestors, survived. It is hardly a coincidence that our genus, *Homo,* appeared on the heels of a mass cooling period approximately 2.5 million years ago. We're in an interglacial period today, a thaw with only the poles covered in ice.

Such periods last about 10,000-20,000 years, and since the current one is 11,500 years old, we may be due for another big chill, if global warming doesn't delay it.[18]

Ever since Christopher Columbus's discovery of America in the fifteenth century, explorers have found identical fossils between continents. For instance, freshwater trilobites that could not survive in salt water appear to have thrived in Europe and North America. Also, the exposed edges of continents like South America and Africa fitted together like a jigsaw puzzle, indicating they were once joined. In 1912, German meteorologist Alfred Wagner said that continents moved slowly, as the plates supporting them drifted. This drifting also caused earthquakes. When landmasses joined, they decreased the total beach area, curtailing access to land from the sea, so sea animals that laid eggs ashore failed to survive and died off. Rain decreased and the inland dried up. Widely fluctuating oxygen levels and surface temperatures caused many species to die, too, in yet another mass extinction. Mass extinctions are increasing in frequency and scope, and this time we may also be responsible. We deforest, overgraze, overhunt, and pollute the air, wreaking havoc on the ecosystem.

As he classified animals and plants in the eighteenth century, Linnaeus was stumped in deciding where to put us, because we looked so much like the apes. Appearances notwithstanding, he reluctantly lumped us together with the monkeys anyway, labelling us primates, or *numeros unos* (first rank). Such a classification matched Linnaeus's Mount Everest-sized ego. He was known as *Princeps Botanicorum* "Prince of Botanists," and he would boast, *"Dues creavit, Linnaeus disposuit,"* Latin for "God created, Linnaeus arranged."

To put distance between us and the beasts, Linnaeus elevated us to *Homo sapiens*, Latin for "wise man," and demoted the chimp to a cave-dwelling man. Sex was never far from his mind, though. This showed in the names he gave different species. One genus of plant he called Clitoria. He also named various parts of a species of clam *vulva*, *labia*, *anus*, and *hymen*, sprinkled his texts with such labels, and others like *Fornicata*, and recorded flower beds as being amorous.[19] God save you if you criticized Linnaeus's self-aggrandizing; he would paste your name onto something disgusting.

LIFE EVOLVING

According to the Bible, we are completely separate, not related to other organisms. God created them before us, each after its own kind. Science, however, finds that all life shares similar DNA. This means we *are* merely branches of the same tree, descendants of LUCA. In humans, DNA is stored on chromosomes in the form of genes inside a cell's nucleus. Genes carry the instructions for building an organism and determining its instinctive behavior. For example, all fourteen cloned horses of the polo player Adolfo Cambiaso had a similar temperament, showing they inherited the behavior.[20]

Today, scientists play God by tweaking genes and growing organs in unfamiliar parts of the body. For instance, when Swiss biologist Walter Gehring activated a gene called *eyeless,* he grew eyes on a fruit fly's legs, wings and antennas.[21]

As cells divide and chromosomes split, their DNA information could be copied wrongly when pairing with another cell, resulting in a mutation. Mutations are a major cause of change of species. Changes may start small but they add up over time. If they improve an organism's chances of survival, its genes will pass into future generations, making the changes the staple of evolution. Such mutations, however, and those that kill the organism terminating its lineage, are few. The majority are neutral, leaving the species unchanged. Scientists then use them as markers to track lineages.

Other evolution mechanisms include sexual selection. Pheromones aside, for many animals, attraction—be it looks, strength, power—is the driving force. The mechanisms also include separation of a group as natural selection weeds out members unsuitable for the new environment. The smaller the group, the faster the change. And for better or for worse, variation occurs when some members of a species fail to reproduce sufficiently to preserve their unique traits, resulting in genetic drift. Also, evolution comes about when a population breeds with another of the same species leading to migration or gene flow. This way, LUCA branched into the tens of millions of species we see today.

Evidently, DNA cannot stop copying itself. Although competition to survive has led many species into an eat-or-be-eaten existence, the flipside, self-sacrifice or altruism, coming to the aid of others at one's own risk, also indirectly preserves the benefactor's DNA. For instance, the male redback spider, by offering itself to its partner for lunch, increases its offspring's chances of feeding and thus survival. Still, many species share, as they do better in groups, finding strength in numbers.

EVOLUTION ATTACKED

Opponents insist evolution is "just a theory," meaning an unproven hypothesis. But a theory is fact, based on evidence. You wouldn't call gravity conjecture, even though some people have never heard of it. A theory may be confounding, but it makes testable predictions, such as Einstein's space-time model. Evolution, defined as genetic variation in genetic makeup of a population adapting to changing environmental conditions over time, is a given. Anyway, change happens, guaranteed. Protesters calling it a so-so story ignore evidence in plain sight. Unable to refute the fact of evolution, opponents cite unresolved questions, gaps in the fossil record to justify reliance on their own theory of divine intervention in creation. New finds, however, poke holes in such arguments. Also, genetic analysis, such as DNA tracking that shows when a species branched from the family tree, discredits their arguments, and fills in the "missing links."

In the 1950s, the Catholic Church grudgingly admitted evolution theory was more than a hypothesis. The evidence was overwhelming. The ancients, doubtless unawares, applied it, but in reverse. For instance, family inbreeding was taboo – Egyptian royalty excepted – and condemned in the Bible. Lv 20:17. Even though they didn't have the technology to explain the cause, they realized that breeding between close relatives could keep harmful traits within the family.

Comparative anatomy offers ample evidence supporting evolution. Many structures, although similar in widely diverse organisms, may have different functions, such as the wings of a bat, flippers of a whale, arms of a human, and on. Also, blinking, characteristic of the animal world, suggests a common ancestor with whom this eye-moisturizing reflex originated. When a body part falls into disuse, it turns into a remnant, or vestige. Such is our tailbone (coccyx, originally a tail – vestigial in some babies), and the appendix. An appendix helps other animals digest food, it is redundant in humans, posing a risk for infection and perforation. At other times, the function of a body part changes, such as the legs of a whale that turned into flippers once the animal forsook land for the big waters.

Small wonder then that as all life is related, embryos of reptiles, birds, and mammals pass through stages in which they resemble one another. In 1812, finding no physical differences between the fetus of a monkey and a human, an author, F. Jacob, declared that the only thing the monkey lacked was a soul.[22] In the nineteenth century, German evolutionary biologist, Ernst Haeckel, taught that our distant ancestors resembled gibbons. Observing different embryos, he believed that as they developed, they repeated the successive stages of their evolutionary journey. Before it began looking like a mammal or a human, the embryo passed through a fish stage with gill-like structures. Also, we share the same body plan, whose blueprint likely originated in a roundworm 600 million years ago, with other animals.

In animals, such a blueprint is executed by Hox genes, a subset of *homeobox* genes. Hox genes regulate other genes that control the development of animals, including insects and flies. They specify when and where body segments form prior and during the growth of an embryo. Hox genes are so similar among animal species that they are interchangeable. As in the *eyeless* gene experiments above, in 1994, when Gehring introduced a mouse eye gene into the fruit fly, eyes grew on its wings, legs and antennae.[23]

Those saying Hox gene order is proof of Intelligent Design also maintain that species are immutable, i.e., they believe that species were created in their present forms. They are blind to a fossil record showing species changing over time. Having nothing to back up their claims, and denying that we are related to all animals, they make contrived, impossible-to-follow arguments. Genesis mentions no such wayward ideas as species changing over time, they insist. But what came first, animals or humans? The two versions of creation in Genesis don't resolve the question, as they contradict one another.

In the early nineteenth century, some people believed evolution occurred when acquired characteristics were passed on to offspring. A French naturalist, Jean-Baptiste Lamarck, believed that a giraffe elongated its neck to reach food in treetops and passed on the trait to its young, a kind of soft inheritance. But without DNA participating, the change would die out and not lead to evolution. Lamarck also viewed the development of various forms of life as an escalator, with simple forms swarming at the bottom, and progressively becoming more complex as they ascended. Like Aristotle before him, Lamarck elevated humans to the penthouse, with a wondrous view of the Earth.[24] Georges Cuvier, a renowned French naturalist, objected to Lamarck's theory. Cuvier had discovered and correctly named many fossils, but – not finding human remains in early deposits – he embraced Catastrophism, the belief that God destroyed the world with a great flood and created new species now and then.[25]

In attacking evolution, Intelligent Design supporters argue that our organs are so complex and refined only a divine power could have created them. Their favorite example of this principle at work is the human eye.

Once again, scientists have shown that the Bible, and God, may not be giving us the full story. Dan Nilsson and Susan Pelger, using computer models at Lund University in Sweden, calculated that small gains in the distribution of light sensitive cells of bacteria could occur through natural variation. Cells that sensed light multiplied, while those that didn't sense light failed to develop and withered away. In 1994, using conservative numbers, Nilsson and Pelger estimated an eye could develop this way within 2000 cycles, or 364,000 generations, corresponding to about 10,000 years of evolution in some animals. So, since appearing around 542 million years ago, an eye could have evolved over and over, as many as 1,500 times in their fish model. Fossils with eyes at various stages of development and existing intermediate steps support their conclusions.[26] As Gehring's experimental induction of eyes onto different parts of the fruit fly showed us, Intelligent Design intervention isn't required. The human eye has major flaws, too, since light traverses nine layers before reaching the sensitive retina, dulling vision. A poor arrangement.

Yet another theory, commonly known as the Watchmaker Analogy, invokes Intelligent Design, based on the premise that a design requires a designer. The argument was formulated and published by William Paley in his book, *Natural Theology,* in 1802. Paley believed order and complexity in the universe required an Intelligent Designer, much as a watch does. But Paley's argument failed to account for a steady progression of timekeeping device designs. The watch took lots of people tinkering through the ages to eventually come up with the current specimens. Over the years, working designs were retained and functionless ones discarded. The original timepiece was the shadow clock, a stick ancient Egyptians stuck in the ground (later obelisks served a similar purpose), and as some people in the tropics still do. The stick and shadow technique were followed by the sundial, which worked well during daytime, but not so well after dark. To tell time at night, people notched candles and knotted ropes. Eventually, these methods gave way to such ingenious timepieces as the hourglass and water clock and, in the fourteenth century, to mechanical and pendulum clocks and watches. Today, the shadow clock comes nowhere near the complexity or precision of smart, digital, solar, or atomic clocks, but as the earliest common ancestor of today's gizmos, it started the research and development process.

In contrast to Intelligent Design, chaos theory (the science of complexity) says that the bigger and more complex a system is, the more likely it is to self-organize, dispensing with a designer. Ant colonies are self-organizing and behave as one functional unit. Living cells of a sponge also show that numbers, not a planner, are behind such order. After a sponge's cells are sieved, they regroup from the suspension as a sponge again. Volvox, another small water animal, self-organizes from individual algae cells. Weather patterns, ecological systems, human societies, and economies self-organize, too. Could this complex phenomenon account for consciousness? Perhaps, since brain activity is infinitely complex.

Some people are unaware they encounter evolution in progress or see evidence of it. Evolution does not have to take millions of years to occur. Cichlids in Lake Victoria, East Africa, multiplied into different species by changing jaws to suit their diet. Also, the depth at which they lived triggered sexual selection. In the relatively short time of 12,000 years since the lake filled up, a single lineage branched into 300 species.[27]

Peter and Rosemary Grant began vacationing in the Galapagos Islands in 1973. Over the next thirty years, studying Darwin's finches, they noticed that the birds' beaks changed shape and size, and their bodies changed size with changes in rainfall. During rainy periods when seeds were bigger, the beaks and body sizes increased, and during drought, when seeds were smaller, the beaks and body sizes decreased.

The fruit fly also shows evolutionary adaptations in a short time. A fruit fly has a lifespan of only two weeks, allowing scientists to observe several generations quickly. Armed with the fruit fly genome, scientists selected and bred specific traits, developing new fruit fly species with a doubled lifespan, for instance; evolution right before our eyes.

We also confront evolution in medicine daily when bacteria and viruses develop resistance to drugs and vaccines, respectively. Bacteria mutate furiously, doubling every twenty minutes on average. As the majority succumb to an antibiotic, a single resistant mutant may "take off," necessitating new drugs to kill it and its offspring. Pesticides such as DDT, used in years past to control malaria-carrying mosquitoes, lost their punch this way. According to the World Health Organization, malaria killed about 619,000 people in 2021, mostly in Africa.

For snail's pace evolution, a ready example is the penguin's wings, which turned into flippers when the bird went to live on the sea. The Australian desert marsupial mole lost its eyesight as the eyes became a burden since the animal spent most of its time underground. And then, isolated on the Pacific island of Mauritius with no predators, the dodo lost flight. Worse, the silly bird (dodo means stupid in Portuguese) laid just one lonesome egg. When people invaded the island with their dogs, hogs, and stowaway rats, the bird was no match. It couldn't outfox or fend off the intruders. People hunted it for meat, and rats and snakes went for its eggs and wiped it out.

In *The Ancestor's* Tale, Richard Dawkins traced our story back to our humble beginnings from the common progenitor – LUCA – as recorded in our DNA. If evolution was a so-so story and everything was created in its present form, how do we account for all the extinct species just shy of 100 percent of all that ever lived?

SURVIVAL OF THE FITTEST

Like River Nile, life evolved, meandering, punctuated by waterfalls (sudden changes), but generally staying the course of least resistance (adapting). As a twisting terrain sculpted the river, nature's erratic conditions shaped life. Species unable to reproduce failed to hold on and fell by the wayside. Instead of the fittest—the brightest, strongest, fastest or most attractive—mainly the lucky ones, those with suitable genes, survived. A designer played no role.

Picture a gazelle on the East African Savanna. During hard times when the grass dries up, nothing, not even its amazing speed, can save it from starvation. However, the gazelle's primary predator, the cheetah, with a horde of other animals – including a million-strong wildebeest throng – to prey upon, would have no problem resisting the drought. Survival of the fittest? Darwin disdained the term. To him, the fittest were those with traits that enabled them to adapt to changing environmental conditions. The environment could change at any time. If we didn't have what it took to adapt, we would be out of here also.

THE BUILDING BLOCKS OF LIFE

If we did not come on Earth in our present form, what form did we have, then? It looks like cells, we were mere cells. As the smallest unit of life, outside a few specialist cells such as nerve cells, most cells possess the complete genetic makeup of an organism and can exist independently. No wonder humans are in a race to clone anything: sheep, dogs, cats, horses, and us. In 2002, a cult in the Bahamas – the Raelians – claimed it had cloned a human baby, but that claim goes unverified.

There is a loud and persistent chorus against copying humans, largely because of ethical reasons. Still, many scientists dream of being the first to make a human. When in 2017 Chinese workers cloned two identical twin-looking monkeys, alarm bells went off, as a human could be next. But those who have been there and done that, caution. The 1996 "creators" of Dolly the sheep pointed to a high failure rate. Dolly also aged quickly, developed arthritis, and died prematurely. Were you to clone a human, instead of being nominated for a Nobel Prize, you likely would be treated to a hearty round of condemnation (and lots of curiosity seekers).

Once life emerges, whether cloned or naturally, in order to survive, or cells to thrive, it feeds on energy. To obtain the energy, the cells convert sugar (glucose) into energy packs and transporters, "batteries" called ATP, or adenosine triphosphate, which they store in *mitochondria*. ATP is used by bacteria and by all animals—fish, birds, humans, and on. During fermentation, yeast also generates ATP, alcohol, and carbon dioxide; the latter gives beer its fizz.

For their energy, plants require sunlight to make sugar from water and carbon dioxide (CO_2) and release oxygen, called photosynthesis. The sugar is converted into the ATP energy packs (akin to batteries). So, the more CO_2 there is available, the more luxuriant the plants grow. In their 1995 book, *What is Life?* Lynn Margulis and son, Dorion

Sagan, theorized that cells fattened themselves by capturing and enslaving smaller cells. After a billion years or so, the cells merged and functioned as one unit resulting in new species. Eventually, they grew into beings our size.

In a later work, Margulis touted a relationship, *symbiosis* (cells beneficially living off one another), as the central force behind evolution, and that mutations and natural selection were only cogs in the gears of evolution. The Russians thought likewise, too. She also maintained that AIDS was caused by a spirochete and not a virus, and AIDS, like syphilis, lived in our bodies symbiotically and could not be cured.[28] It's no wonder she reached this conclusion: although syphilis can be treated with penicillin, patients remain positive for the disease for the rest of their lives. Those with HIV may have to take medications for life, too, even after testing negative. Margulis's suggestions, however, were so controversial that they cemented her reputation as a heretic.

Margulis also suggested that, for mobility, a single-celled filament corkscrew spirochete bored into a cell and formed the tail that propelled it as sperm. Sperm derived energy for swimming from the mitochondria in the tail. Since the tail broke off at the egg's surface, only women have mitochondrial DNA, passing it on to their daughters. By analyzing mitochondrial DNA (mtDNA) of people from different ethnic backgrounds, geneticists traced our ancestry to a woman dubbed the grandmother of humanity, or *The Real Eve*, who walked Africa 200,000 years ago.

After the first cell, LUCA, appeared 3.5 billion years ago, it branched out to the "four corners" of our planet. Along the way, during the age of the dinosaurs, it sprang a branch of shrew-size mammals, our forebearers. Sharing territory with the dinosaurs, the puny mammals hardly had room to breathe before the behemoths left the stage. Our predecessors, then, picked up weight and size and evolved into primates between 65 and 33 million years ago.

After such humble beginnings, how did we become us? Which branch of the family tree is ours?

Such questions are discouraged by the clergy, since the Bible already has the story of our origins, and it cannot be questioned. Doing so used to ruffle feathers in the church and cause a lot of damage to the questioner, especially before the Renaissance. Nevertheless, evidence shows that, indeed, since they emerged, species have changed constantly throughout the ages. Although this may contradict the Bible story, challenging scriptures' veracity, we shouldn't despair, as our basic need for the spiritual experience, and guidance in life – roles we otherwise ascribe to religion – aren't affected in the least.

CHAPTER 6

OUR DEEP ANCESTORS

In the Bible, we were created in several ways. In one story, God made us out of nothing. Gn 1: 26-27. In another, He made Adam from dust. Gn 2:7. In yet another, a potter molded us from clay. Rom 9:20-21, Is 29:16, and 45:9. Finally, a Word (Logos) that became God made all things. Jn 1:1-3. Since then, we haven't changed at all. A closer look, however, throws all these versions of our origins into question. A methodical or scientific approach reveals that we evolved from lowly life over millions of years.

Between 30 to 25 million years ago, after our puny forebears – small, shrew-sized mammals – evolved into primates (much like palm-size, 55-million-year-old Archie, *Archicebus Achilles,* from China), they split into Old World monkeys and primitive apes. Our common ape ancestor then gave rise to gorillas seven million years ago, chimpanzees and bonobos four to six million years ago, and hominins, our line, six million years ago. It means we are more closely related to chimpanzees than we are to gorillas. While we share upwards of 98 percent of our genes with King Kong, we share more than 99 percent with the chimp.

When primates split, the ape line lost its tail. Gone too was the keen sense of smell. To leap from branch to branch, they directed their eyes forward, developing stereoscopic (3-D) vision to judge distances accurately. Their thumb, like the big toe in tree climbers, was opposable for gripping, and with an enlarging brain, they were on their way to complex social behavior.

About fifteen million years ago, the global climate changed and a new type of landscape, the savanna, appeared in East Africa. It became home to many different animals. Ten million years later, when rainfall decreased to the east of the Great Rift Valley, a crisis loomed. Trees became scattered and mostly animals with new modes of locomotion could navigate well over the open terrain. The advantage then shifted to two-legged walking primates, *bipedals,* which then thrived in it. Huge mountains forming to the west of the Rift Valley created a barrier, dramatically reducing rainfall and deepening the crisis. The gorillas and chimpanzees unsuited for the new environment remained behind in the rainy forests to the west, and have not changed for three million years.

To track our journey, we could use a chronological approach. But finds are random, resulting in gaps or "missing links." Humans are sometimes to blame for these gaps. For instance, just like evolution theory contributor Alfred Wallace, Darwin's suggestion that Africa was the place to look for our origins, because apes lived there, fell on deaf ears. Ancient travelers to Africa, such as Herodotus and Diodorus, had also billed the continent as the birthplace of humanity. But in nineteenth century Europe, Africa lacked appeal. A lot was still unknown about it. For instance, pharaohs were considered myth. Even noted philosophers like David Hume or Georg Hegel dismissed Africa as a footnote, without any contribution to civilization. Never mind that an early seventeenth century colonial American plantation owner preferred African workers, who were civilized, to the Irish, who were not. Unlike the Irish or Native Americans, the Africans were good at growing tropical crops, working iron, laying bricks, carpentry, and building houses, among many other skills.[1] Also, the Africans did not get sunburn in the intense sun, because of the melanin in their skin.

Instead, Asia was preferred as our birthplace. After all, it was the home of the gibbon and orangutan (or man of the jungle, the ape with the most human-like features). Ernst Haeckel, a German evolutionary biologist, also recommended it as the place to look for our origins. Haeckel believed humans hailed from a land there that had since sunk in the Indian Ocean, just as Plato's Atlantis disappeared. Other "scholars" believed people migrated from India to Africa and regressed!

Going by discovery order, Neanderthals were the earliest human fossils to be found, the first one in Belgium in 1829, and the second one in Gibraltar in 1848. They were initially not considered prehistoric; that honor went to a skull discovered from the Neander Valley (*tal* in German) in 1856. It was, however, mistaken for that of a northern European savage, or a modern human who had suffered rickets. Then, the tune changed and it was attributed to a Cossack soldier injured while pursuing Napoleon's retreating army. An Irish geologist, William King, finally identified it correctly in 1864. He noted that the fossil's bones were different from those of humans and belonged to a different but close lower species instead. He named the species *Homo neanderthalensis.* Maybe he shouldn't have, for the populace went into a Neanderthal-bashing frenzy. Scientists and the public alike so trashed the species it has yet to recover. Just watch the TV commercials. Neanderthals are usually depicted as dimwitted, lumbering, grunting cave dwellers, and as clumsy brutes. Unfortunately, we tend to belittle those we know little about; the less we know, the worse the put-down. But the Neanderthal brain weighed in at 1,600 cc, a full 200 cc more than today's human brain, although, admittedly, it bulged in the sides to control their heavy musculature.

SEARCHING FOR GRANNY

How do we obtain information about our long-gone ancestors to construct a family tree? Without written records, and with word of mouth that could change in a single pass around a circle, fossils are our best bet. But soft bodies rarely fossilize, so remains are few and far between. Also, fossil hunting requires painstaking graveyard digging and soil sifting—tons of it. One must be on the lookout at ancestral burial grounds for protesters or clandestine grave robbers, too. Were you unnerved by the toiling or the fossil thieves, you might want to give paleoanthropology, the study of human evolution, a wide berth.

Besides the reality of limited artifacts in general, other forces can contribute to the records gap, such as local resistance against disturbing ancestral burial grounds, unauthorized raids of burial sites, carelessness and neglect, inaccessibility, increased fragility over time, and governmental control of the geography where artifacts might be located. For instance, because Mongolia didn't want Genghis Khan's remains found (in deference to the emperor's wishes), his body was buried in a chamber 20 meters underground. To locate it, archeologists have resorted to internet tools, or crowd sourcing, where many people provide input on detailed satellite images posted online. Then small remote-controlled aircraft buzz a promising area with ground-penetrating radar.[2]

Forests and wetlands are poor custodians of the past because they promote decay and don't preserve fossils. Other fossils may be located in dry, hot, dusty, or frigid areas, or in narrow deep caves, while those in populated areas may have been pulverized for medicine, or hauled away as building material. But the hunt for relics is crucial. Filling in the missing pieces in the jigsaw puzzle of life gives us insight into who we are; it should proceed unhampered.

As a find is rarely complete (mainly because of deterioration), forensic experts are left with reconstructing fossils and artifacts from fragments. Scanning techniques such as X-rays and CAT scans help, showing size and shape incredibly accurately, and extracting information invisible to the naked eye. Throw electron microscopy into the mix as well; it may help determine a tooth's age and the type of food its owner consumed by showing growth marks and cuts on the tooth's surface. For signs of very ancient fossils, as in rocks billions of years old, scientists turn to electron microprobe isotope techniques that determine the chemical nature of solid materials. Then they may infer things like intelligence and the user's way of life from tools found at a site.

Discovery of a mere stone blade, tooth, jaw, or finger bone is akin to hitting the jackpot. Besides bringing instant fame, the discovery may win the finder sponsors, awards, and a lecture circuit, justifying the years and fortunes spent toiling. Luck eludes most researchers, however, who then console themselves with poring over finds and debating them *ad nauseum.*

Modern lineage tracing techniques have, to some degree, supplanted fossil hunting. Such techniques avoid gaps and classification by appearances, helping to reveal more about the past than was known only a century ago.

One such method, DNA analysis, links species through genetic markers found in DNA. From the random, harmless mutations accumulated in DNA over time, scientists build a molecular clock. All offspring carrying a specific mutation form a haplogroup, allowing genetic tracking of the group. DNA analysis has broken many an impasse in this highly contested paleoanthropology field.[3] These DNA methods have laid bare inconvenient truths, including the fact that we are also animals, but smarter.

For dating purposes, scientists may count annual tree rings or rock or glacial layers. But by far, the most popular methods are radiometric, such as the decay of carbon-14, which determines the age of fossils or their surroundings.[4]

To track your journey from the first humans, you submit a swab from the inside of your cheek, or a saliva sample. By analyzing mitochondrial DNA (mtDNA) for women and Y-chromosome DNA for men of close to a million people around the globe, the project, Genographic, by *National Geographic,* mapped their migration routes out of Africa.[5] For instance. such results show that my DNA belongs to a 20,000 to 30,000-year-old Bantu sub-Saharan line, hg (haplogroup) E1b1a, similar to Ramses III,'s as discussed in Chapter 2. This group was the final (end) destination of an ancient hg M168, which originated in an East African man about 70,000 years ago and is found in most males in the world. M168 ultimately descended from a man in hg A, dubbed the *Y-chromosome Adam,* who lived in Africa 200,000 to 300,000 years ago and was the ancestor of the vast majority of males on Earth.

Likewise, female mitochondrial lineages found in the rest of the world originated from an African woman dubbed *Mitochondrial Eve.* Her hg L dates from 100,000 to 230,000 years ago.[6] Her descendants, hg L3, left the African continent 74,000 years ago. So, from presidents to the Pope, from the moon landing to deep sea diving, and from the Olympic medalist to the Nobel Prize laureate, we have all been there, too, through our shared DNA.

Cantankerous debates are common among evolutionary scientists, and no radioisotope method is accurate enough to settle them conclusively. For instance, since the volume of carbon-14 varies depending on the extent to which Earth's magnetism deflects cosmic rays, the age of a find may be significantly over- or under-estimated. Or, to support a claim, a scientist may have only a single piece of bone. Worse, definitions could be vague, with more than one possible interpretation. National or personal pride may be at stake, as well. Disputes aside, since finds are random, a fresh find could unceremoniously knock a previously proclaimed "first" off its pedestal. To appreciate the human saga, we should keep an open mind, then. Otherwise, we would be like Galileo's university colleagues, who refused to look through his telescope, lest, instead of a smooth surface, they saw a cratered moon, which contradicted scriptures.

AUSTRALOPITHECINE WHAT?

It was getting late one evening in 1974 and the camp was beckoning. But, feeling "lucky," Donald Johanson and Tom Gray decided to take a last look in a rich fossil area at Hadar in the Afar region of northern Ethiopia. They soon stumbled on a fragment of an arm bone. Up on the slope, more bones. They likely jumped, higher and higher, like kids happy to see their grandma, which would risk trampling the fossils. Oops! They transferred the jubilation to the camp. There, a Beatles' song, "Lucy in the Sky with Diamonds" was playing, giving the fossil its name. At the time, 3.2-million-year-old Lucy was the oldest nearly complete fossil of a human ancestor ever found. She was dubbed the mother of humanity.

Lucy belonged to the mouthful genus Australopithecus, or southern ape. Her skeleton was head to foot 40 percent complete and indicated that she had stood three and half to five feet tall. Her long arms showed that she spent a lot of time in the trees. Being the oldest species yet, she became the most celebrated find to date. But clearly, she was not us. A year later, Johansen found thirty-three more skeletons in the same area. They suggested group social living, earning them the title of "First Family."

In 1978, Lucy was atop the world, having the time of her life, when she was toppled from the "mother of humanity" status. Paleoanthropologist Mary Leaky found 3.6-million-years-old fossilized footprints of a woman and child set in volcanic ash in Tanzania. Like Lucy, the vanquishers were a species of *Australopithecus afarensis*, probably descendants of *Australopithecus anamensis,* a 4.2 million-year-old australopithecine from northern Kenya. The footprints showed our ancestors walked upright more than 3.6 million years ago, long before they developed a big brain. This discovery also discounted the biblical notion that the Earth and we had been around for only 6,000 years. Subsequent discoveries were still older.

So many fossils complicate our family tree. Instead of a single evolutionary trunk, evolutionist Stephen Gould noted that we have a bush with branches leading in different directions. But the fossils help to show our progression from earlier ancestors that the Bible doesn't mention, and that Creationists ignore. Since all lines but ours disappeared, identifying it calls for some sleuthing. For instance, *Kenyanthropus platyops*, or the flat-faced man of Kenya, whose 3.5-million-year-old skull was discovered by Meave Leaky on the shores of Lake Turkana in 1999, also guns for Lucy's esteemed position. *K. platyops'* face looked more like ours.

Also predating the australopithecines was 4.4-million-year-old *Ardipithecus ramidus* (Ardi) from Ethiopia, whose lineage dates back 5.8 million years ago. Ardi was bipedal with small canine teeth and held a position halfway between our common ancestor and the apes. Ardi probably begat *A. anamensis,* who could have split into two: the famous volcanic-ash-strolling *A. afarensis* pair, and *K. platyops.* But Ardi was a baby compared to six-million-year-old *Orrorin tugenensis* unearthed from central Kenya in 2000, or to disputed seven-million-year-old *Sahelanthropus tachadensis* discovered in Chad in 2001.

Australopithecine was small with a flat forehead, a prominent ridge above the eyes, a concave protruding face, and didn't have a protruding chin. The muzzle was pronounced with large cheek teeth. The arms were long, legs short, and the chest apish – shaped like a pyramid. Her 380 cc to 450 cc brain overlapped that of apes, from which she was separated by her smaller canine teeth (fangs) and habit of walking upright.

About three million years ago, *A. afarensis* (Lucy) split into two *A. africanus* species, one gracile and the other robust. As the names suggest, the gracile offshoot was smaller and weaker with a slightly smaller brain and thinner bones than its more robust cousin. It had a mixed softer diet of plants and meat, so its teeth were also smaller. In contrast, its robust cousin had prominent features such as a marked skull crest and enormous cheek teeth operated by powerful jaws to chew leaves, roots, nuts, and seeds. Its brain was only slightly larger than gracile's was. The robust species being a browser (feeding on small vegetation), when the climate changed and its food sources dried up, it died out around a million years ago. At this time, whichever species we directly descended from was still apish and a long way from becoming us. The odds are on the gracile species, which had a wider choice of foods.

WILL THE REAL MISSING LINK PLEASE STAND UP?

For a long time, people searched feverishly for "the missing link," the common ancestor of apes and humans. As happens sometimes, luck comes to those not even in the race. The unsuccessful ones, who could be pillars of society, then opt to make up a discovery rather than admit failure. Such was the motive behind the Piltdown Man hoax. Piltdown

Man was "discovered" in a gravel pit in southern England in 1912. The "finder," Charles Dawson, billed it as the earliest known human from Western Europe. An eminent Scottish anatomist and anthropologist, Sir Arthur Keith, certified Dawson's "find." For forty years, the fraud basked in the limelight, taking in scholars and the public around the globe. It was not until 1953, when it was subjected to new dating X-rays and microscopy methods, that the fraud was exposed: it was a human skull with the jaw of an ape.[7]

Three to four million years ago, australopithecines migrated to southern Africa. There in 1924, a young Australian professor at Witwatersrand University in Johannesburg, Raymond Dart, was given a skull from a limestone quarry at Taung. Judging by its teeth, the skull was that of a three to four-year-old child. Its foramen magnum (the hole through which the spinal cord exited the skull) was located in the center of the base, as with upright walking species. A sediment cast showed its brain looked more like that of a human than an ape. Dart believed it was "the missing link" and called it *Australopithecus africanus,* or Southern ape from Africa.

Further searches of the Taung site and elsewhere uncovered more bones. From these, Dart formed a picture of our early ancestors. They were toolmakers and used tools for hunting animals, he said, not to mention for consuming their rivals. The general public weighed in by any means available, such as via letters to newspapers. Some experts argued that the propensity to kill, coupled with handling increasingly complex tools, stimulated the brain to grow. A cautious few, however, pointed out that africanus's tools were very simple, suitable only for scavenging, not killing. They did not see a savage or cannibal; they believed Dart had misread the facts. Given *Homo sapiens'* propensity for violence though, he was not far off the mark.

Cannibalism was not taboo to our ancestors, either. Apart from preventing starvation, it signified the ultimate defeat and humiliation of one's enemies. It was also believed that cannibalism helped to conquer disease, by allowing the deceased to continue living in the consumer's body. In turn, the consumer acquired the good qualities of the deceased. In ancient Egypt and Central Africa, people ritually sacrificed their king-god and ate his flesh to become one with him.

The idea lives on in the Holy Communion, or Eucharist, where the wine represents the blood of Jesus, and the bread represents his flesh.

The ritual was launched to commemorate the coronation of Ptolemy V, Epiphanes Eucharistos, almost two centuries before the biblical Lord's Supper. The ritual was assigned to the image of Serapis (a merger of Osiris and the Egyptian Apis bull), the god to whom all the Ptolemys, and later the Roman Popes, became vicar. Ptolemy V is also remembered in the Epiphany. Since our ancestors partook of human flesh, genetic studies find that almost all humans have genes designed to provide immunity to certain diseases which can only be transmitted by eating human brains.[8]

Dart achieved instant fame after the British journal *Nature* reluctantly published his report. Then the critics weighed in. Dart's report had clashed with a prevailing but fallacious belief that a large brain came before upright walking and tool use. In the same journal, Keith (he of the Piltdown fraud) dismissed Dart's findings, crediting them to a chimpanzee or gorilla; Piltdown Man supposedly proved it. Adult humans mirrored features of young apes such as gibbons or chimpanzees, he pointed out. He also argued that when fully developed, apes were structurally more advanced than humans. The skull, therefore, belonged to one of these ape species. Dart had also wrongly estimated the skull's age at about one million years old. Keith considered the time too short to the evolution of humans 100,000 years ago; the skull had to be over a million years old. It was actually more than 3.5 million years old, see below.

The popular press had a field day with the young professor. Letters from Creationist groups also poured in from around the world. Religious doctrine forbade mention of a missing link between man and apes, they said; there was none, because God made us in His image. Dart was told that he was doomed and would burn in Hell. (If this story were to break today, Twitter would be all abuzz). Six years later, Sir Keith was still at it, calling Dart's skull that of an ape. The British scientific establishment so shunned Dart that he withdrew from the world stage. At the risk of such rejection, other luminaries, it seems, would think twice before putting their work out. Such concern might have been behind Isaac Newton's secrecy and Darwin's reluctance to publish his book. I envision Darwin resting in Westminster Abbey, scratching his beard at the names Creationists still call him.

But Dart had supporters who eventually helped to vindicate him. One, an enigmatic Scottish country doctor, Robert Broom, was practicing medicine in South Africa at the time. He dabbled in wholesale fossil exports and adored the Taung child, helping Dart to identify it. Broom hunted for similar remains in the limestone caves at Sterkfontein where people searched for "The Missing Link" and for fossils to buy. After only nine days there, a local boy handed him an australopithecine skull cast by a quarry manager and an upper jaw with a tooth.

Broom, a paleontology pioneer, established that austra-lopithecines were credible human ancestors. He showed that Africa, not Asia, was where our early evolution took place. Although he gave his finds many weird names, they all belonged to *A. africanus* and *A. robustus*, descendants of *A. afarensis,* Lucy. The dig where Broom worked proved to be a treasure trove of more than five hundred fossils, including *Homo habilis*, the handy man or tool user.

The discovery that australopiths walked upright came in 1947, confirming that Dart's Taung child was really a half human, not an ape. In a humiliating reversal, Keith admitted his error in a letter to *Nature*. Any lingering doubts evaporated when the Piltdown Man Hoax was exposed. Since we now know when humans separated from the apes, hunting for missing links between our species is no longer a priority.

THE CRADLE OF HUMANITY

In 1916, a new dynasty in the search for our origins dawned when the Kikuyu tribe initiated thirteen-year-old Louis Leakey into their group. Leakey, a son of missionaries working in Kenya, went on to study anthropology in Cambridge. Louis believed the gorge at Olduvai offered a unique opportunity for science to learn about our prehistory. A German scientist had found a skull there in 1913; never mind that the skull turned out to be human. After he left school, Louis returned to Kenya to do research.

The Olduvai Gorge, called the cradle of human evolution, is a thirty-miles long gash in the Great East African Rift Valley in the Serengeti Plains in northern Tanzania. Caches of tools, many made from the 1.9-million-year-old lava found only a few miles away, were preserved there. Louis and his wife, Mary, discovered tools made from many varied types of rock, which pointed to a complex pattern of trade routes. Sweltering in temperatures of 110 degrees Fahrenheit, the

Leakeys also discovered remains of many extinct animals, such as giant pigs and large saber-toothed cats.

Mary Leaky hit pay dirt when she unearthed the skull of Nutcracker Man, *Paranthropus boisei,* in 1959. The approximately 1.75 million-years old skull carried a massive bony crest, flared cheekbones, enormous cheek teeth, and tiny front teeth. Despite its ape-like appearance, Louis Leakey billed it our earliest human ancestor, "the missing link" between the apes and us.

Alongside the find were many scattered tools lying amongst cobbles and cores of rock, the source of thin, sharp-edged flakes of stone. Similar tools found at Gona, Ethiopia, called Oldowan industry tools, reflected activity at least 2.6 million years ago. It was unclear who first used them, since several species, *Australopithecus, Paranthropus,* and *Homo,* lived in the area over time. The anatomy of Nutcracker Man suggested a plant-eater with no use for skin slicing tools, but Louis was adamant about the tools' human origin. The find catapulted Louis onto the world stage. He became a superstar. The National Geographic Society (NGS) sponsored him, published his articles, and afforded him television specials and worldwide speaking tours. In 1962, NGS awarded Leakey the society's prestigious Hubbard Gold Medal.

Louis encouraged his children to hunt for fossils. The training paid off in May of 1960 when at nineteen, his eldest son, Jonathan Leakey, uncovered pieces of a skull of a twelve-year-old child at Olduvai. Since the discovery was the earliest human skull found to date, it torpedoed his father's claim that Nutcracker Man, who was younger, was the "missing link," For the next few years, similar fossils turned up among stone tools and animal bones at the site, all more than two million years old. They had large brains, around 700 cc, and humanlike skulls with thin-bones and teeth. Their tools had improved from Nutcracker Man's natural Oldowan tools to Acheulean tools. The find was called *H. habilis*, or handy man, because it was believed to belong to the toolmaker. *H. habilis* could have evolved from *A. afarensis* or *A. africanus. See Illustration 6.1*

Where did a large brain come from? A big brain appearing millions of years after upright walking means the stimulus had to be more than increased dexterity. Our ancestors were delicate, easy prey. Overcoming predators, starvation, and the vagaries of weather required planning, which required a sharp mind. After separating from the chimps, our ancestors developed brain modification genes, such as ASPM that controlled brain size and boosted it, and HAR1 that helped the cortex to wrinkle and increase its surface area, giving them an advantage.[9] The smaller, weaker brained hopefuls, unable to plan, would have succumbed to the prevailing treacherous conditions and died out without passing along their inferior genes. Kudos to the boost, we survived and multiplied, and from only a handful (hundreds) at one time, took over the planet.

When Louis Leakey died in 1972, his son, Richard Leakey, and Richard's wife, Meave G. Leakey, continued his work. That same year along the shores of Lake Turkana, Richard's Hominid Gang, led by Kenyan Kamoya Kimeu, came across the almost complete skeleton of an eight-year-old boy. The fossil belonged to *Homo erectus*, or Upright Man, and missed only the hands and feet. Believing it was the "missing link," Richard named it Nariokotome Boy for the region where it was found. At a White House ceremony, President Reagan awarded Kimeu the John Oliver La Gorce Medal for "opening doors in the study of mankind."

Nariokotome lived 1.5 million years ago, apparently dying of septicemia (blood infection) from a bad tooth. Had he not died so young, his cranial capacity would have matured to 880 cc. He stood 5 feet 3 inches tall and – had he lived long enough – would have grown to 6 feet 1 inch. His tall stature helped him lose heat efficiently and, without hair hindering sweating, cool him in the hot tropical climate. His physique indicated an extremely active heat-tolerant individual capable of running around in the open. The spine in his chest was S-curved – a uniquely human abnormality known as *scoliosis*. The spinal canal in the chest region was narrower than a modern human's, suggesting he did not have fine control over his breathing muscles; his language skills, therefore, couldn't have been as well developed as ours are. He probably spoke, but in single words or grunts.

Although Nariokotome's brain was the size of a one-year-old human baby, he would have been able to acquire survival skills – including making complex tools, caring for the sick, and controlling fire – by the time he was fully grown. His skull, despite possessing the largest brain at the time, was still quite flat with a prominent brow ridge. His gangly size notwithstanding, he was still only half-human and had a long way to go to become like us. *H. erectus* was likely a later form of *H. ergaster* and the two are generally treated as one.

Two million years ago, *H. erectus* hit the road. He was innovative and solved new challenges along the way, exploring simple tools and modifying them as circumstances dictated. He used fire to prepare meat and supplement plant food. From Eastern Africa, he trekked north through the Middle East, branching out to Eastern Europe and to the Far East into Malaysia and China. In 1991, his 1.8-million-year-old fossil, the earliest ever discovered, was found at the crossroads village of Dmanisi in Georgia, which overlooks the old Silk Road. In Asia, particularly in Indonesia, where Haeckel predicted the human "missing link" would be found, *H. erectus* would top the "ten most wanted" fossils list.

Taking Haeckel at his word, a young Dutchman, Eugene Dubois, left a professorship in anatomy, put on a military officer's uniform, and set sail for the Dutch East Indies in 1887. Strange but true, people would spend their lives and fortunes, putting families and careers on hold, just to find the missing link. Most were disappointed. But not Dubois, one of the lucky few to hit pay dirt.

Dubois gravitated to Java, where orangutans lived. In 1891, after a year of excavating bone fragments from the banks of the Solo River with the help of a team of convicts, he found a flattened skull with a prominent brow ridge. A year later, he found a thighbone. Both fossils belonged to *H. erectus*, dubbed Java Man, confirming Darwin's theory of evolution in Haeckel's and others' eyes. Ultimately, though, Dubois fared worse than Professor Dart, making one wonder whether rummaging through our ancestors' bones also unearthed curses.

Back home, Dubois was initially fêted with honors. The honeymoon did not last long, however, once critics weighed in. "The remains are those of a giant gibbon," they said. Others believed the bones were human. Frustrated, Dubois stopped exhibiting the bones and hid them in the floorboards of his home underneath the dining table. Later, because of similar finds in China, he changed his tune and attributed the skullcap to an ape. That way his would be the only "missing link" between apes and us, since the new discoveries would be human.

PEKING, HERE WE COME

In 1919, a young Canadian physician, Davidson Black, taught neurology and anatomy at the University of Peking. But cadavers of executed criminals came headless. When he objected, the authorities offered to let him execute the convicts himself! A break from his academic duties spared him from carrying out that unenviable task.

In the early 1900s, Munich paleontologist Max Schlosser was convinced China would be a good place to look for primitive man. A fossil hunter in Peking had sent Schlosser some fossils containing a human-like tooth, triggering a stampede of fossil hunters to the country. Black joined in at the invitation of Gunnar Anderson, a Swedish geologist and adviser to the Chinese government.

Over a ten-year period, Black found fragments of human remains at various sites, including forty from the Chikushan site (Chicken Bone Hill) near Peking. Cannibalistic evidence—cut marks on bones—showed they were feast leftovers. Fossil hunting was not only grisly, but it could be trying, too. For one, the Chinese people believed spirits guarded their burial grounds and one could not dig there without permission from authorities. Even when granted, a seal of approval was no guarantee that work would proceed. The locals could refuse to allow access despite permission from authorities. To complicate matters further, the people also might have ground up the bones and sold them for medicine. Despite these challenges, Black found two teeth, which he excitedly showed around the world. Later, he found part of a skull, and called it Chinese man of Peking. Like Java Man, scientists credited the fossil to *H. erectus*. To Dubois' dismay, Black received the dubious distinction as the first person to find "the missing link." To rub salt in the wound, he was made a Fellow of the Royal Society of London in 1932. Because he was a workaholic, Black succumbed to work and silicosis from rock drilling. He died at his desk in 1934, when he was barely fifty years old.

In Africa, *H. erectus* and *ergaster* coexisted with other species of homo, such as *H. rudolfensis* and *H. habilis* around 1.9 million years ago. Ergaster's skull features, such as relatively thin cranial bones and a flat face, were comparatively modern. Ergaster evolved into *H. heidelbergensis,* a massive, stocky individual standing six feet tall and boasting a cranial capacity of about 1100 cc, 200-300 cc more than Erectus. While Erectus used the same tools with little improvement for over a million years, Heidelbergensis, being smarter, crafted superb spears like the one discovered at Schoningen, Germany.

Like Erectus, Heidelbergensis traveled and was capable of some language. He arrived in Europe about 500,000 years ago and died out 300,000 years later, leaving behind a jaw that was discovered near Heidelberg, Germany, in 1906. His other remains were found at Schöningen in Germany, in an underground labyrinth at Atapuerca in Northern Spain, and at Boxgrove in western England. Tools from Boxgrove suggested high intelligence, indicating these humans were fully-fledged hunters who targeted their prey. In Asia, Heidelbergensis may have evolved into *H. neanderthalensis* 500,000 years ago and into *H. sapiens* in Africa 300,000 years ago.

Clearly, there is a lot of evidence showing that we evolved from earlier species and we were not – as the Bible says – created in our present form.

NEANDERTHALS AND CRO-MAGNON

Though the Judeo-Christian scriptures don't mention them, Neanderthals lived alongside humans in Eurasia before they gradually disappeared by 28,000 years ago. We tend to deem Neanderthals unintelligent cavemen. But, despite this unflattering image, they were masters of the Ice Age. Their short statures, about 5 feet to 5.5 feet, bulky joints and bowed legs conserved heat, enabling them to survive -25° F temperatures. This gave them a rather spherical shape, just like people who live in cold climates today. Eskimos, for example, have a low shinbone to thighbone ratio. In contrast, the ratio of people like the Masai, who live in hot climates, is high, as the tall stature helps the wind fan and cool the body in the hot tropical climate. Paradoxically, Neanderthals' noses were large, unlikely to warm their breath.

Neanderthal children were vulnerable; half of them died before their twelfth birthday. A Neanderthal adult's life expectancy was forty-three years, although many did not see forty or even make it out of childhood. Despite their relatively large brains (a five-year-old's brain would have been the same size as that of a modern human adult's brain), they had to catch on fast to be able to survive. As above, the odds were not in their favor.

Neanderthals were also territorial. Their range was a day's hunt, and they hunted in close quarters, as evident from the many healed fractures – sustained in hunting combat – on their upper bodies. At their peak, Neanderthals numbered about 70,000, scattered over Europe and into Siberia, with the heaviest concentration in southern France, where 3,000 lived.

How did Neanderthals think? Despite severe injuries, many of the skeletal remains show healing, suggesting that Neanderthals took care of their sick. They could have tended to injuries using medicinal plants growing close to their caves, and when burying their dead, they might have held symbolic rituals. Burial sites containing pollen suggest that after placing the dead in a grave, they covered them with flowers, adorned them with animal skulls, and sprinkled them with ochre, or haematite (a red iron oxide earth pigment). Such behavior reinforced their humanity, showing they were compassionate and had a sense of grief and hope for a life beyond.

Like us, Neanderthals would have been mystified by nature. While racking their brains for answers, they would have happened on fruits, such as fermented wild berries and grapes, or mushrooms known as *Little Gods,* that induced the spiritual state. In a trance, they could have sensed spirits of animals they admired. Their brains when starved of blood or stimulatory input, like other hominid brains, would have produced near death experiences (NDE), which are related to mystic experiences. It means that they could have had spiritual experiences, too. But, although Neanderthal brain structure was very similar to ours, a sizable chunk of it bulged on the sides, indicating that it controlled their heavy musculature, rather than reinforcing the frontal thinking lobes. Steven Mithen, an archeologist at the University of Reading, England, believed Neanderthals had fragmented minds. Smart they were, but they were unable to daydream and concentrate intensely on a task such as making tools.

Why did Neanderthals die out? Around 39,000 years ago, a group of Early Modern Humans, Cro-Magnon, named for the place where their remains were first discovered in France, arrived in Europe. They were likely a hybrid between Modern Humans and Neanderthals in Asia before 45,000 years ago, and they travelled to the continent from the Middle East across the Bosphorus strait in northwestern Turkey. (As discussed in the next chapter, *Grimaldi*, Anatomically Modern Humans had arrived in Europe a few thousand years earlier than Cro-

Magnon did, so the Neanderthals would have had to contend with two species).

Cro-Magnon had long, fairly low skulls, wide faces with rectangular eye orbits, a poorly defined chin, a large 1600 cc brain, and stood five feet six inches tall on average, three inches taller than Neanderthals. Though physically weaker than the Neanderthal, Cro-Magnon was likely nimbler and smarter and had developed superior tools. Unhampered by the bulky elbow and narrow shoulder socket that condemned Neanderthals to a close-up, hand-to-hand, mortal combat hunting style, Cro-Magnon could throw a spear and kill prey at a distance. He developed communities and exchanged ideas about hunting grounds and new tools. And, as you well know and may sometimes regret, like Modern Humans, they could talk and talk fast, and – like a jet plane flying low – leaving Neanderthals wondering what just happened. Failure to compete with the new arrivals may have contributed to the Neanderthals' disappearance. Being self-sufficient, they did not see the threat as the new species closed in. For 200,000 years, Neanderthals had used about six types of tools, diversifying only towards the end, when they copied the newcomers. Hampered by lack of language and unable to exchange ideas, their ability to forge alliances with other clans was limited. The invaders with superior language skills outfoxed them. When glaciation peaked 35,000 years ago, vegetation dwindled, the land became semi-arid, and large animals disappeared, limiting food supplies. Being resourceful, the newcomers set traps or fished. They also had dogs to help them with hunting.

It appears, telling from stab wounds on their bodies, that Neanderthals fought with the newcomers in their stronghold in southern France. The newcomers had a range of tools, which enabled them to fight better and, in typical human fashion, take whatever they pleased. Forced to compete, Neanderthals fell behind. They sought refuge in enclaves such as the Byzovaya caves in subarctic Russia, where tool kits between 31,000 and 34,000 years old were found, and in the Iberian Peninsula caves.

Isolated, it became harder for Neanderthals to reproduce. Small clan sizes left them at a disadvantage, as their replacement rate could not sustain their numbers. But then again, the different species probably rarely met. Populations in many areas were sparse, only in the hundreds, and Neanderthals may have simply failed to adapt to a changing environment and petered out. By the time Cro-Magnon arrived, Neanderthals were disappearing. For reasons unknown, after two million years of existence, the last *H. erectus* also disappeared in Asia.

Little is known about the Denisovan species, which lived across Eurasia and Oceania between 300,000 to 30,000 years ago. Only bone fragments have been found in a Siberian cave in Russia that was also used by Neanderthals, with whom they split. DNA evidence shows that both interbred with modern humans.

In 2010, a team led by Svante Pääbo mapped the Neanderthal genome at the University of Munich. The team found that a tiny amount of Neanderthal DNA (1 to 6 percent) was like that of modern humans in Asia, Australia, and Europe, but not like those in sub-Saharan Africa.[10] Their mitochondrial DNA, however, was different from ours. Modern Europeans whose habitus resembles that of Neanderthals might have merely adapted to the cold the way the extinct species did, and are not their descendants.

CHAPTER 7

THE ODYSSEY

The deafening roll of a thousand drums
announces Homo sapiens coming.

~Henry Kakembo

When and where did modern humans, *Homo sapiens sapiens,* first appear? Until the 1960s, people assumed we evolved from Neanderthals in Europe 60,000 years ago. But Allan Wilson and his team from the University of California, Berkeley, overturned the idea. After analyzing mitochondrial DNA in samples from women of various ethnic groups, they stunned everyone. In Newsweek of January 11, 1988, they reported that every person on Earth was related to a woman they called "The Real Eve," who walked Africa 200,000 years ago. That magazine issue was the best seller that year.

In addition to genetic studies, fossil evidence identified Africa as the cradle of humanity, the Garden of Eden. In 1967, Richard Leakey found some of the oldest known modern human remains, skulls 195,000 years old, at Omo-Kibish, Ethiopia. He was unaware of their age at the time. But already, Hilary Deacon of the University Stellenbosch, South Africa, had discovered fossils of a modern human, along with advanced stone tools more than 120,000 years old in nearby caves at Klasies River Mouth. Deacon also teased out pieces of a skull with a vertical forehead, lowbrow ridges, and a prominent chin from a layer more than 90,000 years old. Such gracile features were hallmarks of modern humans, contrasting sharply with archaic robust ones like those of Neanderthals and showing that Neanderthals were not our ancestors, as was earlier assumed.

Today, small tribes of hunter-gatherers, the Khoisan, live in these South African parts. Although they no longer hunt in the old ways, their beliefs and lifestyles are fairly traditional. The ashes of burned shells and bones in their fireplaces are like those found at Klasies River mouth. They still speak an ancient language derived from a family of obsolete dialects. Because of 140,000 years of isolation, when their DNA (L0) is compared, it shows their community is the oldest on Earth. Their ancestors likely left behind the fossilized sand footprints on the shore of Langebaan Lagoon about seventy miles northwest of

Cape Town 117,000 years ago. Such finds strengthen the notion that modern humans originated in Africa.

Even today, our birthplace is still considered a mystery by some. Two opposing views exist: a multi-regional theory, and an out-of-Africa theory. The former assumes humans evolved separately on three different continents: from Neanderthals in Europe, *Homo erectus* in Asia, and *Homo sapiens* in Africa. But since we are one species today, how did the three evolve into one? Supporters of the multi-regional theory say the species met often and interbred, trading genes until a common race evolved. To permit the gene flow, such interbreeding required many humans. But the world was too big and ancestors too scattered for such contacts. The world population would have been 10,000 or less at the time. Also, genetic tracking torpedoes the mixed-race theory. DNA studies show that whereas African people are diverse, the rest of the world has a very similar genetic profile. As expected in an older population, DNA in Africans contains a much wider variety of mutations, 199 in all. In contrast, Europeans have 89 variants and Asians 73, showing the African population is the oldest by far. It means everyone living outside Africa comes from a small ancestral population that originated from that continent.[1]

VENTURING INTO THE WILDERNESS

When did our ancestors leave Africa? More than once. In an exodus about 120,000 years ago, a group of *Homo sapiens,* or Anatomically Modern Humans, AMH, hit the road. Why they did so, one can only guess. But wanderlust is in our bones. It helped our ancestors find new pastures during droughts, or evade predatory lions, biting mosquitoes, deadly snakes, and a blistering sun. Likely hugging the Nile, a natural route from the "Land of the Great Lakes," they trekked through North Africa to the Middle East, finding Neanderthals already entrenched there, perhaps going as far back as 185,000 years ago. But because of a deep freeze in the North that turned the region into a terrible desert, these wanderers died out 90,000 years ago. A jawbone found in a collapsed cave on the Israel coast and other remains at Skhul and Qafzeh attest to their failed venture. They were outlived by the Neanderthals, who preferred dry, cold weather to the hot, wet climate in which modern humans from Africa thrived.

Events such as migration or population crashes leave imprints on DNA that could serve as a record of major historical events. Such records show humans in Africa almost died out because of a drought 100,000 years ago. As the deserts expanded and erected barriers our ancestors could not pass, bands became isolated along the coastlines and in small forest communities. They verged on extinction, with less than 10,000 surviving. Eventually, populations bounced back. After the next glaciation dried out the land and expanded the Sahara about 80,000 years ago, people returned to the coastlines of Eastern Africa. There, with their tools and language abilities, they lived off the sea, fishing for their food.

When the fish ran out, about 150 of them packed up to try their luck elsewhere. Sea levels had dropped 150 feet and it was possible to escape the drought by crossing a narrow, shallow Red Sea at the Gate of Grief (now known as the Bab-al Mandab strip), and into present-day Yemen – only ten miles away – where they then headed east. This wasn't the first time they made this trip, though. Stone tools found in the United Arab Emirates show humans were there 125,000 years ago.[2] These earlier migrants also petered out. DNA studies show the ancestors of everyone outside Africa today left the continent starting about 70,000 years ago.

Although the migrants had no clue what lay beyond the horizon, distance and venturing into the unknown didn't deter them. Let's suppose that they walked, hunting and gathering, advancing only four hundred miles a year. In a generation, they would have covered the 10,000 miles to Australia, where a drop in the sea level had created islands.

They rode rafts or boats from Indonesia to New Guinea, ninety miles to the north of the mainland, and hopped the smaller islands to reach the coastline ten miles away. As they travelled, they improved their tools and became better hunters by the time they reached Australia about 50,000 years ago.

Among other groups leaving Africa were people with Mongol features, who, following the earlier emigrants' route, arrived in India 50,000 to 40,000 years ago. The Mongol-featured group eventually migrated to China, where some members continued on to Japan and founded the Jomon culture around 35,000 BCE, and the Ainu culture around 1,300 BCE.

PEOPLING EUROPE

People entered Europe in three major waves and several smaller ones when the weather began to warm up and glacial ice began to retreat 45,000 years ago. The first wave included Khoisan type people (mtDNA U5, strikingly similar to mtDNA U6 of the same age and was distributed across North Africa). They crossed the Straits of Gibraltar and fanned out across Europe and Northern Asia up to Mal'ta in Siberia. They were named Grimaldi for the Prince of Monaco, Albert I Grimaldi, who sponsored exploration of caverns containing their remains near Menton in southern France. Grimaldi were dark and gracile, and like tropical humans, tall, standing over six feet.[3] Like the Neanderthals who were already settled there, Grimaldi were hunter-gatherers, too, but with a complex culture, designing clothes, and making jewelry and fine tools from stone, bone and ivory, and beautiful cave art such as that found at Chauvet Cave in southern France. They left signature steatopygic (curvaceous hips and thighs) Hottentot Venus figurines along their migration routes. Cro-Magnon, arriving 6,000 years later, found them already entrenched.

About 8,000 to 7,000 years ago, a second wave, made of groups of farmers (Y-DNA E1b1b, originally from the Sub-Saharan Horn of Africa 26,000 years ago, and mtDNA H1) migrated from the Middle East across the Bosphorus strait into Europe, and from North Africa across the Strait of Gibraltar into Iberia, displacing the Grimaldi to the periphery. When the Grimaldi returned a few thousand years later, the two groups mixed.

Among the emigrants out of Africa 50,000 years ago were Dravidians and their albino offspring, the latter undoubtedly relieved to leave behind the intense, scorching African sun. On reaching India, because of their fair skin, they gravitated toward the cooler climate to the north. Later, they went through the Khyber Pass in the Hindu Kush mountains to the Central Asian steppes, where the sun's intensity was even lower. According to Thomas Malthus, since populations doubled every generation (approximately twenty-five years), their numbers grew exponentially, outstripping resources in an increasingly arid climate. By the Neolithic Age 12,000 years ago, from a group of ten people, their size would have conservatively mushroomed into the thousands. Being nomads, they moved around a lot, particularly after they domesticated the pony.

In the third millennium BCE, a third wave made of a mixed group, the Yamnaya, or Kurgan (the name of the tumulus or mounds they built over graves), travelled from the western steppes through Russia to Northern Europe, introducing the Indo-European language to the continent. Their arrival contributed to the collapse of Old Europe civilizations, such as that in the Lower Danube Valley, which peaked between 5,000 BCE and 3,500 BCE.[4] Today, their DNA (mtDNA H and Y-DNA R1b) is the commonest in Europe.

By 2000 BCE, the Central Asians had become formidable warriors, especially after breeding horses strong enough to pull chariots. Around 1200 BCE, several waves invaded Southeastern Europe. Some groups went through Thrace in the northeast of Greece to the Aegean area, decimated and brought down the black Mycenaean civilization on the Greek mainland, in towns, and on the islands. The inhabitants, wielding fearsome iron weapons, fought their way overland through the fertile Levant to Egypt. While in the Levant, they routed the Canaanites but spared the Phoenicians, leaving their cities untouched. Earlier, around 1200 BCE, a few of the invaders approaching from the Mediterranean Sea, had attacked the Egyptians at the Nile Delta. The Egyptians called them the *Sea Peoples*, and the name stuck. This second time around, Ramses III stopped them at the border and settled them in Canaan, which was a vassal of the Egyptians. They merged with the Canaanites and become known as the Philistines. Mycenaean pottery showing up in Canaan by the middle of the twelfth century BCE supports their identity.[5]

Why did the Sea Peoples treat the Phoenicians as though they were cousins? Apparently, they knew each other well. Phoenicians had dominated sea trade in the area prior to 1500 BCE. They would have included wheat in their shipments. When prices of wheat went up in the Aegean, western Anatolia (now mainly Turkey), and the Black Sea, areas the migrants came from, the Phoenicians supplied them with cheaper wheat. So, as they rampaged south through the Levant, they would have needed the Phoenicians for food. Canaan and Egypt, being breadbaskets, were stronger magnets to the invaders than was Phoenicia that didn't grow much wheat.[6]

In Greece, the Asian invaders intermingled with those who stayed behind and formed the Hellenes. Other invaders continued westward and displaced or assimilated with the black Italian population, the

Etruscans, and to become the Romans. Roman historian Tacitus (56-118 CE) described the invaders from the East as having fierce blue eyes, red hair and large frames (similar to people of the Pakistan Bhatti tribe). He also noted that they preferred the cold and did not mix with the locals, remaining "pure."[7]

In 300 CE, more waves of invaders comprising millions of people from Central Asia, such as the Germanics, driven West by the Huns from Mongolia, entered Europe in the north, called the barbarian invasions or *Volkswanderung* (migration of people). The Germanic peoples such as the Goths, Vandals, Angles, Saxons, Frisians, and Franks, drove the current inhabitants, the Celts and others, to the fringes. Under the leadership of King Attila (called the scourge of God) in 476 CE, the Huns destroyed the Western Roman Empire and plunged the continent into the so called 300-year Dark Ages. The Eastern Roman Empire continued as the Byzantine for another thousand years.

In the fifth century, Germanic tribes, including the Saxons, the Angles, and the Jutes from Germany, Denmark, and Holland begun to arrive in Britain. The island's inhabitants, the Britons, invited them in for protection against attacks from Scotland and Ireland after the Romans withdrew. To stave off Viking raids, the tribes merged their seven main kingdoms, Northumbria, Mercia, East Anglia, Essex, Kent, Sussex and Wessex, and created England. *See Illustration 7.1*

In the sixth century, the Mongols displaced into Europe the Slavs, the most numerous and linguistic body of people, from their original home in the Central Asian steppes. Then starting in the seventh century CE, the Turks, the last waves to come, followed, pushed west by the Mongols, until just before Genghis Khan arrived in the twelfth century CE.

In the seventeenth century BCE, Aryans, the ancestors of the Hindi today, migrated south from the Central Asian steppes to Iran and then to India. There they displaced the inhabitants, the Dravidians, now their distant cousins, to the center and south of the sub-continent. Genetic distance maps by Sarah Tishkoff and Cavalli-Sforza show white Europeans and pigmented Asian Indians, the Dravidians, through their albino offspring, are "closely related, alone together, separate from all other humans, like two peas in a pod."[8] Another group, the Zhou, migrated east to China, took over the black Shang Dynasty, and integrated peacefully with the indigenous Mongol population in 1064

BCE. The Zhou likely introduced there the now extinct Indo-European Tocharian language. Today, 2,000 generations later, about 45,000 years, lives a man in Kazakhstan, Niyazov, the direct descendant of the original Central Asia emigrants from South East Asia. His Y-DNA P1 (or P-M45, branches Q and R) is found in many Europeans, Asian Indians, Russians, and Native Americans.[9] Haplogroup P1 originated in Southeast Asia or East Asia.

RACE

Why is race such a loaded word if, according to the Bible, we are all related, descendants of God through Adam, Enoch, and Noah? Lk 3:23. Genetic studies have supported this relationship. Buddhists go even further, saying that everything in the universe is related and that we are all connected. We distinguish race mainly by skin color, which depends on the amount of the brown pigment, melanin, in the skin. Lack of melanin largely results from a mutation of a recessive gene, OCA2, or oculocutaneous albinism type II (albinism from Portuguese *albino* for white, Spanish *albo* and Latin *albus*). This mutation causes the commonest of about seven types of albinism. Those affected typically have white skin, blue eyes, and pale blonde to golden strawberry blonde hair (a tyrosinase gene mutation results in red hair and blue eyes for those with type OCA1 albinism).

To prevent sunburn or skin cancer such as melanoma, light-skinned people need protection from UV light, hence the preference for cooler climates than Africa. While people with the OCA1b variant may tan, those with OCA2 may develop only freckles or moles. Eyeglasses may be needed to correct visual abnormalities caused by lack of pigment in the iris. So, despite belief to the contrary, lack of vitamin D, or lower UV light in temperate climates such as Europe, did not cause skin color to change from "black" to "white."[10] Grimaldi lived in Europe for thousands of years, but their skin remained dark or swarthy unless they interbred with incoming Central Asian migrants. The Nenets, who live in the Russian arctic, farther north than any other people, are swarthy with black bristly hair, too. Also, some Eskimos are dark skinned. So, the name "Cushite" – the Latin term meaning burned faces, which was the name Greeks used for the old Ethiopians, or all Africans – suggesting intense sunlight caused skin color to turn black in Africa –

isn't entirely correct. Otherwise, light-skinned people living in hot sunny zones such as the tropics would become dark, too.

The concept of race, then, is problematic, as albino children (but not their offspring) would be classified differently from their black parents and siblings. Grouping people by ethnicity rather than race, as Japan does, or by culture, as Yuval Harari suggests in *Sapiens,* would be more appropriate.[11] Also, the Table of Nations in Genesis 10 does not mention races, only nations. It seems the word *race* as generally used may be a misnomer for *class,* for the *haves* and *have nots.* Such social class systems have bedeviled humanity for eons.

The idea of race originated in the United States, and was intended to distinguish white indentured servants from black indentured servants and African slaves in the seventeenth century. The two groups used to rebel together against their inhumane treatment by plantation owners. During Bacon's Rebellion in Virginia in 1676, in which the groups participated, the capitol, Jamestown, was burned down. To prevent future uprisings, it became expedient to separate the groups. But, since the nascent United States Constitution clearly proclaimed that all men were created equal, to justify slavery, blacks were demoted to heathens or sub-humans, and therefore, like animals, had no rights, thus justifying taking away their voting privileges and land.[12] Unlike Brazil, however, most black slaves in the United States and the Caribbean came from Europe. Many were prisoners of war from Scotland, Ireland, and the 1618 Thirty Years' War in Europe. Others, like their white counterparts, were indentured servants seeking a better life abroad but couldn't afford sea fare. For the journey, they sailed together with convicts and murderers, as Britain was wont to dispatch its unsavory characters to the colonies.

In fact, up to 20 percent of blacks from Europe weren't slaves. Some owned slaves themselves. According to "The Journey of Man, A Genetic Odyssey," by Spencer Wells, Y-DNA E1b1a is the most common haplogroup among black males in the United States with a frequency up to 60 percent. In Brazil, where most slaves came directly from Africa, it's only 4 to 19 percent, indicating the two groups came from different places. Haplogroup E1b1a and E1b1b are virtually the same, and E1b1b is common in Europe. Fouad Zakharia, and team demonstrated in their study, "The term 'African American,' was it a mistake?" that the genetic architecture of African Americans was distinct from that of Africans. Their conclusions strongly suggested that the term was indeed a misnomer.[13]

According to the Emory University history project, "Voyages: The Trans-Atlantic Slave Trade Database," the total number of slaves imported from Africa to the United States was only 305,326. In contrast, over 5,000,000 were imported to Brazil. Interbreeding in the States then would not solely have been responsible for the high percentage of E1b1a in North American Blacks, particularly as intermarriage was strictly forbidden by law. Penalties for violation were severe, including banishment or thirty years of servitude by the children, if they survived.

One must, then, look to geography and proximity to explain the great numbers of blacks in Europe, some of whom eventually ended up in America, many of them as indentured servants. Applying genetic distance measures to various populations, Cavalli-Sforza found the distance between Sub-Saharan Africa and Europe the smallest between continents. This is not surprising, as the two continents are contiguous at the Sinai Peninsula and very close at the Strait of Gibraltar. This proximity would account for the influx of human species from Africa into Europe starting with *Homo erectus.*

DISCOVERING AMERICA

America was settled from different directions over time. Stone tools and fire pits in Serra da Capivara National Park in central Brazil indicate people lived there 50,000 years ago. Skeletal remains and tools, and rock paintings depicting giant armadillos and mastodon hunting allow researchers to track their movements northward from South America as ice retreated.

In 1971, an 11,500-year-old skull with female Negroid features was discovered in Northeastern Brazil, and over forty similar remains were discovered in a nearby cave. The skull was dubbed Luzia. The remains probably belonged to people who arrived in Brazil from Africa riding the powerful South Equatorial Current. The same current also could have carried the five African fishermen washed up ashore there in a storm after only a few weeks at sea in 1996.[14]

Monkeys may likewise have made the same journey thirty million years ago, drifting across the water on large clumps of vegetation and soil. Contrary to what some archeologists say, the skulls in

Northeastern Brazil were unlikely to be those of Polynesians from the Pacific or Aborigines from Australia. These places are twice as far away from Brazil as they are from Africa, which is only 1000 miles away, and – unlike Africa – lack direct sea currents to carry boats there. The intimidating mighty Antarctic Circumpolar Current flows eastward, driven by strong westerly winds. Old Australasians' genes do not show up in people along the Pacific route to South America either, negating the theory that the migrants took that route.

Luzia's features are similar to those of a teenage girl with negroid features, Naia, whose submerged remains were found by a team of scientific divers as they explored a deep Yucatan cave in 2007.[15] Naia, who lived 12,000 to 13,000 years ago, shared mtDNA Haplogroup D1 with some Native Americans. Nuclear DNA should be even more telling, for instance, about her admixture.

Both the skulls referred to above resemble the modern people of Africa, Australia, and the South Pacific, and not modern Native Americans. Naia's skull also resembles that of 9,000-year-old Kennewick Man, once thought to be Caucasian, whose remains were discovered in Washington state in 1996. Such skulls likely belonged to the immigrants who made up the bulk of the early native population. Following Christopher Columbus's arrival, up to 90 percent of these early immigrants perished at the hands of settlers. Some were given smallpox-contaminated blankets, and others were scalped. According to Historian Professor Roxanne Dunbar Ortiz, the bloody corpses left behind were known as "Redskins," the source of that name.

By studying artifacts such as stone tool kits, living quarters, human footprints, or coprolites (fossilized faeces) found at excavation sites, archeologists posit that people arrived in the Americas by several routes. Some migrated overland from Asia to Alaska across the Beringian land bridge exposed during the glacial maximum when water was sucked up by glaciers over 13,000 years ago. Others travelled by boat along the northern Pacific rim, then turned southward along the coast. Also, groups from the Iberian Peninsula likely sailed along the edge of an ice cap to Newfoundland. And, beginning 50,000 years ago, emigrants from West Africa could have crossed the Atlantic to Brazil.[16] Without written records, memories of where people came from faded quickly. As mentioned earlier, Native American origin stories say that their ancestors migrated or emerged from the ground.

Life was challenging for the immigrants. To survive, they used their wits and planned ahead. Hunters, for example, learned to fish.

"Use of tools showed they could pose the what-if questions," Ian Tattersall with the American Museum of Natural History at the time, said in *The Evolution of Man* TV documentary. "They were capable of symbolic reasoning, which allowed them to imagine situations they had never seen or that may never have existed. They manipulated those situations in their minds, revealing their behavior potential."[17]

As glaciers retreated and retracted, a new America emerged. The giant mammals –mammoths, mastodons, and saber-toothed cats – vanished. Most of the Clovis people, named for their spear points and long thought to be the first Americans, also disappeared. Dry, dusty conditions following the Younger Dryas climate shift when the Ice Age made a brief comeback and plants and animals died in 11,700 BCE – probably contributed to their disappearance. An asteroid strike also may have helped. But this time, though, the newcomers – already skilled hunters – survived a looming food crisis and thrived.

CONQUERING THE WORLD

Emigrations out of Africa included the Anu, 20,000 years ago, whose remains were found by European explorers on every continent. About 4,000 BCE, Cushites, modern humans from the Highland regions of central Africa, hit the road. They settled the great river valleys, founding the Sumerian, Elamite, Xia, and Harappan civilizations, taking over former Anu trade centers, including those in the Americas.[18] People permanently settled Britain and Ireland about 12,000 years ago and New Zealand about 800 years ago, finally occupying the whole planet.

WERE THE WANDERERS FULLY HUMAN?

People question whether humans coming out of Africa were fully developed in the same way we are. Were they capable of symbolic thought, art, music, storytelling, and religion, activities that distinguished us from the beasts? To understand these early people and gain insight into what went on in their minds, scientists searched for traces they left behind. They looked at the relationship between bones and tools and the people who made them. Archeologist Christopher Henshilwood, studying 90,000-year-old tools from caves in South

Africa, said these people showed advanced thought and planning. For instance, they obtained ochre from a source twenty miles away. With a sophistication not found in Europe until 50,000 years later, they used the rock to make different kinds of tools, such as a double-edged piece for piercing leather to make clothing. This meant people did not merely use any rock they found lying about. They thought about what they had to do, made plans, and skillfully executed them.[19]

Henshilwood also noted that shellfish found in their caves showed these people consumed a wide range of foodstuffs, including fish. "Fishing required cooperation between people in a group. To catch fish, they had to have a template in their minds, a plan of how to do it," he said. Layers upon layers of ashes in the caves more than 100,000 years ago revealed that they had prepared their food over hearths – hearths they would have had to build, intentionally, to cook. So yes, our ancestors were capable of complex thought.

Were they capable of symbolic thought such as art? Art found in the South African caves was created using carefully stored ochre more than 100,000 years ago. The red permanent iron oxide pigment, was also probably used to paint their bodies with, as the Khoisan still do. Shamans, or Wisemen, also used ochre to record trance experiences on rock, depicting their sojourns into the spiritual world. This early art graces rock walls in South Africa, such as the Drakensburg rock shelter on the Great Escarpment. Paintings were usually of animals placed on top of one another, particularly the eland antelope. This motif repeated in French cave and Australian rock art. Archeologists and psychologists believe that the stories behind these paintings could be a way into understanding the minds of the ancient people.

Shamans painted the animals they saw around them. In some way in the drawings, the animals appeared to interact with the rock face, which, according to archeologist Thomas Dawson, was like a membrane between the real and the spiritual world. The animals' spirits would reside beyond the membrane. The artist used specific natural features in the rock to enhance a painting. For instance, he drew an eland stepping out of a crack in the rock face and himself with an eland head and hoofed feet, holding the tail of a dying animal, thus indicating the two were one, the essence of a mystical experience. Then he made a handprint on the wall, as though reaching through the rock face into the spiritual world yonder.[20] Similar handprint signatures adorn rock art in France and Australia.

The African continent has over 8,000 prehistoric artifacts, half of them in Tsolido, Botswana. One of the earliest artifacts, small stone plaquettes, 27,000 years old with images of animals, is found at the Apollo 11 shelter in Namibia. Another, a decorated 70,000-year-old ostrich eggshell, lies at Diepkloof Rock Shelter. Out in the open, such art was vulnerable, and much of it likely fell to the vagaries of weather over the years. The oldest rock art in Europe - found in a cave at El Castillo, Spain - dates to around 40,000 years ago. Another can be found in a cave at Chauvet, France, dated 33,000 years ago. Australian Aboriginal rock art is very old, too, over 40,000 years. Ochre is used as the dye in all these paintings.

Similarities among rock paintings on all three continents are uncanny. In many, animals are covered with dots and grid patterns as they appear in trance experiences and hallucinations (see Chapter 10). In the French paintings, figures of men are rare and hidden. They appear to be in extreme pain, indicated by spear or arrow-like lines radiating from their flanks. One figure is called the Wounded Man. By appearing to show pain in the kidney and stomach areas, the paintings depict a Khoisan shaman's experience during an hours-long, trance-inducing dance. The Wounded Man, then, is a shaman. During the ritual, the shaman bleeds from the nose, and falls forward like a dying eland. Afterwards, he relates experiencing taking on the animal's power. This comforts the people and, through the mind-body link, boosts their health.[21]

In a cave at Blombos, South Africa, Henshilwood also found a 90,000-year-old piece of red ochre with a diamond mesh pattern. The intricate design signified early art and intelligence, indicating symbolic thought. Such artifacts strongly suggest that the small groups of people leaving Africa beginning more than 60,000 years ago were fully human. Their brains were bigger than any other species matched to body size. Though physically weak, they were innovative and capable of solving problems, each danger helping to forge a sharper mind.[22] They were also deeply spiritual, so on striking out, they were fully prepared to settle the whole world. They were us.

KINGS, QUEENS, AND GODS

Fifteen thousand years ago, the weather warmed up briefly. The world population was about one million humans, and if people remained hunter-gatherers, it was going to be hard for populations to grow. During the Younger Dryas event, the ancients realized they needed to save food for a bad year. With main staple food yields running out, people became resourceful. About 10,000 years ago, from the Sahara to the fertile Sumerian riverbanks, and from the Indus Valley to the Yangtze River valley in China, they started growing crops.

At the time, the Sahara was lush. People grew sorghum, millet, rice, ensete, and sesame by 6,000 BCE. They domesticated cattle and made pottery. Ethiopia joined in by 5,500 BCE, and by 5,200 BCE, the Fayum, the Fertile Crescent to the North in Egypt, was growing wheat and barley and raising sheep, goats and pigs. In a first, in 2,000 BCE, the Egyptians started ploughing, yoking a bow-shaped stick, or hoe, to cattle horns. Farming spread into Europe from the Middle East with the Linear Pottery Culture around 5,500 BCE and was underway in the Gulf of Mexico by 5,100 BCE. People began working metal and crafting tools, too, the advent of technology.

Farming enabled trade, cooperation, order, and complex societies. The first state organization at the time emerged 5,200 years ago or earlier at Qustul in Nubia. Jean Champollion, the Rosetta stone wizard, said Qustul could have been moved to Hierokapolis in Upper Egypt in 5867 BCE, the time he established the Narmer Palette was made.[23] At the time, occupied by the Anu, Egypt had forty-two provinces (*nomes*), each belonging to a different ethnicity. By relocating the capitol to Memphis, King Menes, also called Narmer, symbolically united Upper and Lower Egypt. In Memphis, the first recorded dam was constructed by 4,000 BCE. Over time, kingdoms grew into empires, then into dynasties. Kings and queens ruled them as representatives of the gods.

WHENCE CIVILIZATION?

Where did civilization begin? That is a tricky question. But, as understood to be social or political organization with division of labor, commerce and technology, civilization didn't come from aliens out of space. Extra Terrestrials did not build the Pyramid of Giza, as some, Egyptians included, say. Nor, contrary to what others believe, did it start with the people of Atlantis, as this Plato's island has never been found (the Olmecs, discussed below, may have a clue). The answer could depend on one's origins or funding, tossing objectivity to the winds.

Civilization and religion have gone hand-in-hand for ages. According to Yuval Harari, the global political order was built on monotheistic foundations.[24] The Catholic Church played a major role in shaping it for Western civilization, and in Rome, the emperor was subordinate to the pope. Religion's influence on government shows, for instance, when government officials are sworn into office on the Bible, or witnesses are sworn in before giving testimony in court. In the Sahara, the earliest examples of writing, generally considered a hallmark of civilization, in addition to keeping merchants' records, preserved religious doctrines and recorded obituaries.[25] Also, in Egypt, order was the will of the gods such that after Ra organized the world from chaos, his daughter, Maat, goddess of balance, harmony, and order, prevented it from relapsing. Maat's consort, the god Thoth, "invented" writing. As Hermes Trismegistus, the Greeks credited Thoth with authoring all works of science, religion, and philosophy. Hence the name for the Egyptian script, *hieroglyphic*, Greek for "God's Words" (*hieros* means sacred, and *glypho*, inscription).

Historians like Herodotus spoke to witnesses of events in the places they visited and wrote about. But, at the turn of the nineteenth century, scholars unaware of such records tended only to look for evidence confirming their beliefs, ignoring or downplaying that which contradicted them. Called *social rigidity*, it increased as they aged, and they became more inflexible and less likely to abandon their narrow approach. So, those coming across the ancient Greek writings claimed their authors knew not what they were talking about when they said the Greeks learned all their traditions from Egypt, or that the Egyptians had black skins and frizzled hair. Thus, they failed to recognize Africa's primacy in the advent of civilization.

To early Europeans, Africa seemed too backward to contribute to civilization. As elsewhere, African civilizations had risen and fallen, becoming victims to time (Mapungubwe, South Africa) or abandonment (Blue Swallow Nature Reserve 200,000 year-old gold metropolis, South Africa).[26] The latter was likely behind the legend of King Solomon's mines and that of Atrahasis from Sumeria. Other civilizations didn't stand a chance against ruthless invaders (Meroe, Egypt, Sudan), slow decline (Axum), and, more recently, political shenanigans (Great Zimbabwe), and were forgotten.

Scholars, for the sake of obtaining funding for research, would present their work in a form to please sponsors. In 1916, for example, Henry Breasted revised his school textbook, *Ancient Times,* to read that the Egyptians belonged to "the great white race" instead of "a brown-skinned race."[27] Others coming upon the Twenty-Fifth Nubian Dynasty identified the Cushites as Caucasian and the pharaohs as "white." In 2013, a Swiss DNA company, Igenea, after watching microsecond TV snippets, announced that, based on bits of King Tut's genetic profile (despite his dolicocephalic skull), 70 percent of British men and 40 percent of Europeans were genetically related to the king. DNA Tribes, a Virginia genetic ancestry testing company, countered by showing that the king's ancestry was mainly Central African (*see* Chapter 2 for expanded discussion on this topic).

Sumeria in western Asia could not have been the cradle of civilization, despite nineteenth century scholars' claims. Archeologists, anthropologists, historians, and others have yet to find evidence pre-dating that from Africa. After working the first Meroitic cemetery on record, British excavator Leonard Woolley considered Sudan a backwater, saying civilization there was an isolated phenomenon with no importance to culture or art elsewhere. Instead, he attributed such influence to Mesopotamia, where he found many artifacts at the Ur of the Chaldees. Such discoveries, however, showed the Sumerians responsible for civilization came from the Nile River valley.[28] "In their literature, the Sumerians called themselves 'the black heads,'" writes Runoko Rashid.[29]

Relying on Franz Kugler's astronomical calculations, the chronology of Viktor Christian dated the Sumerian Ur Dynasty to between 2,600 BCE and 2,580 BCE, at least a millennium or so after Egyptian history began, based on the over 6,000-years-old Narmer Palette. According to C. C. J. Bunsen, such areas owed their civilizations to the Hamitic family, mainly the Cushite branch.[30] Indeed, "The ancient Ethiopians were the architectural giants of the past..." Thomas Maurice said, "It was they who built the tower of Babel or Belus, and raised the pyramids of Egypt."[31]

In the Bible, Nimrod, Ham's grandson and the mythical founder of Mesopotamian civilization, was a Cushite. Ham was the ancestor of black people occupying ancient Ethiopia (Cush), Egypt (Mizraim), Libya (Put) and Palestine (Canaan). Gn 10: 6-8. According to the Babylonian Talmud Bible and Islam's interpretation of Noah's curse against Ham's son, Canaan, Ham designated black as in Kemit (*Km-t*, Hebrew *Kham)*, a name the ancient Egyptians called themselves. The Sumerians also stayed in touch with the Egyptians from the outset. Their early texts mentioned Magan and Mcluhha, which an Assyrian inscription said were old names of Egypt and Ethiopia.[32] So, as Bunsen wrote in *Philosophy of Ancient History,* the Hamitic family was the fountainhead of civilization.

As these people from ancient history left Africa, they took their know-how with them to India, Mongolia, China, and Japan in the East, and to continental Europe, Ireland, and Iceland in the north. In Europe, they mixed with earlier immigrants from 45,000 BCE in the Aurignacian era.

Awestruck by a 3,000-room Labyrinth at Tanis, Herodotus billed Egypt the cradle of civilization, which he estimated was more than 10,000 years old.[33] This civilization, according to Diodorus Siculus, first-century BCE Greek historian, originated in Nubia. Siculus quoted the ancient Ethiopians as saying they were the oldest people on Earth and Egypt was a colony of theirs.[34]

Tracing the roots of modern culture, mincing no words, Massey said, "… Egypt for the mouthpiece and Africa as the birthplace."[35]

CIVILIZATION SPREADS TO THE EAST

Iranian legends say a race of black men with short, wooly hair occupied the region from ancient Persia to India. In 2,500 BCE, they founded the Harappan civilization in the Pakistan Indus Valley that flourished until 1900 BCE, when they migrated into India. Their presence is recorded in names such as the Hindu Kush Mountain range and the Iranian province, Khuzestan, meaning the land of Khuz or the Cushite. They are today's Dravidians.

The presence of prehistoric and historic blacks in China and Japan can be teased from folklore, names of dynasties, and their founders, and writing such as Manding script on the shang bones. People from the fertile African crescent would have reached China by way of India, establishing the Xia and Shang/Yin dynasties. In 2005, DNA analysis by specialist Jin Li and team confirmed the African origins of the Chinese.

Emigrants carried their gods along with them. On the Indian Central Plane, the Dravidians renamed them Brahma, Shiva, and Vishnu, or his incarnation Krishna, after the Egyptian trinity. Krishna and Kali were black like their worshippers, while the first three clearly carried African concepts. Siddhartha Gautama Buddha (likely named after an Egyptian priest, who fled crazed Cambyses II) and Confucius were relatively recent additions to the list of priests or gods, who clearly showed African features in their early portrayals.

THEY CAME BEFORE COLUMBUS

It boggles the mind that ocean traffic between Africa and the Americas took place over 20,000 years ago, but the discovery of Luzia Woman's skull and others like it support this. People then were restless and inquisitive, much like us today. Whereas some crossings could have been exploratory with families, or search parties, others were likely accidental, stranding sailors on foreign territory.

For the treacherous journey, people were building boats such as the Nigerian wooden Dafuna canoe as many as 8,000 years ago, and the Egyptians invented the sailboat by 7000 years ago. The sailboats were used when visiting Punt, the land of their ancestors, 4,000 years ago, and they achieved speeds up to nine knots.[36] Even papyrus reed boats were up to the task of crossing the Atlantic Ocean as the *Ra II* demonstrated in 1970. The Portuguese found Ethiopians with maps and charts showing routes extending to the new world long before Columbus went there.[37] Caribbean Indians had stories of trading with black people. Their wares included spear points made of a yellow metal called guanin, a mixture of gold, silver, and copper in the same ratio as the metal produced in African Guinea. The word *guanin* can be traced to Vai from a group of Mande languages of West Africa.[38] This reinforces the idea that trade between Africa and America took place in antiquity.

When Columbus returned from the West Indies in 1493, instead of going to Spain, whose monarchy had sponsored him (actually, Ferdinand II's finance minister, Luis de Santangelo, provided the funds), Columbus rerouted – by design or storm – and docked in

Portugal, where a beaming King Don Juan welcomed him. Nonetheless, Columbus was guarded. Had the king not tried to kill him to prevent him from setting out on the journey to the west? Don Juan had heard about mariners and traders from Africa carrying merchandise into the southwestern waters below the equatorial line, but had ignored the information. Now, looking at a triumphant Columbus, he was beside himself with grief (and possibly regret) for not sending an expedition. Lest God dispatched his soul to Hell if he killed Columbus, the king spared his life, opting to vie for territory with Spain instead.

Going by a map of the Americas drawn at Don Juan's request and based on information obtained from African mariners in Guinea, traders traveled regularly between the two continents. The Africans most probably rode the North Equatorial Current to Central America and returned via the Gulf Stream that, at Spain divides into the North Atlantic Drift and the Canary Current going south into the Guinea Current off the African coast. This would explain Native Americans found living in Berber country in North Africa.[39] To get to north Brazil, West Africans would have travelled by the Canary Current to the Equatorial Counter Current that bathed the shores on both sides.

Sertima extols eleven helmeted realistic negroid stone heads (now over seventeen) found at sacred Olmec sites, such as Trez Zapotes, La Venta, San Lorenzo, and Vera Cruz in Mexico. The first statue was excavated by Matthew Stirling from Trez Zapotes in 1938. Made of a single block of basalt, it measured eight feet in height and eighteen feet in circumference and weighed over ten tons. It matched in style and conception the heads of Nubian blacks found at Tanis, the maritime Egyptian harbor that served as the Nubian kings' Egyptian capital (Zoan in the Bible). The following year, Stirling discovered four more similar colossal domes in a ceremonial plaza at La Venta, a site that later turned out to be the holy center of the Olmec and the seat of their civilization. Of the statues, one of the heads wore ear plugs, curved with a cross resembling the Egyptian fertility cross, a symbol of regeneration. Their helmets were also similar to the leather helmets worn by the Egyptian-Nubian military during the era of the Rameses pharaohs. Also found at the plaza was the earliest pyramid ever built in the Americas, arranged in a north-south axis, "as all Egyptian and Nubian pyramids" were placed. Elsewhere, American pyramids resemble Imhotep's Step Pyramid with similar measurements and the same north-south axis orientation.[40] Pyramids were also oriented toward the star Sirius. As with the Egyptians, the massive building blocks, weighing between two and sixty tons, didn't have to be hauled from eighty miles away. It was impossible. They would have been made on site, as described below. The Olmec stella 5 of Izapa, "The Tree of Life," like a memento, depicts the three pyramids of Giza with the temples in front.

The Olmec statues are dated from 900 BCE, or much earlier (the helmeted ones from 800 BCE) to 400 BCE, or less. These dates are imprecise, as many statues were moved. When the Egyptians landed in Mexico, they could have added to the colossi above, as some heads resemble those found at Tanis.

The statues, like those with negroid features found at Durham, Cambridge, and Benwell in England, represented priest-kings or gods, and were worshipped. Next to the stone heads at La Venta lay numerous clay pieces (terra cottas) and portraits in jade and stone with kinky hair, broad nose, and generous lips, complementing the African features of their builders. In 1974, at the 41st Congress of Americanists in Mexico, a skull expert, Andrzej Wierciński, reported finding a prevalence of negroid skulls of a continental African type at Olmec sites, supporting pre-Columbian African presence there.[41] As the earliest monoliths show negroid features, and gods look like their worshippers– the makers - the founders of the Olmec –civilization - were negroid.

Africans migrated from Nuwba, an area comprising modern day Sudan (Nubia in Latin), Ethiopia, Kenya, and Uganda, to Mali, Dogon country. Then, as their predecessors had done for millennia, riding the North Equatorial current, they reached America 5,000 years ago. They called the new land Utla (Cushite for vacate), and later, because it was divided into north and south, used the plural Atlan, from which the names *Atlantis* and *Atlanta* derived, which solves the origins-of-Atlantis puzzle. The Olmecs (Olmec stands for "rubber people") introduced the rubber tree, indigenous only to Africa, to the Americas. In Africa, they used the sap to make rubber capes, coats, and shoes and were the first to add rubber soles to shoes. They also made large rubber balls and used them during ceremonies in stone arenas. This gave the Spaniards the idea of soccer.

In 727 BCE, Egypt was in decline and foreign kings were calling the shots when a Nubian king, Piankhy, swept in and expelled the invaders. He did it with such ease that people in surrounding towns at Memphis "...fled headlong and no one knew where they went," probably into the hills.[42] Piankhy restored Egypt, launching the Twenty Fifth Egyptian Dynasty, which lasted from 760 BCE to 656 BCE. Sometime during this dynasty, most likely during the reign of Piankhy's son, Taharka, a Phoenician ship set sail looking for iron. As the ship rounded North Africa, it veered off course, and fell into the ocean's powerful currents.

The crew and passengers, including Hittites, reflected the Nubian-Egyptian-Mediterranean milieu at the time. Since 1,085 BCE, the Nubian military was the most powerful militia in Africa and had become the main force behind power factions in Egypt.[43] Although the Phoenicians were mercenaries and vassals of the Egyptians then, the two kept a close relationship. The Phoenicians operated a profitable maritime trade in metals and other commodities and the Egyptians traded with them for centuries, using their ships to transport trade. The Egyptians also profited from the tributes the Phoenicians paid them. It wouldn't come as a surprise, then, that the two would become stranded together in a foreign land. Next to the colossal negroid stone heads on the plaza at La Venta stands a slab of stone etched with a figure with an aquiline nose, thin lips, a long beard, and wearing turned up shoes, elements found in Phoenician culture. A Phoenician traveling mascot, the god Melkart, was also found at an Olmec site at Rio Balsas in Mexico, further supporting the Phoenicians' presence there.[44]

The Egyptians profoundly influenced the Olmec culture. A cluster of traits shared by both civilizations shows that similarities between the two couldn't be coincidental. For instance, the double crown signifying the union between Upper and Lower Egypt adorned by an Olmec dignitary speaks to such influence in royal circles. Other similarities include the royal flail in an Olmec painting, the use of purple to denote high rank as in Egypt, and the pharaohs' artificial beard appearing on the usually American hairless chin. Also, feathered fans, or sunshades, in paintings in the pyramid of Las Higueras are almost identical to the Egyptians,' as is the parasol, or ceremonial umbrella, an emblem of royalty in both cultures, that in Mexico is traditionally said to have come from the east, or the other side of the water.[45]

Mexican practices like circumcision and the ceremony of the opening of the mouth of the dead were also Egyptian, as was the Mexican royal incest practiced among siblings to keep royal blood pure, which started after the arrival of the Nubians. The Egyptians had abandoned it a few centuries earlier when Ramses II sent for a Hittite bride, but it never ceased in Nubia, and the Ptolemys revived it in Egypt. By regularly killing one another for the Egyptian throne, however, the Ptolemys ran out of eligible mates. Cleopatra was the last. Common characteristics were also shared by Ra and Horus and American Mayan sun god Kinick Ahau Ixhel.

Egyptian scripts also turned up in ancient America as well. Hieroglyphic writing was established there no earlier than 650 BCE. The Micmac, a tribe of Native Americans living in the eastern provinces of Canada and closely related to tribes in Maine, used many hieroglyphic signs in the Lord's Prayer that were similar to the Egyptians', suggesting a contact with the Egyptians or others related to them.[46] Language similarities can also be a giveaway. *Ra,* in Egyptian, Mexican, and Peruvian, was the sun, and its hieroglyph was similar in the first two languages.

From their Olmec center, the Africans' influence spread throughout North America. Containing secrets of regulating the calendar, the Davenport stele, dated at around 800 BCE and now housed in the Putnam Museum in Iowa, has inscriptions in Egyptian, Phoenician (Iberian Punic), and Libyan. Stone portraits of people with Nubian, Phoenician, and Iberian features or clothing have also been found in Davenport fields. In Mexico, the African visitors are remembered in oral traditions for their black skin color and giant size; they stood an average of a foot and a half taller than the Mexicans.[47] Totemism, the idea that a group, or clan, is protected by a certain animal it is named after, is practiced by Native Americans as well as Africans. In Uganda, the Baganda do not marry within their own totem group, or clan, or hunt the clan's totem animal for food.

The Olmec people likely blended with the indigenous populations and succeeding emigrants. Their traces are found in the Hopi Indians' ceremonial dance, identical to the Dogon Bado dance, complete with symbols and spirit names.[48]

Centuries later in 1311 CE, Abubakari II, King of Mali, temporarily ceded his throne to a deputy, his brother Kakou Musa I, and set sail with a fleet of two thousand boats to search for knowledge in the world yonder. That Abubakari reached Mexico is not disputed. But when the king left Mexico fifteen years later, he must have succumbed to the merciless high seas and was never seen again. His brother, now Emperor Mansa Musa, expanded the empire and became the richest man in history. His worth in gold would have been the equivalent of trillions of dollars in 2020 prices. While passing through Egypt on his way to Mecca, he handed out the precious metal like candy.

Approaching from the east, Abubakari appeared to the Aztecs to be the Sun's messenger, Quetzalcoatl, the plumed serpent. He showed up at a time when the cyclical god was due to return from the land of the rising sun. As the ships approached, the natives recognized the god's emblem in the outlines of a great golden bird, the serpent-slaying eagle, emblazoned on the king's parasol.

In fact, Quetzalcoatl was the counterpart of the Dasiri of Abubakari's ancient Manding Mali tribe. The Dasiri's' sacred animal was also a snake. Ceremonies on both sides of the Atlantic, such as penance with whips and bloodletting with thorns, were also similar, indicating that they shared beliefs.[49] The Africans brought the god with them when they initially migrated to America.

The falcon-like parasol emblem, then, represented the African snake hunting "secretary bird" that made its home south of the Sahara. First Dynasty Egyptian Pharaoh Menes adopted a logo – the winged disc symbol of two intertwining snakes, denoting the union of Nekhbet (Upper Egypt) and Wadjet (lower Egypt). Nekhbet was the vulture (represented by a bird) and goddess of the northern kingdom. Wadjet was the serpent, goddess of the southern kingdom. In their endless clashes, really territorial tussles, the bird as the hawk, or falcon, represented Horus, and the serpent represented Set, who was usually shown cowering in a hole. Over the hole would be a pole capped with Horus's falcon head ready to flatten the snake's ugly face as it popped out. In real life, the long-legged secretary bird sneaks up on the slithering abhorrence, and stamps on its head.

Today, the secretary bird is emblazoned on the South African coat of arms. It's the national emblem of Egypt's neighbor, Sudan. Its cousin, the eagle, is the national symbol of both Egypt and the United States and appears on stamps in many countries.

THE QUEST FOR KNOWLEDGE

Ancient Egyptians excelled in mathematics, astronomy, literature, law, boat building, mechanics (a model of a glider from the fourth or third century BCE, number 6347, sits in room 22 of the Cairo Museum), and architecture. Their architectural skills are still on display in the pyramids, huge temple hypostyles, Ramses II's Abu Simbel temple, and obelisks (many still standing) weighing up to 1,000 tons. Egyptians built in–stone early - Djoser's Saqqara Pyramid complex around 2611 BCE,–for example - and the Greeks consulted them when constructing the Parthenon. The Moscow Mathematical Papyrus, dated to about 1850 BCE, and the Rhind Papyrus at Luxor, dated to 1650 BCE, contained trigonometry and algebra, what some have called Aha calculus, testifying to the Egyptians' mathematical skills.

Egyptians were foremost in medicine, too. The Edwin Smith papyrus recorded results of injuries to various parts of the brain from nearly five thousand years before Broca. The papyrus listed guidelines for treatment of brain injuries as well as injuries to upper parts of the body. The Ebers papyrus in the Memphis Temple Library shows they associated heart function with circulation or pulses. For their research, Hippocrates and Galen consulted and credited the annals of Imhotep, the Saqqara pyramid builder, also found in the library.[50]

Since only priests and their students sworn to secrecy were privy to academic or special knowledge, we cannot tell all that the ancient Egyptians discovered. Were it not for the Greeks, most of this knowledge would have died with the ancient Egyptians. But then, instead of the Egyptians, the Greeks are given credit for the discoveries. For instance, 1,500 years before Archimedes, the Egyptians had worked out the surface area of the sphere, recorded as problem 10 in the Moscow Mathematical Papyrus,[51] and the volume of the pyramid's frustum recorded in the same document as problem 14. For a long time, people puzzled over how the Egyptians built the pyramids, or made, transported, and erected the obelisks and huge monuments. In the documentary movie, "The Great Pyramid K2019," director Fehmi Krasniqi,[52] shows how the pyramids could have been made on site by mixing concrete, as minerologist Jose Davidovits does in the movie. They mixed limestone with a white clay called kaolin. Then, they mixed together caustic soda (lye) dissolved in water with the limestone and clay, along with a little bit more water, to form a paste, which they poured into a mold as we do to cement. After a month, the paste hardened into a white stone block. Obelisks and monuments were made from granite melted with solar lenses and molded before it cooled.

The Egyptians derived the value of pi using it in circle and sphere measurements, today known as problem 50 of the Rhind Papyrus.[53] Archimedes only refined its value. That he invented the endless screw to raise water while in Egypt, where it is still used, seems too coincidental. The Egyptians also developed measuring units and constants, such as the meter and kilogram based on the one-centimeter size or weight of a drop of water, the Royal Cubit, and Phi, also known as the Golden ratio. They called the latter "The Addition Sequence" millennia before Fibonacci brought it to Europe. According to Krasniqi, since the constants were universal in nature, the Egyptians believed they were sacred and, to remain connected to nature themselves, they integrated the constants liberally in the pyramids, temples, and so on, "to do as Gods do." The Egyptians' obsession with geometry is on display in the incredibly precise measurements inside the pyramids and their chambers.

Pythagoras was unforgiving if students disclosed his teachings, even though he broke his own oath of secrecy to the Egyptian priests, as mentioned in Chapter 2, by attributing the Egyptian discoveries to himself. Such Greek "inventions" led Clement of Alexandria to exclaim, "A one-thousand-page book will not be long enough to cite the names of my fellow countrymen who have used and abused the Egyptian science!"[54]

The Samaritan Simon Magus, like many Greek intellectuals, also trained in Egypt, where he met Philo Judaeus, the Alexandrian born Jewish philosopher and likely writer of John's Gospel. Later, Simon became the first Bishop of Rome (see Chapter 13). The Hebrew Jehoshua Ben-Pandira, also thought to be Jesus, spent some time in Egypt with a Jewish ascetic sect resembling the Essenes, the Therapeutae, considered the first Christians. Philo, himself a follower of Pythagoras and a chief priest of the Eleusinian mysteries, wrote in a book, On the Contemplative Life, that the Therapeutae resembled a Pythagorean community and were found in many parts of the inhabited world.[55]

About 1,400 BCE, long before Alexandria was built, elements of Egypto-Phoenician civilization crossed into Greece, including agriculture, iron working, and the alphabet. The Greeks say the alphabet was introduced by Cadmus the Phoenician.

Besides arts and sciences, it seems the Egyptians were physical training enthusiasts, as well. At Beni Hasan, the walls of early Twelfth Dynastic tombs show pairs of wrestling figures in various moves, one dressed in black and the other white. On entering the Grand Lodge of Luxor, a neophyte followed ten rules of engagement, which were comparable to those governing Shaolin Temple boxing for novices. These rules appeared in China in the eighth century BCE, indicating that the Shaolin form of martial arts and its ethical standards were Egyptian.

WRITING

Improving on numeric symbols to record very large numbers, the Egyptians were writing using hieroglyphics by 3400 BCE. But in *Guns, Germs, and Steel,* 1998 literature Pulitzer Prize winner Jared Diamond indicated this writing appeared rather suddenly, suggesting it started elsewhere.[56] Topping the list of potential sources would be the Egyptians' ancestors and neighbors to the south, the Cushites. The Sumerians were also Cushites and originally used the Mande African script to write Uruk before they changed to Cuneiform.

Overall, the world's largest and oldest collection of ancient writing systems were scattered across Middle Africa, the homeland of the Dravidians, Egyptians, Sumerians, Niger-Kordofanian-Mande and Elamite speakers. These systems, invented by Cushites, started in East Africa and spread throughout the lush Sahara by 9,000 BCE. The oldest inscriptions, at least 7,000 years old, were found near the Kharga oasis in western Nubia. They show similarities with later inscriptions such as Vai, a descendant of old Mande, used in the highland region of the Fertile African Crescent in the Western Sahara at Oued Mertoutek. These inscriptions, over 5,000 years old, contain merchant records, religious doctrines, and obituaries.

The systems were originally syllabic, with hundreds of phonetic sounds that were condensed to between twenty and thirty key signs over time. The Cushites' descendants such as the Meroites, Phoenicians, Ethiopians, Egyptians, Sumerians, Elamites, Dravidians, Xia Dynasty Chinese, and Olmecs derived their alphabets from these key signs. Not only is Cushitic syllabic script matched by signs in

Olmec, but also by Cretan, Indus Valley Harappan, and Chinese Shang Dynasty Oracle Bone writing. All of it could be deciphered using Vai. Sudan's Meroitic writing was deciphered using terms from Manding, Kushana and Dravidian. Despite the Sumerians abandoning the Manding script in favor of cuneiform writing, Henry Rawlinson employed Oromo (Galla), a Cushitic language spoken in modern Ethiopia, and Mahri (a south Semitic language) to decipher their writing. It showed that the old Saharan script could be read using recent Manding and Dravidian linguistic data.

Many Egyptologists were shocked to learn that hieroglyphic writing in *Medu Neter,* the script of the 4000 BCE Narmer Palette, was in use in Qustul, Nubia, two centuries before the Egyptians began using it. At the time, the Egyptian border ran through Asyut, 365 miles north of the current one established in the Eighteenth Dynasty. Symbols representing Nubian and African interior plants and animals point to the hieroglyphics' source. To decipher the Egyptian writing, Jean Champollion used Coptic, a language closely related to Western Sahara Manding.[57] Also, since the Olmec spoke an aspect of Manding (Malinke-Bambara) language used in West Africa, their hieroglyphics and syllables – the earliest confirmed writing in America dating back to 900 BCE – could be read using Vai signs.[58]

According to Diodorus, writing was confined to priests in Egypt but was widely used in Nubia. By the time the ancient Nubian capital relocated from Napata to Meroe in the third century BCE, this writing had advanced to an alphabetic script, while Egyptian, even in its hieratic (priests' form) and demotic (people script) phases, was still hieroglyphic.

In the fourth and third centuries BCE, Sumerians, who are professed to have invented writing, were using a system of record keeping like that in the Sahara, Iran, the Indus Valley, and China.[59] Around 3,000 BCE, they changed to accounting systems consisting of simple hieroglyphic pictures and numbers written on cuneiform clay tablets. The earliest of these examples are records of mainly agricultural yields. But in 1998, German Gunter Dreyer found clay tablet inscriptions from 3,400 BCE in an Abydos tomb in the Nile Valley recording linen and oil deliveries. Comparable writing, other than simple merchant accounts, did not appear in Mesopotamia until the eighth century BCE. Unlike Sumerian writing but like modern hieroglyphs, the Egyptians' writing contained sound signs by 3,400 BCE.

STAR GAZING

When Acadians conquered the Mesopotamians in the 1700s BCE, they found Sumerians producing astronomy books and signs of the Zodiac for the calendar. But they were not the only ones or the first to do so. Living and breathing the skies, the ancients, such as the Dogon in the Western Sahara, were also familiar with the stars in antiquity. The Dogon apparently had uncanny knowledge of astronomy and mathematics. They believed they came from a planet – invisible to the naked eye – that orbited Sirius 50 trillion miles away. Such a star, a white dwarf called Sirius B, was identified in 1970.

The Egyptians had also long recognized constellations like – Ursa Major - the Great Bear – and mentioned it in Pyramid Texts over 4,000 years ago. They were aware of precession, that star positions changed over time. Today, King Djoser's statue sits in a serdab, a small stone enclosure on the north side of the Saqqara step pyramid, gazing at the empty spot where the star – Al kaid in the Great Bear, Ursa Major's "thigh," – twinkled in 2800 BCE, 190 years before the pyramid was built.[60] Robert Schoch, a Boston University geologist, analyzed the Lascaux French cave paintings and concluded Taurus was known in about 15,000 BCE, and its origins were known even earlier.[61]

The Narmer Palette, writes Audrey Fletcher, is a sky chart that shows Egyptian familiarity with precession by the beginning of the dynastic period in the fifth century BCE. It depicts the Lord of precession and creation, Hu, or Harakhate of the Horizon, who made the world in 14,000 BCE. The Egyptians celebrated the occasion as the First Time, or *Zep Tepi*. On the front, the palette bids farewell to the Gemini twins running across the bottom and welcomes the new age, Taurus (represented by the two bulls above Osiris, or Orion, the giant), in 4468 BCE. In front of the giant, with papyrus growing on its back, sits the Celestial Sphinx, Hu, characterized as the constellation Eridanus. Eventually, Hu sank below the horizon, and his secret as the Giver of Life and Lord of Precession would soon be lost and usurped by Osiris. The stick, or bow, held by the falcon standing on the papyrus, represents stars usually shown with Orion.

On the back, a bull smashes the Zodiac, knocking three teeth out of a cogwheel, denoting the three elapsed eras of the Zodiac since the First Time.[62] As the sky events on the palette appear in the order of the constellations at the dawn of Taurus, this dates the beginning of the Egyptians to around 12,000 BCE in Leo at the end of the last Ice Age, the moment when, for them, the Zodiac began. So, as Plato and a slew of modern astronomers starting with Jean-Baptiste Biot maintained, and as the Dendera zodiac, now displayed in the Louvre, shows, the Egyptians (or their southern ancestors) understood precession of the equinoxes and had long before begun mapping the skies.[63] The Greeks then acquired their vast astronomical knowledge from the Egyptians. *See Illustration 7.2*

By 4325 BCE, counting moon cycles, the Egyptians created a calendar consisting of 360 days a year with twelve months of thirty days each. They named a month for each of the twelve major gods. Later, they tacked on five days and gods, making the year last 365 days. They knew the civil year was a quarter of a day shorter than the solar year. But, because the gods didn't like anyone tampering with order, or "Maat," instead of adding an extra day every four years as the Gregorian calendar does, they waited 1,460 years, called a Sothic cycle, for the disparity to correct itself. But it caused the civil year to get out of sync with the solar year until then. This disoriented the Romans, so they changed the calendar.

The Olmec calendar mirrored the Egyptian calendar, including the five "useless" days. And like the Egyptians, the Olmec gave the months Zodiacal names. Some marked special events such as the birth of animals (Aries, Taurus) or the coming of the rains that brought a lot of fish (Pisces).

According to Robert Bauval, the Giza pyramid layout reflected the skies at *Zep Tepi,* the First Time, or the beginning of the Egyptians in Leo in 11,541 BCE, 9,000 years before the monuments were built. At the time, the Nile aligned with the Milky Way in the south, appearing to flow from heaven.[64] The event was commemorated in conjunction with the thirty-year *heb-sed* festival, or royal jubilee, at Saqqara.[65] But on studying Nabta Playa, an ancient site in the southwestern Egyptian desert, former NASA physicist Thomas Brophy suggested the Giza arrangement was a Zodiacal clock bracketing the period.

Nabta Playa was occupied on and off from around 9,000 BCE to 3,200 BCE, depending on the rains. By 6,000 BCE, settlers there had created a calendar circle, a stone rink aligned with the position of the three stars in Orion's belt at summer solstice. Brophy believed that this site might have been the center of it all. At the time, Egypt was occupied by the Anu. And, as Egyptologist Flinders Petrie noted, the Sphinx's Nubian sculptural form reveals the makers' Central African origins.[66]

Legend has it that at sunset, 8:27 p.m. July fourth in 16,115 BCE at Giza, the Celestial Sphinx god Hu, comprising Eridanus and Lepus the Hare (known to the Egyptians as Orion's chair), gently began expelling breaths, "Hhhhoooo…," called the Word, throughout the night. The first breath created the soul of Osiris, or Orion, in front. More breaths until dawn slowly created the rest of the heavens. Before midnight, however, with the Bennu bird Constellation Phoenix hovering behind him, Hu straddled the southern horizon gazing east at what would be the place of his final Act of Creation. Twelve hours after sunset at exactly 8.27 a.m., on July fourth at dawn, as the sun rose directly due east, Hu expelled the last Word. Twelve hours later at precisely 8:27 p.m., the sun set in the west, the only time this conjunction of Eridanus with equal night and day occurred in the 25,800-Great-Year Zodiac cycle, which ended with Taurus. A new Great Year began with Aries in 2308 BCE.

Audrey Fletcher dated the Egyptian First Time to 14,000 BCE. But, for dusk to dawn and dawn to dusk to occur exactly twelve hours apart, using the Starry Night Pro computer program, I pushed the event back by 2,000 years, to 16,000 BCE in Scorpio.

Since Egypt was occupied by people other than the Cushites in 16,000 BCE, the idea of the First Time probably originated elsewhere, most likely in the Western Sahara. There, the conjunction of equal night and day came at exactly eight o'clock. Memory of the event would have been preserved in folklore and constructions such as Nabta Playa, where a circular stone sky map depicted Orion's position in 16,500 BCE. To Schoch, the circle froze the event in time. New circles built on top of old ones between 20,000 and 7,000 years ago passed the star arrangement down through the generations.[67] Most likely, then, 16,115 BCE was the First Time, *Zep Tepi*, commemorated by the Egyptians. Even Bauval admits that "the Nabta Playa sky map could be some sort of memorial of an important event, perhaps a beginning in the history of those sub-Saharan herders...."[68] Studying Egyptian priests' data, Herodotus and Manetho dated the beginning of time close to this period, around 17,000 BCE.[69]

So, in the beginning was the Word (Jn 1:1-2), and the Word was with God, and the Word was God, or Hu, aka the constellation Eridanus, represented by the Celestial Sphinx. Settlers along the Nile coming across a Sacred Mound surrounded by the Waters of Nun (the Flood) at Giza, sculpted Hu into the Sphinx, with a face like theirs.

At the dawn of Taurus, Hu had already sunk below the horizon and was remembered only as the Lost Secret, or The Secret of Precession, by the initiated intellectual elite. Hu would become the Lost Word, represented by his terrestrial counterpart, the Sphinx at Giza, or "Harakhate (Hu) of the Horizon," its name in the New kingdom. Orion's Belt, or god's Word (the three pyramids) stood in the background. Today, Hu is worshipped by the Eckankars, headquartered in Chanhassen, Minnesota.

As the Sphinx's symbolism representing the god Hu faded from cultural memory, people transferred their loyalty to another lion, Leo, which rose due east with the sun in 11,541 BCE. But unlike the Celestial Sphinx, Leo faces west. Also, as signs of rainwater erosion indicate, the Sphinx appears to be much older than the pyramids. Religious sets of rings made of huge 16-ton stone pillars and dating from the tenth millennium BCE at Göbekli Tepe (Guh-behk-LEE TEH-peh), or Potbelly Hill, in southern Turkey demonstrate that such massive stone constructions were possible in antiquity.[70] Cave art in France oriented toward the sun, and artifacts from South Africa and Siberia recording moon and star patterns about this time or earlier, support that our ancestors were avid observers of the skies, reproducing those patterns here below on Earth.

All told, as people have lived the longest in Africa, the way we are today began there. The Meroitic script gives us insight into the lives of those early ancestors, revealing that Isis, Osiris, and Amun were worshipped in Nubia before they were worshipped in Egypt. Farther west, the Dogon still worship Amun, or Amm, who has a ram's head, and a Jackal-headed god similar to the Egyptian god of the dead, Anubis. Egyptian gods, then, came from the south. According to Massey in *A Book of the Beginnings II*, "Ethiopia and Egypt produced the earliest civilization in the world, and it was indigenous." He argues, therefore, that all the world's gods came from or through Egypt.[71]

GOLD SANDS

In all, Egyptian civilization spanned thirty centuries. It beat out the Mayan, which lasted twenty-nine centuries, for first place (China's continues uninterrupted after thirty-three centuries, and could be even older; its written records date back 4,000 years). The Egyptian civilization peaked in the eighteenth Dynasty, when Egypt became a superpower. Its rule stretched from Nubia to the Levant, and from Mesopotamia across Syria to prosperous Colchis on the Black Sea. Colchis practiced circumcision like the Egyptians and Ethiopians, showing Egyptian influence reached there. Going by Herodotus, although Colchis remembered Egypt, Egypt didn't remember them.[72]

Egypt's Eighteenth Dynasty began when Ahmose drove the Hyksos out of Egypt back to the Levant in 1,550 BCE. It reached its zenith during the reign of Amenhotep III (whose name means "Amun is pleased"). Amenhotep III was called the Napoleon of the ancient world; he won all his foreign campaigns and expanded the empire. Egyptians did not colonize other lands, for dying abroad meant they could not go to the hereafter. From afar, how could they prepare for the afterlife, particularly the pharaohs?

Egypt was stable during Amenhotep's reign and throughout the dynasty. The country's coffers overflowed with treasures from the pharaoh's foreign conquests and trade. Gold poured in from Nubia and east of the Nile, as if blown by sandstorms. One young king, Tutankhamen, or Tut as he's fondly known, boasted of building ships plated with the shiny metal to illuminate the Nile. Foreign kings jostled for some of the gold when building their palaces, too.

Because they believed in a happy everlasting afterlife, pharaohs prepared for it in grand style. They built vast tombs in the Valley of the Kings, lavishly filled them with furniture and daily essentials and packed them with lots of gold for a want-free eternal rest as gods. They inscribed walls with embellished accounts of their foreign conquests and adorned ceilings with intricate eulogies to the gods and beautiful celestial scenes. A royal butler rewarded in gold in turn dutifully splashed the imperishable metal in his tomb. Naturally, thieves helped themselves to the unprotected treasures throughout the centuries, confident their owners would never miss them.

Early Egyptian tombs were simple, called mastabas (Arabic for bench) before they were eclipsed by pyramids. Both the mastabas and the pyramids were easy to loot. So, pharaohs were buried inside unscalable cliffs in the Valley of the Kings. The massive pyramids then served only as the abode of the king's *ka,* or the life force that left the body temporarily at death, and his *ba,* or personal spirit, which roamed about or accompanied the king when he ascended to Heaven as a full-fledged god and became a star. Still, the hidden tombs were pillaged, most likely by workers – except one. For thirty-three centuries, his tomb remained undisturbed, as though meant to give us a glimpse of Egypt's fabulous golden era. No wonder the occupant, Tutankhamen, became a legend.

When Howard Carter discovered the tomb in 1922, he found Tutankhamen's mummy lying peacefully in the innermost of three coffins. These in turn rested inside a yellow quartzite sarcophagus decorated with a goddess on each corner, their outstretched, winged arms surrounding it. Above the goddesses ran vertical columns of inscriptions, primarily spells chanted to protect the king, surmounted by a horizontal line containing more spells. In turn, the sarcophagus was enclosed in four gilded wooden shrines.

The innermost coffin was made of 245 pounds of solid gold, worth nearly eight million dollars at 2023 prices. Tut's face was covered by a gold mask weighing 21.5 pounds and inlaid with lapis lazuli, obsidian, carnelian, quartz, turquoise, and feldspar. Next to the sarcophagus stood a throne covered in sheets of gold and silver inlaid with semi-precious stones, faience, and carnelian red glass. On the backrest panel, a queen wearing a pleated robe anointed Tut with perfume. The colors of the chair blended to create a vivid effect when combined with the gold and silver. Also, in the sanctuary were statues of gods cast in gold and electrum, decorated with lapis lazuli, and all kinds of precious stones. Within the wrappings of the entombed mummy lay a dagger with a blade made of meteorite iron.

In total, despite robberies (there were some), the tomb contained over 5,000 items of clothes, jewelry, perfumery, weapons, chariots — many covered with gold — an arsenal of weapons, and dried food. About 430 shabtis, small servant figures, were ready to wait on the young king, one for each day of the year and then some, when he awoke in the afterlife. Because Tut died at age nineteen, his tomb was modest compared to what other pharaohs, like Ramses II, stocked in theirs. Ramses II lived ninety years, long enough to prepare hundreds of rooms on several levels and stuff them with priceless treasures, all eventually looted. For four hundred years, though, the Egyptians wallowed in the yellow stuff.

Safe to say, civilization, and empire and dynasty building began in Africa and spread mainly through Egypt. Ancient Egypt's legacy is ubiquitous today, represented in the sciences, architecture, arts, literature, religion, on money—the US dollar bill, and more. Also, besides the Bible, the Catholic Bishop's stick (which resembles Tutankhamen's hook), the Pope's miter (which represents the white crown of Upper Egypt), and the five-pointed star decorating Christmas trees (which represents Horus's star, Sirius) are all rooted in early Egyptian civilization.

So today, with our planet choking in pollution, bedeviled by stress, climate change, deadly viruses, terrorism, wanton shootings, the threat of nuclear war, fake media, and growing insensitivity, you may wonder if modern civilization isn't sputtering. Some would argue that human civilization peaked 3,500 years ago during the incredible Egyptian Eighteenth Dynasty and that we have been on a downhill trajectory ever since.

Now that we are at the end of our historical odyssey, we should have a good idea about our true origins and makeup, even though our purpose and the essential nature of the creator still elude us. As spirituality is universal, invoked through practices such as religion, and as such practices vary by region and culture, our number of deities and religions is overwhelming. Until recently, when travel and communications became much easier, such deities were only known locally. The Bible mentions several, presided over by El Shaddai, the Supreme God, after He unseated El Elyon, god of the Canaanites, from the throne. How do we find El Shaddai? Church fathers don't know and give vague answers, such as "He's everywhere." Like the mystics, e.g., Kabbalists, Buddhists, Sufis, and so on, these ecclesiastical leaders are unable to describe their deities, because they merely experience them, and not see them (more on this later). As a result, many creeds insist on blind faith as the foundation of their theology. To learn about the One and Only God, then, unless we are "lifted" into Heaven like Enoch, we have our work cut out for us.

HENRY KAKEMBO, M.D.

CHAPTER 8

GODS GALORE

Truth unmasked blinds as Llaima's snowcap at sunrise. [1]

~Henry Kakembo

We are in a jam when it comes to establishing the nature of our creator. The Bible distinguishes Him from other gods by name only, no descriptions. But since I was born a Christian, and not because He is better than the others, I will focus on the God of Abraham, El Shaddai, who is also the God of the Jews and Muslims.

Before creating us, God said, "Let us make man in our image," Gn 1:26-27, the plural "us" indicating there were other gods – Elohim – as mentioned earlier. To become the only God of the Jews, He rebuked and fought other gods repeatedly until He overthrew El Elyon, the Canaanite High God, and took his place. [2] But, like their names, gods changed with time. We cannot be completely sure, then, that we're focusing on El Shaddai specifically.

Originally, as was indicated earlier, the god of Abraham was El Shaddai and that of Moses was Yahweh (Jehovah, Adonai, or Lord). While El Shaddai was mild, open, and friendly, taking on human form like other gods, Yahweh was hidden, distant, and terrifying. The gods of Isaac and Jacob were also different. [3] Eventually, though, all of them likely merged into Yahweh. As His real name, YHVH, was not pronounced outside the Temple, substitutes abounded, such as *Aloha, the Existing One, Lord, Supreme Being, The Unknowable*, and *Ultimate Reality*. Other names used for God were Aristotle's *Unmoved Mover*, and the Roman philosopher Plotinus's *The One*. Jewish mystics, the Kabbalists, called Him *Ein-Sof* (for endless or infinite), and Isaac Newton, the *Great Mechanick*. *Allah*, also the God of Abraham, had ninety-nine names in the Quran. But, as with the Hindu and Egyptian gods, scholars of ancient history believe that all these different names of deities represented one reality, probably a solar deity like *Amun*, the Egyptian king of the gods, who, like *Yahweh*, was hidden and unknowable. Most likely, however, this reality could be the power, or the Universal Force, responsible for the universe.

PROOFS OF GOD

To establish whether the God I'm zeroing in on is or is not a figment of the mind, let's look at the evidence. Mainly, people find God through religious experiences, or revelations. Gautama Siddhartha Buddha, Paul the Apostle, Saint Augustine, and Pascal all offered their own experiences to the existence of a divine power. In surveys in Britain, Australia, and the USA, 31 to 48 percent of respondents admitted to having religious experiences, which actually were spiritual experiences in which they sensed a divine presence. In Britain, 5 percent of the respondents had an inner feeling of union with the universe,[4] but none of these people could describe their experiences or God.

In his runaway bestseller, *Why God Won't Go Away,* Andrew Newberg wrote that interpretation of such experiences depended on one's cultural beliefs. Whereas Nuns experienced a tangible sense of closeness to God, Buddhist monks sensed a Universal Force or spirit. According to Michael Argyle, the nature of the experience depended on the trance stage. Lower stages produced vivid images, such as sensing God, while the highest (or deepest) stage produced a feeling of the individual merging with God and the universe, and the three becoming one, as the British respondents above felt, called *Unio Mystica.*[5] But are these experiences of true reality and not a brain trick? In his 1917 essay, "Mysticism and Logic," British philosopher Bertrand Russell called these experiences intense illusions that imparted no true insight about the universe.[6] In contrast, Buddhists consider the visions and such insights natural to humanity rather than supernatural phenomena.[7] Since Russell's essay was published, profound similarities between physicists and those of mystics about their views of the universe and reality have come to the fore.

Because explaining such events defies common sense, belief in supernatural phenomena seems illogical. If such a belief is hereditary, however, in that it passes from generation to generation, it means that it somehow aided survival and evolution selected it. The talent to induce the experiences such beliefs are based on, then, is stamped in our DNA, as Dean Hamer indicates in chapter 13, and the feeling God exists is hereditary. Call it the faith instinct. Acknowledging the futility of justifying faith, Eastern societies discourage describing God or the Universal Force.

Because iron-clad proof of God's existence is unavailable, some people based their belief on ontological or cosmological arguments, Pascal's Wager, or Intelligent Design (ID).

The ontological argument was introduced by a former Archbishop of Canterbury, Saint Anselm, in 1078. He based it on the nature of being or existence, saying that God was best understood as "something than which nothing greater can be thought." People interpreted this to mean God was a perfect being, the most perfect idea possible. Therefore, as one characteristic of "perfection" was "existence," God must exist. A seventeenth century Frenchman, René Descartes, considered the father of modern philosophy and the source of the maxim, "I think therefore I am," agreed with Saint Anselm. To Descartes, God was a supreme and perfect being. The evidence was in our sublime human consciousness. But critics of Descartes, philosophers no doubt, countered that one could not infer the extramental (outside the mind) existence of anything by analyzing its definition (God as a supreme and perfect being).

In the seventeenth century, after a religious experience convinced him that God existed, Blaise Pascal, a French mathematician and inventor of the syringe and hydraulic press, argued that since daily life itself was a gamble, belief was a choice we should make because it was safer to bet that God existed than to bet against the idea. If you were right, you won; wrong, you lost nothing. This approach to belief is known as Pascal's Wager. Until the nineteenth century, wary of incurring the wrath of the church, philosophers chose to believe. In support of his equally dubious mathematical Bayesian Arguments in *Probability of God,* Stephen Unwin called Pascal's alleged proofs empty.[8] These wise men, fully aware of the horror stories of the punishments meted out to heretics, opted to toe the church's line and live another day.

The cosmological arguments as summarized by thirteenth-century Catholic philosopher Saint Thomas Aquinas included *the first cause* and *argument from design.* The first cause stated that the universe and us had a cause itself uncaused, God. But the argument leaked whenever a mystery was explained. For instance, we now know catastrophes like earthquakes are not havoc wreaked by an irate God on His disobedient children. So, most people no longer believe God does the heaving that causes earthquakes.

Small wonder then that in some parts of the world, particularly in Europe, God's importance has waned. The time when people believed in God and spirits by day and nymphs, goblins, and demons by night is long gone. Still, in a Pew Research Center poll in 2014, 89 percent of adults in the United States believed in God. In China, where religion was once banned, a similar Pew poll in 2005 had found that only about half the population considered belief in God either very important or somewhat important in their lives. Today, instead of answers, for most people belief in God provides mainly comfort (*see* Chapter 10).

The Intelligent Design (ID) argument cites order in the universe as a primary proof of the existence of God. Its proponents focus on the eye, which, like a watch, they insist could not have evolved on its own. They don't accept that, like the watch, natural selection developed the eye in umpteen stages, obviating ID. If ID were responsible, the world would be perfect. Instead, as in product manufacturing, evolution discarded rejected defective or non-functioning parts, retaining only those that worked, again obviating ID.

But, like the eye, a poor design for clearest vision, many things are imperfect. "If order supported ID, what of disorder?" Scottish philosopher David Hume, asked in *Dialogues Concerning Natural Religion* (1750-1776).[9] Instead of ID then, natural forces like gravity or electromagnetism could account for our existence. In a corollary of the anthropic principle introduced in chapter 5, physicist Steven Weinberg said, "…we observe conditions that permit our existence precisely because if the conditions were any different, we wouldn't be around to observe them."[10] In a way, he answered philosopher Gottfried Leibniz's question, "Why is there something rather than nothing?"[11] This question is related to those posed in the introduction to this book. Why does the universe exist? How did it come about? Electrons, the four forces of nature, and virtual particles were unknown in Leibniz's time, and – despite our expanded knowledge since then – debate still surrounds the issue.

If God exists, how does one explain evil? Why would an all-caring, all-powerful God let His children suffer? Could evil be the penalty for Adam and Eve's "Original Sin," as Saint Augustine believed? Are famines, typhoons, and earthquakes retribution? Why doesn't God chase away the Devil and save us from temptation and Hell? Since He's all-powerful, not even Satan can withstand His might. Some say suffering is the price we pay for free will. Make the wrong choice, you'll pay.

Alternatively, perhaps suffering is the motivation that helps one to develop and grow character, called *soul making theodicy*. According to Charles Darwin, however, natural selection is an indifferent mechanistic process that bypasses God altogether, thus making Him powerless over evil. The outlook is even worse in Hinduism, where gods also suffer because evil and suffering are part of reality. But, as we will also find in the final chapter, because of His nature, God is powerless to turn evil away, or prevent pain and suffering of His children.

Unable to test these proofs, we cannot fault them either. According to Christopher Hitchens, author of *God is not Great,* since such arguments lack supporting evidence, we also need none to dismiss them. [12] Regardless of the belief, a belief is not fact until proven. Beliefs change, too. But for the devoted, proof is expendable for them to hold on to their faith. Their experiences in the mystical world suffice.

SEARCHING FOR GOD

Since religions are based in spirituality or the mystical experience, it's no surprise that worship worldwide shares common features, making different gods, symbols of the Glory sensed in the experiences, represent the same reality. The god of the West, known as Yahweh or Allah, seems, however, to have gone AWOL. He is not in Jerusalem, as the Romans razed the Temple in 70 CE. Might He be the white-bearded old man straddling the skies with Jesus on His right side? Jet planes and spacecraft have hurtled through the sky and telescopes peered through it into space, and found nothing. All they see are planets, stars, rarefied gases, and dust. Of the risen Christ? No sign.

In biblical times, God appeared to people. Once He had dinner with Abraham near the great trees of Mamre. Gn 18:1-8. Then in Midian, He spoke to Moses from a burning bush that was not consumed. When Moses approached, God said, "Moses! Moses!" "Here I am." "Do not come any closer…" Ex 3:2. Why can't we see God today? Is it perhaps because we only visualize Him in our image? Some mystics say He does not really exist, calling Him "Nothing." At the turn of the twentieth century, Hermann Cohen, a German Jewish philosopher, thought God was an idea formed by the human mind, a symbol of the

153

ethical ideal. And according to philosopher and psychologist William James, evidence suggested God lay primarily "in inner personal experience."[13] "Who created Him?" some pointedly ask, as Stephen Hawking did in his video, *A Brief History of Time*. Others are more pragmatic. For example, true to character, when asked if there was something special, perhaps even divine, about the human soul in a January 13, 1997 *Time Magazine* interview, Microsoft tech company founder, Bill Gates, who lived and breathed data to minimize glitches in computer programing, said, "I don't have any evidence on that." Feeling that God watches over us all the time, we conjecture He must be everywhere, omnipresent. So, if God exists and we are created in His image, it means that we should also be omnipresent. But this doesn't add up unless you're a Buddhist monk or a mystic at one with the universe.

A MAN-MADE GOD?

Since we lack proof of God, did we invent him? Understanding human views of things to be human creations, including knowledge, Xenophanes, a sixth century BCE Greek philosopher and poet, , thought we did. He said that although learning brought us closer to the truth, ideas remained ours. So, if cattle, horses, and lions could paint, they would depict their gods as themselves, implying that gods generally looked like those who worshipped them, since we made them in our own image.[14] Ludwig Feuerbach, a nineteenth century German philosopher agreed. He explained that the reason God was personal with human attributes was because God and gods were our creations.

Is God a human projection reflecting our desires and fears? Perhaps. Karen Armstrong suggests that the word *God* is a symbol of an indescribable reality beyond, an experience everyone feels differently to answer the needs of one's own temperament. If Armstrong is correct, a universal image of God is impossible. So, various cultures envisage God differently. Next to the sun, people have worshipped the moon, stars, planets, mountains, animal spirits, and themselves. As all gods were considered equal in the ancient world, it was normal for people to worship each other's deities, and to adopt them.

The sun has been tops as a god for millennia, in Egypt as Ra, Amun, or Horus, in Greece as Zeus' son Apollo, and in Rome as Mithra. In paintings, sculpture, and other artistic depictions of Mithra, Horus, the Greek god Serapis, and various Christian saints, the halo surrounding the head is clearly the sun's aura.

Timothy Freke and Peter Gandy wrote in *The Jesus Mysteries* that church Father Clement of Alexandria called Jesus "The Sun of Righteousness." In fact, a third century mosaic from the Mausoleum of the Julii underneath present-day St. Peter's Basilica in Rome shows Christ as Sol Invictus (Unconquered Sun), the Sun God. "Jesus was a disguised form of the solar god of Egypt," wrote Gerald Massey.[15]

If we erred and worshipped false gods, the all-powerful, all-knowing, and ever-present God would surely have reprimanded us as he did the Israelites during the Exodus. God demanded complete loyalty from the Jews, then. He slapped a plague onto them for worshipping a gold calf, the sign of the previous age of Taurus the Bull. The new age now was Aries the Ram. Ex 32:35.[16]

But this God is just like us. He can be angry and jealous of other gods. He even fought the other gods to become the only God of the Jews. Also, God kills. Didn't He wipe out wicked humans with a flood, sparing only a righteous few like Noah and some innocent animals? And didn't He send the "Angel of Death" to slay every Egyptian firstborn son in order to free the Israelites from bondage? All these sound-like things humans also do.

And just like us, according to the Canaanites, God had a wife, Asherah, mother of the Gods, and a consort, Sophia, queen of Heaven, who Philo said was mother of the Logos. God, then, had a son, Jesus, by Mary, Joseph's betrothed. God's other children, the angels, befriended earthly women, siring the Nephilim, giants that He destroyed in Noah's Flood. Gn 6:2-7. With time, God mellowed. His angry, vengeful image softened, and His prophets, like Isaiah, Amos, and later His son, Jesus, preached a message of love and compassion. This message became the backbone of the theology of the three major Western religions. But sometimes it expressed itself in surprising ways. For example, during the 1099 Crusades, Christians massacred Jews and Muslims in Jerusalem with abandon. A century later, the Inquisition, the judicial investigation of heresy under papal direction, allegedly resulted in close to five million witches and heretics being burned at the stake. Saddam Hussein's atrocities pale in comparison.

Creating gods is human nature. According to French philosopher François-Marie de Voltaire, had God not existed, it would have been necessary to invent Him.[17] Karl Marx believed that expressing faith in a supreme being with a cosmic purpose was the sigh of the oppressed creature, the opium of the people, and it helped them to deal with suffering.[18] According to Karen Armstrong, "…people created gods to meet a purpose." In times of crisis, gods served as a crutch to help them cope with inevitable death. Once their purpose was fulfilled, they were quietly dropped, and new ones created for a different purpose. Although people invented gods to account for nature's mysteries, in the Hindu religious ideas from the Rig-Veda odes, nobody, not even the gods, understood the mystery of existence.[19] Also, Eastern religions see the Western God as too limited. Because the Western God was seen as a being whose image was patterned on humans, He seems like a king when speaking to the prophets. Eastern philosophy believes a god should be indefinable and not subject to speculation.

Alas, today, there are no reports of tangible theophanies (God appearing to humans) and Epiphanies (God as a human) because of His nature, as we see later. Even mystics merely report sensing or uniting with God, not seeing Him in person. In lieu of a description of God, they report losing a sense of self and experiencing rapture, bliss, and ecstasy.

WHO IS JESUS?

If gods are our creations, symbols representing our spirituality, what about Jesus? Was Jesus a personification of God's son, or God himself? Muslims call him a prophet, actually *The Messenger,* as explained in the next chapter. *Jesus*, as the anglicized name of Joshua, means "savior"; *Christ* and *Messiah* (*Kristos* or *Christos* in Greek) mean "anointed." Instead of names, therefore, they are titles also given to other gods like Horus, Attis, Mithra, or Dionysus in the ancient world. They don't denote an individual historical figure. This Christian God, then, is nameless, and His name could refer to any of these other savior gods and more.

Consider this: the name *Christ* originated in Egypt, where, as KRST or Karast, it meant "covered in embalming oil," or "anointed." KRST described Horus, whose emblem it had been for thousands of years. According to Flinders Petrie, a similar Indian name of the Hindu god

Krishna also hailed from Egypt through the Cushites. The name Krishna comes from the Sanskrit root "krst," meaning Christ, and was formerly spelled "Christ-Na" in early English.[20] Krishna's story mirrors that of Horus.

Jesus never claimed He was God. "Why do you call me good? No one is good – except God alone," he said. Mk 10:18. In the fourth century, Arius, a popular Alexandrian presbyter from Libya, also taught that Jesus was not the same substance as His father, while Arius's supporter, Bishop Nestorius of Constantinople, believed Jesus was half divine and half human. Their controversial positions led to street fights in Alexandria, and to end the mayhem, Constantine convened the Council of Nicaea. Arius was unfazed when the Council ruled Jesus was divine, and his followers, the Arians, survived him by 400 years. Even today in England, for example, some people still consider themselves Arians. Nestorians, too, are alive and well in the Middle East, India, China, and Mongolia.

During the turbulent times of the first century CE, Jews yearned for the promised Messiah, a warrior king from the line of David, to liberate them from Roman rule. But a Messiah had existed in the mystery religions for millennia. Horus and Osiris played this role in Egypt. To dupe stubborn Messianic Jews and dagger-carrying Sicarii into worshipping the Roman Vespasians instead of God, historian Josephus Flavius (aka Josephus ben Mathias) created a new religion based on a typological application of the story of Jesus as it relates to the story of Joseph in the Old Testament. Both had miracle births, both had twelve brothers or disciples, both were betrayed and sold by Judah or Judas, and both began work at age thirty.[21] Their stories were similar to those of the inclusive Greco-Egyptian religion of Serapis Christus, also a Messiah and Savior, just like all the Ptolemys, thus marking the beginning date of Christianity. In a letter to Servianus in 134 CE, Emperor Hadrian lamented that some Egyptian Christians worshipped Serapis and called themselves bishops of Christ.[22] Yet, the emperor built a Serapis sanctuary for himself at Tivoli and – for his inauguration – made a coin depicting him on one side and Serapis on the other. Such similarities are unsurprising since the New Testament fulfills prophecies from the Old Testament, which is Egyptian wisdom poetry and literature.

The wealthy family of Julius Alexander Lysimachus, the ruler of Alexandrian Jews, was supremely positioned and powerful enough to influence the budding Christian faith at the time and shape its theology. The family shared a long-standing relationship with the Roman Flavians and the Judean Herod. This relationship was traceable to the household of Antonia, mother of the Emperor Claudius. Antonia employed Lysimachus as her financial overseer around 45 CE. Scholars believe it was Lysimachus's younger brother, Philo Judaeus, an eminent Gnostic Jewish philosopher, who developed the doctrine of the Word (Logos), and merged Judaism and Greek wisdom to compile the New Testament, giving John's gospel its philosophical tone. Otherwise, who of the disciples knew the sacred Pythagorean number *vesica pisces* below?

Facts about Jesus, including his portrayals, birthday, life story, and death circumstances, have changed with time. According to Arthur Weigall, Jesus's birthplace, generally said to be in Bethlehem, was originally Tammuz's (Osiris's) shrine, leading many to confuse or conflate these gods with Jesus.[23] A prophecy had foretold Bethlehem as the birthplace of the savior of the Jews. Mi 5:2-6. The salvation, however, was not *from* the Romans but rather from the Assyrians in the eighth century BCE. In fact, this place name may also refer to that of Horus at Heliopolis in Egypt. Heliopolis, being the winter grain store, was the House of Bread, "Bethlehem" in Hebrew. It represented Virgo, signifying grain harvest time.[24] Some archeologists doubt Bethlehem existed in Judea in the first century CE.[25] Also, since the census was held to count Roman citizens only, Joseph (not a Roman citizen) would have had no need to be counted.

According to *Encyclopedia Biblica,* the city of Nazareth in Galilee, where Joseph and Mary lived, didn't exist at the time Jesus was supposedly born. Nazareth appeared between 70 CE-135 CE, long past the time when Jesus would have already risen.[26] Jesus's bewildered followers believed he came from Galilee not Bethlehem. Jn 7: 41-42. Regardless of where he might have been born, Jesus couldn't have been born during Herod's rule (Mt 2:1), as this king had died four years earlier in 4 BCE and he wouldn't have ordered the slaughter of all first-born Egyptian male babies. Jesus also couldn't have been born during the rule of Quinirius, governor of Syria, which began in 7 CE (Lk 2:2), a discrepancy of ten years in His birthdate. According to Massey, the crib and manger, *Apta* in Egyptian, were hieroglyphic signs of the sun's birthplace that were co-opted to serve as the physical representation of Jesus's birth environment.[27] During early Christianity, no one really knew when or where Jesus was born.

Jesus' crucifixion is also problematic. Irenaeus, Bishop of Lyon, whose business it was to defend the four gospels, believed Jesus was not crucified at thirty-three, three years after one entered manhood according to Egyptian custom, but rather lived to be an oldish man. Archeological evidence of thousands of crucifixions the Romans allegedly carried out in the first century has not materialized. The only possible finds of crucifixion from the period consist of two bone artifacts, unconfirmed but possibly human heels, one found in Jerusalem and the other in Northern Italy near Venice.[28] Various Jews and Christians dated Jesus's death between 103 BCE and 76 BCE. There is evidence to suggest that Jesus's identity (or at least the story of his death) has been confused with that of the Hebrew Jehoshua ben Pandira, who was born in the reign of King Alexander Jannaeus around 102 BCE. Jehoshua studied in Egypt with the Essenes for many years. On returning to Palestine, he went about teaching, healing, and performing wonders. One Passover when Jehoshua was about fifty years old, Jewish authorities were incensed and hanged him on a tree for sorcery. Some people believe Jesus's name and story were derived from Jehoshua's.

Jews in Israel weren't taken in by the idea of Vespasians as gods, particularly Titus Flavius as the Messiah. Unlike the Osiris-like gods, the Jews' savior was to be a king and not die yearly like the others. Above all, instead of being a pacifist like Jesus, the Jewish savior would be a warrior, and such a savior had not yet arrived. They therefore considered Titus Flavius a false Jesus, as they did the many other wannabes. Wannabes claimed to be the promised Messiah. Over time, their stories grew by order of magnitude so that they too came by virgin births, performed miracles, and resurrected yearly. For instance, God announced Apollonius of Tyanna's miraculous birth. Apollonius preached a message of peace and love, raised the dead and, in the end, was lifted alive into heaven by a choir. Other reports say that he vanished from Domitian's courtroom, or more exciting, disappeared in the temple of the beautiful goddess Dictynna. As many Christian texts were being written in 132, another Messiah, Simon Bar Kokhba, who was believed to be descended from King David, fought with the Romans and liberated Judea for two years. So beloved was he that coins were minted with him on one side and as the Son of the Sun on the other. After the Romans defeated the Bar Kokhba revolt, they scattered the Jews far and wide. Bar Kokhba's fate remains unclear; he either died of a snake bite or was beheaded and the head presented to the emperor, Hadrian.

To sever Christianity's ties to its "pagan" roots and simplify its message for the masses, the church embodied the mystical Christ. Paul had not envisaged a human Jesus. He knew best, of course, being one of the earliest to write about Him. Like the Gnostics, Paul promoted a symbolic Jesus. Following a vision while travelling to Damascus, he wrote, "…God was pleased to reveal his son in me." Gal 1:15-16. God had chosen him to make known the mystery "…which is Christ in you." Col 1:27. For Paul, Jesus had no definite beginning. According to Massey, this was because in ancient Egyptian religion, Christ was reborn in heaven every 2150 years at the dawn of each new Zodiac age. Then, again, He was reborn yearly at the spring equinox like all other gods in the region.[29]

Zodiac signs change with the dawning of a new age. In 2,410 BCE, the sign changed from Taurus to Aries. Jesus, then, was depicted as a shepherd with a lamb. According to Gerald Massey, when the equinox entered the age of Pisces in 255 BCE, His symbol became the fish.[30] At a loss to describe what Jesus looked like, early Christians portrayed Jesus as a fish for most of the first five centuries. Others depicted him as Serapis Christus complete with a beard. On a Roman coin, according to Sertima in *African Presence in Early Asia,* Jesus had African features: black with woolly hair, full lips and a round nose. Jesus also resembled 660 BCE Jewish Assyrian captives from Palestine. Today, He is represented in statuettes of the Black Madonna and Child in churches throughout Europe. Such statuettes can be found in a Vatican Chapel, or enshrined in the Church of St. Maria in Cologne, Germany, a replica of that found in the Basilica della Santa Casa in Loreto, Italy. He is represented by a wooden carving of a Black Nazarene on a cross made in Mexico in the sixteenth century. Its replica was brought to the Philippines in 1606, where it is borne in processions three times a year. A similar image is worshipped in Panama.

Jesus may also be represented by blonde, blue-eyed Swedish actors in movies. So, His image has changed with time and resembles the people worshipping Him.

To fulfill his role as the "fisher of men," Jesus helped his disciples miraculously catch one hundred fifty-three fish, a number sacred to the Pythagoreans. Jn 21:11. Pythagoras had apparently predicted this event centuries earlier. This number is the ratio of the height to the length of a fish produced by overlapping two similar circles, called "the measure of the fish" or *vesica pisces*. Pisces was Horus's symbol, too. When it became a Christian emblem, Christians became little fishes, or Pisciculi.

As with Horus's birth, Jesus's birth was astronomical, based on the sun. No wonder that outside of altered biblical texts, Paul never mentioned Jesus's virgin birth, miracles, or parables. Paul's approach to salvation, or union with the spirit Jesus, was Gnostic, as were his writings. To him, Jesus did not come as a person, but came instead in the likeness of sinful flesh. Rom 8:3. The church, to counter and suppress such heretical views and promote an historical Jesus, forged many of Paul's letters. Among the fraudulent epistles were the Pastoral letters to Timothy and Titus. The writer of Hebrews remains anonymous.[31]

No doubt Jesus's character and story were borrowed from those of other mystery gods. For example, the story of Jesus compares with that of Dionysus, the son of Zeus, who was born of a virgin in a stable. Just as Jesus did at a wedding (Jn 2:1-11), Dionysus likewise preached, turned water into wine, and performed miracles. He did not walk on water, though; this antigravity feat was saved for Jesus. Before Dionysus, however, Pythagoras and Osiris had also "prevailed" over the waters.[32] Such stories were really about the sun crossing the river in the sky – the Milky Way – in May.[33] Dionysus incensed authorities by calling himself the son of God. His fate, at least in some stories, was crucifixion. Guess what? He woke up three days later and appeared to women weeping for him.

Also, Jesus resembled the Roman soldiers' god, Mithra, primarily known as Pator or Peter in the East. Mithra's story, however, predated the Jesus story by thousands of years in Persia and by seventy years in Rome. Like Jesus, Mithra, was born on December 25, in a cave (as was Jesus in some stories such as Justin Martyr's), and three Magi came to see Him. Like Jesus, Mithra was "the light of the world" and "the good Shepherd." He raised the dead, had a last supper with twelve disciples, and ascended to heaven on Easter. In 321, since Mithra was worshipped as the sun god, Constantine moved the Sabbath from Saturday to *Sunday*. Inscribed in the ruins of a Mithraic temple underneath the Vatican are the words, "He who will not eat of my body and drink of my blood, so that he will be made one with me and I with him, the same shall not know salvation."[34]

Such parallels between the gods confused early Christians. Some depicted Jesus as Orpheus holding a lyre, or Apollo driving a chariot. Embarrassed by the confusion, Father Justin Martyr, a third century church apologist, accused the Devil of sending imposters ahead of Christ to muddle followers. This is still the official explanation today. According to Freke and Gandy, the ambiguity arose because Jesus was a spirit that descended into man. So, the Gnostics taught that "the true Christian experienced *Gnosis* or mystical 'knowledge' for themselves and became a Christ!"[35]

In his lectures Massey said, "...the history in our Gospels is from beginning to end the identifiable story of the Sun God and the Gnostic Christ who never could become flesh." The Christian's habit of turning toward the east to bow to Christ supports the notion that the cult was astronomical, and, as in other religions, was based in mythology. Pictures of the Holy child with a radiant halo, the sun's aura, also support this argument. The obscuring mist lifted when the Egyptian hieroglyphs were deciphered.[36] It's no wonder that writers around the first century didn't mention Jesus. To the historian Josephus, Titus Flavius was Jesus returned. Acknowledging other religions' contributions to his faith in a 1950 paper, Martin Luther King wrote, "Christianity was truly indebted to the mystery religions."[37] Indeed, other anointed gods like Attis, Mithra, Dionysus, and so on also contributed. Serapis became a European savior god when Constantine was baptized into the Greco-Egyptian religion. Eventually, by merging other saviors into himself, Constantine established modern Christianity.

THE HOLY GRAIL

When Christians are asked whether Jesus was married, as Dan Brown's novel, *The da Vinci Code*, seems to suggest, they answer categorically, "No!" Nevertheless, the question is superfluous, even though God and His sons, the angels, Zeus, and others had families. To the apostle Paul, Christianity's founder and Jesus's strongest advocate, Jesus couldn't marry, since He was simply an idea, a high priest in a mythical realm. "If Jesus were on Earth," Paul said, "He would not be a priest." Heb 8:3-4. Like the Gnostics, Paul believed salvation was in becoming one with Christ. Eph 2:13. And according to the Gnostics, the spirit Jesus married his sister, Sophia, God's daughter; he wouldn't have married a mortal like Mary Magdalene.[38] They were not equals. Mary Magdalene, too, it appears, was the embodiment of Isis-Hathor. Both their feast days were fixed on 22 July, when Sirius reappeared around the first century.

In the novel *The DaVinci Code,* Sir Leigh Teabing, fearing he might be on a wild goose chase, remarked, "Nobody is saying Christ was a fraud [that the story was a hoax] or denying that He walked the earth..."[39] Teabing was in denial, having searched for the Holy Grail (Mary Magdalene) for fifteen years, spending a lot of time and money. Since Jesus and Mary Magdalene were mythical, the search for the bloodline, sarcophagus, and chalice or Holy Grail was indeed a wild goose chase.

Our story, as told earlier, is completely different from that of the Bible. To find out why, we should find out more about its writers. Where did they get their information, and what role did Egypt, where the Book was compiled, play?

HENRY KAKEMBO, M.D.

CHAPTER 9

WHO WROTE
THE BIBLE?

Who fused Egyptian religion with Jewish traditions into a Holy Book? As in the conflicting Genesis accounts of creation, the process involved borrowing, merging, and filling gaps. Ambiguous passages were clarified or interpreted. Different writing styles and inconsistencies give away the Bible's many contributors. Coming from different regions and times, the writers introduced discrepancies when writing about distant events. For instance, according to scientist and religion writer Michael Magee and Bible scholar Russell Gmirkin, Jericho would have fallen twelve centuries before Genesis was written in the third century BCE.[1]

Stories attributed to God reflected the authors' personalities, prejudices, and agendas. For example, Isaiah, being an aristocrat, saw God as a king. On the other hand, Amos, a shepherd, saw God as empathetic with the suffering poor. The Southern Kingdom, Judah, was the favorite over the Northern Kingdom. Scribes painted a grand glorified picture of Judah. But in the time of David and Solomon (said to be approximately between 1010 BCE and 930 BCE), Jerusalem was little more than a village and Judah was a sparsely populated rural region. According to archeologists Neil Silberman and Israel Finklestein in *The Bible Unearthed*, despite the scorn poured on it in the Bible by the Southern Kingdom, the Kingdom of Israel in the north, which emerged during the Omride kings' era, was the greater of the two and "…grew to be among the richest, most cosmopolitan, and most powerful in the region."[2]

Differences in style may not be as obvious in translations as they are in the Hebrew texts. And despite widespread belief, Moses did *not* write the Torah or the law, as claimed. Dt 31:9. He would have died centuries before it was penned, as we see below.

NOAH CHEATS DEATH

Because of constant threats and destruction from flooding, myths reflecting flooding were common worldwide. Noah's story of The Flood, or Deluge, was no exception. According to Finch, it has elements from the Nile Valley. Noah ("Nuach" in Hebrew), is equivalent to the Egyptian word Nu-akh, meaning "the floodwaters that irrigated and fertilized cultivated lands."[3] As for the ark, Khufu's 143-foot, 45-ton solar boat sealed at the foot of Khufu's pyramid was also called Nu Akh (*Nu* stands for water and *Akh* for fertile field, garden, or irrigated lands). The boat, built over 4,500 years ago, was a replica of the boats used during the flood season. Such evidence suggests that Noah's story originated in Egypt.

To rid the Earth of sinners, and giants called Nephilim, children of angels (all angels are male, by the way) by earthly women,[4] God drowned all wickedness in a big flood. Gn 6:1-7. Bible scholars recognize several writers in the story. In the Documentary "Hypothesis," German scholar Julius Wellhausen, using *source criticism*, identified four different writers in the Torah: J or Yahwist, who called God Lord or Yahweh (often erroneously translated as *Jehovah*); E or Elohist, who called God *Elohim*; D or Deuteronomist, who wrote the Law Code in Deuteronomy; and P or Priestly, who sounded like a priest. E would have written the second creation story of Adam and Eve, and P would have written the first. Like the Jefferson Bible, P "purged" stories of angels, dreams, talking animals, and other fantasies. Drawing on the Enuma Elish epic, he produced a grand version of the text, glorifying Yahweh. So, unlike Adam, who came from dust, God made man in His image out "of nothing" or "ex nihilo." In 325 CE, the First Nicaea Council of Bishops sided with this more palatable version of creation from nothing, giving the Adam version short shrift. Anyway, regardless of what the writers may have believed, the Bible version of creation was and still is wrong. Archeological evidence and DNA analysis both conclusively show that humanity evolved in Africa.

P's account of the Deluge was interwoven with J's. According to P, God instructed Noah to make the Ark, a boat 450 feet long, 75 feet wide and 45 feet high. Noah brought aboard his family and *two* of all living creatures, male and female. Then *underground springs burst forth and the floodgates of heaven opened. The water rose for 150 days* (five months), and receded in the seventh month, grounding the ark "upon the mountains of Ararat." To test for dry land, Noah sent out a raven several times. Gn 8:7.

But according to J, after Noah loaded the boat with *seven pairs* of all clean animals and *two* of all unclean animals, *rain fell for forty days and nights.* Noah then opened a window and sent out a dove to check the water level. The dove went back and forth a couple of times and on the third time, returned with a fresh olive leaf in its beak. About two weeks later, it flew out and did not return. Noah pulled back the Ark's covering and found the land dry. Gn 8:13. Where the Ark settled, J did not say. In gratitude, Noah offered the Lord sacrifices on the altar. Smelling the pleasant aroma, the Lord vowed never to destroy life on Earth with a flood again and set a rainbow in the clouds as a reminder. Gn 20-21. That's when He decreased life expectancy to 120 years. Gn 6:3. Before then, people lived up to nearly a thousand years. But Enoch didn't die at all. He walked with God. And when he was 365 years old, he was no more. God whisked him away to live in Heaven. Gn 5:23-24.

Extensive searches have failed to turn up sediment deposits of a worldwide flood. And you might wonder: how could a 40-day rainfall (roughly five times the water in the oceans) submerge mountains and continents? Also, where did all that water go that rose to twenty-two feet above the highest mountains? Gn 6:20. Also, loading a boat with millions of species at the rate of one per second would require thirty years, not seven days. And, most damning, a boat nearly the size of the Titanic built entirely of wood cannot maintain its shape in water and would sink. The three hundred-foot wooden *Wyoming* collapsed and sank in 1909, taking all fourteen hands down with it. In 2014, so, it wouldn't sink, the next boat, a full-size replica of the ark, was built on dry land at Williamstown, Kentucky, to give curiosity seekers an idea about the size of Noah's boat.

IMMORTALITY AND THE RULER

Other stories that were even earlier than Noah's Deluge wafted about in the region. In Greece, Zeus planned to destroy humans in a flood because they were inherently corrupt and they liked Prometheus, benefactor of humanity, who had stolen fire from the god for them. Zeus vowed to punish them and sent a bride, Pandora, to Epimetheus, Prometheus's brother. Pandora, the first woman created, was the keeper of a box or jar she was told never to open, but opened it anyway and accidentally released the world's evils and pain, hence the idiom. Only hope remained in the box.[5] Zeus's son, Deucalion, and his wife, Pyrrha, were obedient to the gods, so Titan Prometheus, forewarned them about

167

the impending deluge. They built a boat and took on board their children and a collection of animals and survived nine days and nights of a flood. After the water receded, the boat settled safely on Mount Parnassus.

Such stories could also be related to the *Epic of Gilgamesh,* a "Babylonian story of a flood," passed down orally and recorded on twelve cuneiform tablets in 1900 BCE. Gilgamesh was a Mesopotamian tyrannical king around 2,700 BCE. After losing his dearest friend to a mysterious death, he sought to live forever. He looked for a wise old man, Utnapishtim, who had survived "the great flood" and knew immortality. According to Utnapishtim, a god – Ea – had warned him about the flood. He built a huge boat and took his family and all kinds of animals aboard [during] the flood. Then when the boat grounded on Mount Nisir, he sent out a dove, a swallow, and a raven to test for dry land.

After the story, the old man asked Gilgamesh to look in the sea for a plant that bestowed eternal youth. Gilgamesh dived into the water and brought it out. Unfortunately, a serpent snatched it away and Gilgamesh died a miserable death. Snakes gained the immortality Gilgamesh had sought, renewing themselves by simply shedding their skins. Also, in the Garden of Eden, the sly snake tricked Adam and Eve into partaking of the forbidden fruit from the tree of knowledge, thus passing their immortality on to the serpent. Humans have died ever since.

Several scholars have concluded that the biblical narrative, and others like the Greek and Gilgamesh stories, were adapted from the epic of *Atrahasis* about a local flood, or one earlier elsewhere, likely Egypt. In addition to irrigation and cultivation, Noah's name signified both survival of a flood, and vine growing through which Noah got drunk. As Nuach, Finch writes that the name is identical to the Egyptian "nuch," meaning "drunkenness," a reference to Noah's drunken episode. Initially, drunkenness was likened to the spiritual state, the source of the term "spirits" for alcoholic beverages. Egyptian priests used the alcohol for religious ceremonies only, but the alcohol became common and it fell into disfavor.[6]

On the front of the Narmer Palette, a bird carries a twig above the Sphinx, denoting the end of centuries of floods from the Waters of Nun.

This was the first record of the Flood myth, the source of Noah's deluge story.

A CATASTROPHE INDEED

Catastrophists believe that sudden events such as Noah's flood shaped the Earth and explain large-scale changes in its crust, such as the Grand Canyon. But the origin of the Grand Canyon gorge is no mystery. A meandering Colorado River aided by volcanic activity, winter ice at rock tops, and plateau uplift, gouged out the spectacular ravine – 277 miles long, up to 18 miles wide in spots, and one mile deep – over the past six million years, and it continues to do so. Despite this compelling evidence, creationists from the Creation-Science Research Center in San Diego direct searches for the ark's relics in the Middle East, giving Egypt, where Khufu's solar boat, the Nu-akh, rests, a wide berth. Exemplifying the two steps forward one step backwards progress of science, a creationists' book, *"Grand Canyon: A Different View,"* by Tom Vail, claiming the Deluge curved the Grand Canyon, went on sale in American National Park Service stores in 2004.[7]

Creationists argue that God deliberately gave the universe the "appearance of old age" to explain away the time difference between Bishop Ussher's 4004 BCE world creation date and the much older geological records identified by eighteenth century British geologist, Charles Lyell. Why God would do this, they don't say. Others strike a compromise and adopt the Day-Age theory. Instead of seven days, they allow hundreds of millions of years for the process of creation.

But, even the creation stories in Genesis weren't original. According to Godfrey Higgins in *Analcalypsis* I, the Fall of Man was depicted on the walls of Ramses II's Abu Simbel temple in Nubia in the thirteenth century BCE. Colonel James Tod's *History of Rajputana* located the same myth in India.[8] Also, the story was painted on the walls of Ramses III's temple at Karnak in the twelfth century, predating the Israelites. Here, the pharaoh prayed for wisdom to rule his kingdom and God sent him a woman, Shanu, who handed him fruit from the Tree of Knowledge. The scene was repeated in the Enki and Ninti Sumerian creation story, and again in Genesis centuries later. Very likely, however, through a hint in Rom 9:20-22, the Genesis story was related to the creation of man out of clay from the Nile on a potter's wheel by Khnum. Since writer P borrowed from the Enuma Elish epic, creation would have played out in seven days, instead of hundreds of millions of years, torpedoing the Day-Age theory compromise.

THE AUTHORS

The Torah was written over time and not chronologically, which might explain the contradictions" and "doublets," as in Mt 5:29 and 18:9, or Mt 5:32 and 19:9. Legend dates the composition of "The Law Code," or Law (*tôrâ* in Hebrew), to the seventh century BCE. Jeremiah assigned the task to his scribe, Baruch. Then to redeem his "lost" people, King Josiah's high priest, Hilkiah, conveniently "discovered" the Book of the Law in the "Temple" and read it in public. According to Magee in "When Was Exodus Written?" to mollify a defiant captive people returning from the Babylonian exile, and unite them with those left behind, their Persian overlords would have introduced the Law, imposing the message of one God on the Jerusalem priesthood.[9]

Legend has it that in the 3rd century BCE, Ptolemy II Philadelphus assembled seventy-two priests from Israel to translate the Hebrew texts and Law Code into the Greek Bible, or *Septuagint* (Greek for seventy). Their instructions were to work at the Alexandrian library and complete the translations in seventy-two days. Their use of Greek dialect, Alexandrian koine, however, shows that the translators were local. Rewriting the Persian story of the Lawgiver, they added Moses (meaning *born of* or *son of* in Egyptian) to gain favor with the Jewish priesthood so Judah could act as a buffer against the Ptolemys' rivals, the Greek Seleucid kings of Syria. The writers expanded the Law Code into what became Deuteronomy, and they added Exodus, Numbers, Leviticus, and Joshua, creating the Pentateuch (literally Greek for *five volumes*). The second book, Exodus, would have been based on the return of the Jews from Babylon. But according to Manetho, it tells the story of the expulsion of the Hyksos, "men of ignoble birth," from the eastern parts [the desert] of Egypt, back to the Levant in 1550 BCE. Even though it continued the Moses story, for the books in the Pentateuch to remain five, Joshua gave way to Genesis written around 273 – 272 BCE.[10]

Magee also notes that The Law was translated twice, first from Imperial Aramaic - the *lingua franca* of the Near East at the time - into Greek, and then into a Paleo-Hebrew script not used since the sixth century BCE. Moses was unknown before the Septuagint.[11] Legend places him in the thirteenth century BCE during Ramses II's reign. But Deuteronomy mentions Israel throughout before such a country existed. And, whereas the Sabbath commemorates the seven days of creation in Ex 20:11, it honors the flight of the Jews from Egypt in Dt 5:15. Also, was Moses' father-in-law Reuel, Hobab, or Jethro? The in-laws are mixed up. Moses had one wife, Zipporah. And if Moses wrote the Torah, how could he record his own death in Dt 34:5-6? Was it because he was a foreteller and could see into the future? Prophets, however, are problematic. They defy logic. Otherwise, if they could see into the future, wouldn't they make a killing in the stock market? But none are here today. Muhammad was last. Perhaps there was no more work for them.

To give the Law Code authority, it was attributed to Moses. In the fifth and fourth centuries BCE, Governor Nehemiah (with or without a priest, supposedly Ezra) enforced these laws to reintegrate the exiles from Babylon with the people left behind, to purify their ethnicity, and to preserve Jewish identity. He outlawed marriages between Jews and non-Jews and dissolved existing ones.

In all, many authors contributed to the Old Testament. Some texts were borrowed. For instance, Psalm 104, attributed to King David, matches the Hymn to Aten from 1,350 BCE. Other texts written in an Egyptian composition style focused on an historical event involving a particular king. Joseph's story in Egypt, according to Middle Eastern archeologist Donald Redford, was a novella containing regional wise man motifs.[12] It appears that if one wished to document the factual history of Israel, one would need to step away from the biblical, scriptural structure, eliminate references to God, and double check the details with outside sources. Stories lacking verifiable proof, such as Noah's Deluge, or the legendary First Temple, would surely not pass muster.

The Old Testament was completed in the first century BCE or later. Scholars long believed that Rabbis at the Council of Jamnia, about 90 CE, rejected the Greek version for the *Tanakh,* insisting almost exclusively on texts in Hebrew. Enoch's five books appeared between the third century BCE and the seventh century CE. The only complete copy in existence, 1 Enoch, was found in Ethiopia, where it's canonical. Written in the Ge'ez language, it depicts Enoch with Ethiopian features on the cover. His influence is visible in the numerous images of black angels in a riot of colors decorating Ethiopian churches, like the church at Lalibela in Axum hewn out rock—by angels no less—in the twelfth century. Hebrew texts found in the Qumran caves and the Samaritan Torah match the Septuagint, suggesting the Septuagint was the source. It's not clear when the Old Testament was canonized. The Jamnia council most likely never took place. [13]

All twenty-seven books of the New Testament were written in Greek. The assumed authors, the disciples, were ordinary folk, who, like 90 percent of the population, would have been illiterate. Unschooled in the Greek language, they couldn't have written the gospels. Also, they would have been too old. John's gospel was recorded in 90 or 95 CE, when John would have been about one hundred years old. It is unlikely that he managed to live that long, making it equally unlikely that the actual authors of the New Testament books would have known Jesus personally.

Numerous clues within the text itself also point to the likelihood that the NT authors were not original disciples of Jesus. Consider the following:

> •Unfamiliar with the Hebrew Bible, the NT authors referred to the Septuagint, mistranslations and all.
> •Mark, crafting the first gospel somewhere after 70 CE, apparently did not know Palestinian geography. Although the sea of Galilee lies south of Tyre, he had Jesus travel north from that city to Sidon to reach it. Mk 7:31.

Church historian Bishop Eusebius of Caesarea believed Mark brought Christianity to Egypt and wrote the gospel in Alexandria, supporting Father Clements's belief that the gospel was written there. According to 1919 German scholar Karl Schmidt, Mark likely gathered existing smaller stories to which, for instance, Matthew and Luke had added the Nativity account to promote the argument for an historical Jesus. Like many other stories, the Nativity Scene was not original. It was engraved on the walls of Amun's Temple at Luxor around 1,500 BCE, showing the divine birth of the infant-pharaoh Amenhotep III, an avatar of Horus, in the fourteenth century BCE. Such similarities caused Massey to write that gospel history had been "written before" from beginning to end.[14] Massey also argued that "the myths of Egypt supplied the mysteries of the world."[15]

Scholars don't know any of the authors of the Gospels. They were anonymous, and the names were arbitrarily assigned, thus they were titled *the gospel according to* so and so, not *by* so and so. Paul was probably the only historical figure, but seven of the fourteen letters attributed to him were forgeries to conceal his Gnostic leanings.[16] According to Father Clement, that of Mark might have shown similar views in a "secret" gospel he wrote for the initiated, the Gnostics.

Because the gospels of Mark, Matthew, and Luke are very similar, they are called the *Synoptic* Gospels. Their authors could have borrowed from a fourth, called "Q" or *quelle*, German for "source." As the Q document has never been found, scholars recreated it by teasing texts out of the Synoptics and Thomas's Gnostic gospel and called it the *Lost Gospel: Q*. In the Q gospel, Jesus is not a real person. He does not have a virgin birth, perform miracles, or die. Instead, as in Thomas's gospel, he is a talking head with no history, no real existence, and no miracles. However, the other two gospels could be revisions of Mark, as Christianity scholar Richard Carrier also said in an interview.

As we find out below, Josephus, and the Alexandrian– Philo Judaeus – who developed the doctrine of the Logos (or Word) clandestinely with the Flavians – helped to create the gospels, particularly the gospel of John. They dated the gospels back to 30 CE to make it appear as though Jesus's predictions of the Jewish war around 70 CE had come to pass.[17] Philo was a Platonist and wrote beautiful Greek.

Since the Bible was written by many authors over hundreds of years, how could they fact check to stay on message and create a book to be trusted? Of course, they couldn't have. There were no facts. Only created stories. It's noteworthy, however, that many texts, mostly those that differed from church doctrine were revised, falsified, rejected, burnt, or lost. So, what's the Bible's story?

THE BIBLE, WORD OF GOD?

If compiling the Bible took centuries by scribes unknown, how can it be the word of God? God would have written it in one fell swoop and passed it down to us mortals, as the Lord Ahura Mazda did with the *Book of Law* to the Persian prophet, Zoroaster. When Hilkiah read the

Book of the Law in public, Deuteronomy hadn't yet been written. Perhaps the writers were inspired. But by whom or what? Pitting the Bible's timeline against historical events could help to provide an answer. Since it has no dates, however, verifying the Bible's historical claims is a daunting task, regardless of whatever position the church maintains. Contemporaneous societies around the time of its stories, such as the ancient Egyptians, have dates. Also, stories about well-known people, like some kings of Israel, and actual events, could have become so embellished that they became legends.

THE OLD TESTAMENT

Before the reign of Darius II, from 423 BCE to 405 BCE at the earliest, the Jews and their Temple did not exist. According to traditional rabbinic sources, the Temple stood for 420 years before its destruction in 70 CE, so, it had to have been built during the reign of Darius II's grandson, Artaxerxes III. According to *Seder Olam Rabbah's* chronology based on the Bible, it was constructed in 350 BCE, reason enough for historian Herodotus and others such as Aristotle not to mention it.

Before the mid-thirteenth century BCE, nomadic shepherds frequented the Palestinian hills in the Levant, bartering goods with the Canaanites, and later the Sea Peoples, who arrived in the valley below in the twelfth century BCE. Eventually, as Canaanite cities and economy declined, the bartering ceased, and people migrated into the hills. There they mixed with the nomads, settled, and became farmers. An Egyptian stone inscription – a stele – from 1,207 BCE written by Merneptah, son of Ramses II, reading, "Israel is devastated, its seed is no more," was the first indication of the Israelites' existence, and that the king had devastated them. The stele did not say the Israelites were a country yet. Despite historian Flavius Josephus's claim in the essay *Against Apion* that the Jews were very ancient, Israel did not exist as a nation before the ninth century BCE. Not until the second century BCE did Greek historians and poets start describing Jewish history, more than a thousand years after it supposedly started.

Religion was used as a powerful tool by rulers to control the masses eager for information about the creator. As Magee mentioned earlier, to pacify the Israelites during captivity, their Persian overlords wrote *The Law of Moses,* imposing their religion of a universal god, Ahura

Mazda, on the priesthood that was contemptuous of the many-god worshipping population they left behind in Judah. Referring to the story in Manetho's *Aegyptiaca,* written about 285 – 280 BCE, Magee, like Russell Girmkin, believed that Exodus was composed later in Alexandria. He also dated Isaiah, Ezekiel, and Numbers to this period, as their authors mentioned Zoan (cognate with Tanis), which flourished in the Ptolemaic period. The appearance of Talmai (Ptolemy) in the same verse with Zoan (Nm 13:22), was the clincher.[18] Likely, Exodus contained sections that ultimately went into Genesis. Genesis chapters 1 through 11 are clearly borrowed from Berossus's *History of Babylonia*, which was written between 290-278 BCE. Speaking to this late date of the Bible, although Ur, Abraham's birthplace, was ancient, it did not become known as Chaldea until around the eighth century BCE when various tribes from the west migrated into the region, 1200 years after legend says Abraham was born. According to the Bible, Abraham's story was figurative (Gal 4:21-24), so perhaps we don't really need to assign a date to it. Also, Enoch's story in Genesis first appeared in *The Book of the Watchers,* one of the earliest forms of the Enoch legend, in the second or third century BCE, and before it was incorporated in 1 Enoch.[19] So Genesis cannot be older than Enoch's story. Since Deuteronomy mentions the first four books of the Torah, it must necessarily have been written after they were. And as Magee noted, the table of nations references third century BCE geopolitical relationships in Genesis 10.

Because humans evolved in Africa, creation stories can only be guesses about those origins. For instance, the river that watered the garden of Eden and divided into the four branches, Pishon, Gihon, Tigris, and Euphrates in western Asia, has not been located. In Ife, Western Nigeria, a Yoruba myth mentions a garden, Edina, where Olodumare, the local god, created the Earth. Ife, however, doesn't have such rivers, even though its religion bears some resemblance to Christianity with its belief in a trinity.

To win over the Jewish priesthood from their Syrian enemies, the Seleucid kings, the Egyptians, Ptolemy Philadelphus, and scholars such as Manetho, translated, rewrote, and embellished the Jewish texts, likely the earliest form of the scriptures we have today, according to Magee. Come the second century BCE, however, after the Seleucid kings wrestled Israel from the Ptolemys, they reworked the texts and could have added events such as the evil Pharaoh and the plagues as a way of vilifying the Egyptians.

When they finally drove out the Syrians, the Maccabees found the scriptures tattered, scattered, or lost, and restored them. They filled in gaps, created new stories, and merged different versions of similar events, thus contributing to the doublets and contradictions in the Book. Joshua's drama would have remained intact, as it was a novella with a single source. Daniel was written nearly last, mainly in Aramaic, in 165 BCE, with Ecclesiastes not far behind, according to Magee. Full of reflections, such as "Everything is meaningless" or "There is nothing new under the sun," Ecclesiastes appears to contain Epicurean and Stoic philosophy, its last two verses spuriously added to soften its defeatist tone:

> Now all has been heard; here is the conclusion of the matter: Fear God and keep his commandments, for this is the duty of all mankind.
> For God will bring every deed into judgment, including every hidden thing, whether it is good or evil. Eccl 12:13-14.

In the fourteenth century, John Wycliffe translated the Vulgate, or Latin, texts into the first English version of the Bible. William Tyndale wrote a second version from Greek and Hebrew texts in the sixteenth century. As in any translation, meaning would be lost or corrupted. The original texts being unavailable for comparison, versions differed. Also, a Hebrew word could have several interpretations, resulting in mistranslations. For instance, although the Hebrew Bible says a young woman [*alama* in Hebrew] would conceive and give birth to a son (Is 7:14), the Greek version uses *parthenos,* originally meaning girl but over time taking on the connotation of "virgin." Mt 1:23. Yet, how could Joseph marry a woman who was expecting someone else's baby? A pregnancy out of wedlock was taboo, and the woman was usually stoned to death. The switch from a young woman to a virgin appears deliberate, intended to fulfill the Bible prophesy of Jesus as the Messiah, born of immaculate conception. But if Mary was already expecting, then a husband was not necessary for her to conceive and Mark, the earliest gospel, doesn't mention one. Below, we find that gospel revisions endeavor to show that Joseph was not Jesus's father.

Also, even though Joseph may be traced to the line of King David, he couldn't be related to Jesus. A woman, then, was considered merely a vessel for the man's "seed" with no contribution to a child's DNA. The story of Mary's marriage to Joseph was meant to indicate fulfillment of an Old Testament prophecy that the Messiah would be a descendant of the king. 2 Sam 7:12. Nevertheless, Jesus's genealogies in Matthew and Luke diverge completely from David to Joseph.[20] His immaculate birth was not exceptional either, as many gods of old were born by virgins.

Other factors, such as translation, also contributed to the Bible's questionable veracity. Some versions, riddled with errors, circulated at the time, and could have influenced translations into other languages. For instance, Ex 20:14 of the Wicked Bible read, "Thou shalt commit adultery," "And ye said, Behold, the Lord our God hath shewed us his glory, and his great asse," instead of "greatness." (Compare this to Dt 5:24, KJV.)

Until a couple of centuries ago, questioning the Bible was suicidal. If you wondered publicly whether Enoch was the son of Cain and father of Irad (Gn 4:17-18), or the son of Jared and father of Methuselah (Gn 5:18-21), or whether Joseph was the son of Jacob (Mt 1:16), or the son of Heli (Lk 3:23), Spanish Inquisitor Leon of Castro would answer you with a sword. In 1576, speaking of the Vulgate, he warned that nothing could be changed, "be it a single period, a single little conclusion or a single clause, a single word of expression, a single syllable or one iota." Despite Leon's edict to the contrary, a Coptic version had existed in Egypt since the second century. Yet, those who ignored Leon's advice, some going as far as translating the Book into other languages, were accused of heresy and executed. After his death, Wycliffe's body was exhumed and burned. Tyndale fared even worse. He was imprisoned for sixteen months, retried in 1536, strangled, and his body burned at the stake. This was all in vain, however, as his translation lives on in the King James Bible.

WAS MOSES HISTORICAL?

In Exodus, God is said to have dictated only the Ten Commandments to Moses. Going by the timeline, though, Moses couldn't have written the Torah, as mentioned earlier. The Bible says, "Since then, no prophet has risen in Israel like Moses…." Dt 34:1.

Deuteronomy was written after Moses's time. If Moses was the first prophet, how could he have known about those who came afterwards? If the Bible timeline is accurate, Moses would have lived centuries before the Edomite kings listed in Genesis. Gn 36:31-39. They couldn't have denied him passage through their land during the exodus. Sites like Kadesh Barnea, Heshbon, and Gibeon weren't even occupied at the time. Importantly, laws credited to Moses are similar to the seventeenth century BCE Babylonian Code of Hammurabi and others. In turn, these were likely based on the Egyptian 147 Negative Confessions from the book of the dead.

Was Moses historical, then? The story about him floating down River Nile in a basket is similar to that of Sargon of Akkad, a third millennium BCE Mesopotamian king. Moreover, what happened to Moses on Mount Sinai (if you can locate it) was not unique. Laws of the Greek mythological god Bacchus, called "the Lawgiver," were written on two tablets of stone. And one day as the Persian prophet Zoroaster prayed on a high mountain, the Lord appeared before him and handed him the "Book of Law." Zoroaster brought the book to the King and people at the foot of the mountain. The Persians could have introduced this story when imposing The Law on the Jews after their return from their Babylonian exile.

Were Moses and Abraham human? One can hardly imagine living 120 years as Moses did with any semblance of a sound mind, but Abraham's sojourn of 175 years is truly a stretch and Noah's 950 years is merely a dream. None of the tens of thousands of cuneiform tablets from the archives of the ancient city of Mari specifically refers to Abraham. Failing to find evidence for these figures, or the Exodus—the Egyptians have no records of it—John Loftus, a former preacher, wrote in *Why I Became an Atheist* that Jewish history "…rested on myths, legends superstitions."[21]

Because Bible events tally poorly with history, it seems Joseph's Egyptian stay, the Exodus, and accounts revolving around figures like Abraham and Jacob were based on earlier Canaanite traditions and written late.[22] For instance, when Joshua supposedly conquered the Canaanite cities in 1407 BCE, the area was under constant Egyptian rule. Also, cities like Jericho, unoccupied since 1550 BCE, had no sky-high walls, if any. Moreover, even if there had been walls, Joshua would have already torn them down before the time of Moses and the

Exodus, dated to the thirteenth century during Ramses II's reign. The Israelites do not mention the Egyptians, as Israel wasn't a nation yet. Like resettled Jericho, cities that fell did so to the Sea Peoples. And since Pithom was a fort built by Pharaoh Necho around 608 BCE, and Ramses, a city constructed by Ramses II in the 1,200s BCE on top of Avaris, the main center of the Hyksos until 1,550 BCE, Moses couldn't have existed before the third century BCE, when the Egyptians invented him. Ex 1:11.

THE ISRAELITES

In the third century BCE when the Bible was written, it seems that the scribes were unaware that the Levant was governed by the Egyptians from 1550 BCE to 1100 BCE.[23] Natufians, or Negroids, the ancestors of the Canaanites, Phoenicians, and Samaritans, migrated there during the Stone Age around 9,600 BCE, building Jericho, one of the world's oldest settlements. Their elongated (dolicocephalic) heads, like those of Bantu and Egyptians, link them to Africa. Obsidian and shellfish found in the area point to their origins near the Nile, and malachite beads found in the area point to Congo. In Congo, a highly developed culture at Katanga and Ishango – by Lake Albert on the border with Uganda – was crafting fine harpoon points 70,000 years ago. Similar harpoon cultures later appeared sequentially along the Nile in Sudan, Egypt, on up to Palestine. Indeed, Robert Graves traces Canaanite origins from Uganda through Egypt.[24]

In 1020 BCE, Hebrew, a Semitic subfamily of Afro-Asiatic languages from Northern Africa, was not yet a codified language. It took hold about three centuries later as groups in the hills mixed with incoming Canaanite sources. Called Israelites today, Jacob's (Israel's) descendants were only identified as Jews, or *Yehudi,* the name of his fourth son, after their return from the Babylonian exile in 538 BCE. The land was then known as Yahud (Judea), which, after translation through several European countries, morphed into *Jew* in a comedy, *The Rivals,* in England in 1775. Thereafter, it was used in the New Testament in the King James Bible. The name, though, includes many groups, and some would prefer to be called Israelites rather than Jews. For their origins, DNA studies show that Israelites are genetically similar to Palestinians, Syrians, Lebanese, and Saudi Arabians.[25] So, conflicts among them notwithstanding, they are basically one family—

with Abraham their patriarch. Approximately 13 percent of Jewish people carry the only recognized Semitic hg J1. Twenty-eight percent of Ashkenazi Jews carry the sub-Saharan hg E1b1b. Famous E1b1b carriers include Albert Einstein, Lyndon Johnson, the Wright Brothers, and Hitler (whose DNA of sub-Saharan origin ironically differs from that of Aryans or pure white people). Napoleon Bonaparte carried its variant, E-M34.

Although, as with many legends, the Bible may contain a kernel of truth, it's mainly the product of political gamesmanship as nations sought to control Judah or gain its support against rivals. Likely politics played a part during the canonization of the Old Testament. The Maccabees' Books were left out, perhaps because of rivalry between the priestly class: the Sadducees to whose family they belonged, and the descendants of the Pharisees, the rabbis, who might have had the last word at the debates. Books considered to be outlandish, of questionable authorship, or blatant fabrications (depending on the standard used), were relegated to the Apocrypha or rejected outright. Still, we get to read about the sun stopping for Joshua, so that's something.

Enoch's authenticity as an inspired work was discredited at the Council of Laodicea in the fourth century CE. Though the council may have demonized 2 Enoch, leading to the disappearance of his books, early Christians, who had initially lauded the books, were rejecting them by the time of the meeting. Interest in the books revived in the late eighteenth-century after James Bruce, a Scottish traveller tracing the origins of the Blue Nile, brought a complete copy of 1 Enoch from Ethiopia to Europe. Early Christians considered Enoch's books scripture and his influence on the New Testament was substantial. Of the over one hundred parallels between the New Testament and the Enoch books, the use of "the Elect One" for Jesus stands out. The original Greek texts read, "This is my Son, the Elect One. Hear him." Lk 9:35.[26] Both Peter and Jude borrowed heavily from Enoch, too.[27] And by influencing the Old Testament, as indicated above, Enoch, it seems, predated Genesis. Besides the "taking" of Enoch by God appearing in Genesis 5, writer P references him when he mentions "eating animals" and "bloodshed" in Genesis 9, and Noah's Deluge. Such observations support the view that Genesis was written in the third century BCE, when the Enoch books were first written.[28]

Compiled in the third century BCE, the Old Testament is clearly the work of humans. Its stories were Egyptian, translated word for word into Hebrew.[29] Tel Aviv archeologist Professor Ze'ev Herzog, finding many accounts fictional, wrote, "Israelites were never in Egypt, did not wander in the desert, did not conquer the land in a military campaign and did not pass it on to the 12 tribes of Israel...."[30]

THE NEW TESTAMENT

Is the New Testament "gospel truth"? Following the publication of Dan Brown's novel, *The Da Vinci Code*, interest in its history and Christianity was piqued. But, had it not been for the Roman Emperor Constantine, Christianity would be very different today, if it were to exist at all. When building their armies, Roman emperors recruited soldiers from different groups. Constantine recruited from worshippers of Serapis Christus, Mithras (the Roman soldier's sun god), and Apollo (the Roman sun god), courting them by adopting their faiths. And, perhaps, to recruit the obstinate Christians, in the Edict of Milan in 313 CE, Constantine allowed them to worship their god without fear of persecution.

Like the pharaohs, Roman emperors were God's envoys on Earth. Constantine's commemorative medallion showed him as Mithra Sol Invictus. According to Ian Wilson in *Jesus: The Evidence,* Constantine was represented wearing a helmet with a Chi-Rho monogram, the symbol of Christ, and a leaping Sol (sun) chariot below it.

To win over Serapis's followers, Constantine gained recognition in the Egyptian Secret Order. Raphael's fresco in the Vatican captures his baptism into this Greco-Egyptian religion (also found in Italy and Turkey at the time) by Egyptian Bishop Sylvester. In gratitude, Constantine made Sylvester head of the Roman Church. This also made Sylvester supreme over Constantine himself. Little is known about Sylvester, even though he became a saint, suggesting the church suppressed his story, as it did with those it found unflattering.

In his documentary film,[31] Canadian award-winning investigative journalist Simcha Jacobovich reported that Constantine erected an arch in the center of Rome to celebrate his victory over General Maxseutius. The arch was flanked by the Flavians in front of their Coliseum. At the top of the arch, Constantine wrote, "Divinely Inspired," suggesting inspired by the sun god, Apollo, whose statue towered in the foreground.

If Constantine championed Christianity, his arch didn't show it. Instead, he filled the arch with pagan symbols, mostly of Mithra, the god of the soldiers and Roman elite. Christian symbols were absent. Of the vision on the cross, nada. This event was first recorded by Eusebius of Caesarea, Constantine's historian, thirteen years after the battle. Constantine invented it to give bishops convening for the first Nicaean Council in 325 the impression that he was a prophet, the literal embodiment of Jesus returned. Flavius Josephus had done the same for Titus Flavius when he presented him as the Savior, the Messiah returned as promised. From there on, Constantine passed himself off as the Savior.

On a colossal Apollo column in Constantinople, a city Constantine made home and later re-named after himself, he replaced the sun god's face with his own, adding a halo –the sun's rays – around his head, as Emperor Nero had done. He merged Apollo and Mithra with Jesus, and then merged the two into himself. Two centuries after his death, a mosaic portrait in the Archbishop's Chapel in Ravenna, Italy, shows he replaced Jesus with himself by dressing God's son in the emperor soldier's uniform. Eusebius had already told his followers that the emperor was the voice of Christ on Earth.[32]

By placing his arch in front of the Colosseum amid the Flavians, the same Flavians who destroyed the Temple in 70 CE and created the Jesus story, and by taking their name of Flavius, Constantine gave the impression that he was a Flavian and was as divinely sanctioned as they were, but better. Even though he was an avatar of the Christian God, Constantine worshipped other gods, too, mainly Mithra, the Roman elite and soldiers' god. Anyway, to him, all gods were the same, differing in name only.

To dispel any doubts that Constantine was the Messiah, he was buried surrounded by twelve coffins representing the apostles.[33] Unbeknownst to most, that's the Savior Christians worship today.

Before Constantine, Christians did not have standard texts. Rome had Mark, Greece had Luke, Antioch had Matthew, and Ephesus had John. In 325, hoping to forge some uniformity, Constantine summoned the first Council of Nicaea, attended by 318 bishops. The Council did not start from scratch. By 170 CE, Irenaeus, Bishop of Lyon, had listed the current four gospels, claiming they were obviously the authentic ones, as they corresponded to the four corners of the universe and four principle winds. Earlier in 150 CE, Marcion of Sinope had compiled a canon consisting of Luke and ten Pauline Epistles.[34] Marcion omitted anything Jewish, as he believed the Jewish God, always angry and punishing people, was different and inferior to that of Jesus. Leviticus 20 lists the sins and punishments, mainly death. Like the Gnostics, Marcion was unaware of the Incarnation or Resurrection. Along with many other texts circulating at the time, his canon did not pass muster.

There are over forty similarities in the same sequence between the gospels and Josephus's books, *Wars of the Jews,* and *Antiquities of the Jews,* based on the Old Testament. The books were called gospels, Greek *evangelion*, meaning "Good News of Military Victory," as the Romans celebrated their conquest of Judea. To dupe Messianic movements, the Jewish Zealots and their rebellious splinter group, the Sicarii, into worshipping Titus Flavius as the Messiah returned, the Romans created a religion centered on this belief.[35]

Titus wrote himself into the Gospels as the Son of God—Emperor Flavius Vespasian—who crushed Galilean towns, encircled Jerusalem, and razed the Temple in 70 CE, true to Jesus's prophecy forty years earlier in Lk 19:37-44. Josephus had backdated the gospels forty years, when Jesus would have been thirty years old, to make it appear as though Jesus's predictions had come true about the war, word for word. It meant that Jesus had returned as promised, but now as Titus Flavius.[36]

The Flavians featured prominently in the early Catholic Church, too. Titus's sister, St. Flavia Domitilla, was the first saint. Domitilla's son, Titus Flavius Clement, became the fourth Pope of Rome.

The gospels were Roman propaganda and cast the Jews in a bad light. They clearly favored Rome at the expense of the Jews, as in the saying give to Caesar what belongs to Caesar. Mt 22:21. Christian literature made the Jews responsible for Jesus's crucifixion. For instance, in the parallel New Testament story, the Gospels replaced the Israelites, who tempted God in the Old Testament, with the devil.[37] Going forward, the Flavians suppressed and destroyed any literature that could expose their conspiracy to tempt the Sicarii into worshipping Titus as the Messiah. So, texts like the Dead Sea Scrolls and the Nag Hammadi papyrus manuscripts were hidden to preserve them.

At the Council of Nicaea, the bishops were deeply divided over Jesus's identity. Those like Arius, the Alexandrian presbyter from Libya, and Eusebius of Nicomedia, disputed the claim that Jesus was of the same essence as the Father, basing their argument on Jesus's own words, "...the Father is greater than I." They interpreted this to mean that as a son, He came after the Father from nothing, and had a beginning. Jn 14:28. Thus, He couldn't be coequal or coeternal with God. This claim also put him in direct conflict with the belief in the concept of the Holy Trinity. Arius believed Christ was like the man-made god, Serapis, and not a true god. In the heated debate, Arius lost to his strongest critic, Athanasius, the contentious Coptic Alexandrian pope, who was exiled five times. The Nicene Creed refuted Arius word for word, in effect saying that Jesus Christ was the Son of God, begotten, not manmade, of one substance, the same as the Father. But, if Jesus was made of the same substance as the Father, didn't that make Him a god, too? Nevertheless, despite deep divisions and disagreements in the meeting, texts referring to Serapis and other cults were ruled heretical and burned. Constantine decreed that anyone harboring such writings be executed. Arius was deposed and exiled.

In 331 CE, Constantine commissioned fifty Bibles. Eusebius of Caesarea selected eighteen books, and added them to the Hebrew scriptures, as they fulfilled its promises of a savior, the Messiah. The current twenty-seven books of the New Testament were listed by Athanasius in 367 CE. After merging the gods into himself, Constantine was worshipped as Jesus outside Rome. Within the city, however, the Romans prayed to Mithra, performing the Communion in caves below the first Christian churches. Constantine was baptized again on his deathbed, this time by the Arian bishop, Eusebius of Nicomedia, to boot.

An uneasy Christian church, differing from earlier religions in name only, embarked on a campaign to eradicate signs of them, imposing the new religion on the empire. Removing hieroglyphic symbols cut in stone in chambers, monuments, temples, or the Serapeum proved impossible, so they were covered with stucco instead. When the plaster aged and peeled, though, there were the symbols again for all to see.

Considered the first Christian writings, the Gnostic Gospels were among the books ultimately rejected and excluded from the Bible. As in Marcion's canon, and later Nazi Christianity, the God of Jesus had nothing in common with the God of the Bible. He was more like the god of the Mystery Religion of Osiris, or Dionysus, experienced mystically by initiates. Other books omitted included the Gospel of James, Jesus's brother, which was deemed too frivolous, and Barnabas, too anti-Jewish. Paul's Epistles, too well known, were altered instead.

As in Freemasonry, mystery systems required an oath of secrecy. But with membership restricted, sometimes to only those good in mathematics, the masses flocked to Orthodox Christianity. There they heard captivating and amazing things about an historical Jesus whose nature was easy to grasp, the more miracles and Christian relics the better. To meet demand, the number of available relics grew dramatically. John Calvin, the leader of the Swiss Reformation, joked that relics from the wooden crucifixion cross could fill a ship. From Ptolemy I to Constantine the Great, it took Christianity almost six hundred years to become the official religion of an empire, the Roman Empire.

GNOSTICISM

Gnosticism, reflecting Greek origin ideas, especially those of the Platonists, taught that attaining wisdom was the gateway to salvation. Gnostics also taught that people should shun the material world created by an intermediary of God, and instead embrace the spiritual world of Jesus. The Gnostic gospels came to light when a local peasant, Muhammed al-Samman, discovered papyrus manuscripts hidden in jars at Nag Hammadi in the Egyptian desert in 1945. The manuscripts included the Gospels of Thomas and Philip. Thomas's gospel recorded Jesus's teachings only, as Thomas was not aware of Jesus's virgin birth, the stories of his youth, or his miracles. Though he reputedly touched Jesus's wounds, Thomas said nothing about the crucifixion or resurrection. Gnostics believed in Jesus as a Divine Mind, and thus they did not believe that He could die. Since all Gnostic texts were based on Christ, Gnostics were considered Christians.

In Judea, too, people did not believe that the Messiah who had come to liberate them could die as a common criminal. The apostles accused the Sanhedrin high priest of hanging Jesus on a tree, Acts 5:30, as though God had accursed Him. Dt 21:23. This story closely parallels that of the Nazarene magician, Jehoshua (Joshua or Savior) Ben-

Pandira, who at the time many believed was the promised Messiah. Ben-Pandira suffered this ignoble death around 70 BCE.

Gnostics were strong rivals of Orthodox Christianity. For them, one came to salvation on one's own and did not need the hierarchical system of bishops. You could be a high priest one day and an ordinary person the next. Also, women could become priests, not only men. Gnostics taught that the key to salvation lay in knowledge of who you were: According to gnostic belief, Jesus brought the knowledge that everyone has a divine spark that needs to be set free.

Gnostics understood Jesus's teachings intuitively through mystical experience. While he spoke in public in simple terms – often in parables – in private Jesus would break out into riddles with his disciples. Hence, the Gnostics saw Jesus not as a teacher, but as a divine *revealer.* "For many who are first will become last," He said talking about unity with the divine, "and they will become one and the same."[38]

In contrast, Orthodox Christians taught that Jesus had lived among us and died for our sins. No wonder then, that in 180 CE in his work entitled *Against the Heresies,* Bishop Irenaeus, a strong opponent of the Gnostics, let loose a tirade that drove the sect underground. There, unknown, their gospels languished until Muhammed al-Samman discovered them in 1945.

THE "PAGAN'S" ROUT

Despite Pope Theodosius's declaration that Christianity was the only religion permitted in the Roman Empire, and an edict shutting down all pagan temples in 391 CE, Gnostic teachings still made up 20 percent of the New Testament. These included sayings from Thomas's gospel and seven of the thirteen letters attributed to Paul portraying a mythical Jesus. Mainstream Christians simply merged Gnostic ideas with the cults of Osiris and Isis and the wisdom of the Greek philosophers. As early Christians didn't have temples and worshipped in homes, they seized the Pagans' shrines, like the Serapeum at Ephesus, and built churches atop them. Saint Peter's Basilica towers over a former Mithraic Temple, and Notre Dame (Our Lady) in Paris (*Par Isidos* or Place or House of Isis) was erected over the site of an Isis Temple.[39] In 354, because the birthdays of Osiris, Horus, and Krishna and sun gods like Mithra and Attis all fell on 25 December, Bishop Liberius of Rome decreed that Jesus's birthday would also be celebrated on the same day. Before then, Christians, not knowing when Jesus was born, had not observed his birthday.

Because on 25 December, the sun slowly turned around and began ascending northwards from a three-day winter slumber, or *sun stop,* the solstice – Christmas – was a celebration of this return or *annual rebirth.* The event was known as Yule Tide, from Anglo-Saxon *Iul,* meaning "wheel turn," since the sun turned around and began climbing northward to revive crops and save lives from hellish winter and bring heavenly summer. Not surprisingly, because of its pagan roots, Christmas was not celebrated widely in the United States until it was declared a federal holiday in 1870. Groups such as the Puritans initially banned it, while Jehovah's Witnesses still don't observe it. Easter celebrated crop renewal and animal births in the pagan world as well.

Following the adoption of the 325 CE Nicene Creed, many people were left disgruntled. To counter the persistent Gnostic view of a mythical Jesus and instead portray Jesus as a real person, scholars found that the story of Christ's resurrection was added to the gospels. Lk 24:28-43; Mk 16:9-20. Such gospel "corrections" – known as pious fraud – were common. Church fathers considered it a virtue to deceive and lie for the glory of the faith. Church historian Eusebius of Caesarea admitted to promoting only that which might contribute to the church and suppressing whatever could disgrace it.[40] Such additions were essential, as there was no evidence an historical Jesus ever existed.

None of twenty-seven history writers around Jesus's time mentioned Him, and the Flavians made sure any literature that could expose their machinations was destroyed. Renowned historian Josephus, who wrote about many events from Galilee, also wrote about noted people and "little [ordinary] people," such as wonder makers like Theudas, and criminals in *The Antiquities of the Jews,* but not a word about Jesus. He couldn't, since he promoted Titus Flavius – the son of Flavius Vespasian, the historian's adopted father – as the Messiah, the son of "god." Not until three hundred years later did church propagandist Eusebius of Caesarea "help out" by inserting a brief biography of Christ into Josephus's work. Other modern scholars, like historian Richard Carrier, found that Jesus lived in one of the lower celestial spheres, and declared that any texts saying Jesus was historical were forgeries inserted later.[41]

No wonder then, that the later gospels differed from earlier versions. For instance, earlier gospels said about Jesus's birth, "The child's father and mother marveled at what was said about him" (Lk 2:33), but later versions replaced "father" with "Joseph," implying Joseph was not Jesus's biological parent. And in Lk 2:48, when as a young boy Jesus went missing at the Temple for three days, Mary said, "Son, why have you treated us like this? Your father and I have been anxiously looking for you." Newer versions changed the last sentence to, "We have been anxiously looking for you."[42]

Incidentally, since we can't find Heaven, where did Jesus go when he ascended? Maybe we are looking in the wrong place. Jesus had told the Pharisees, "The kingdom of god is within you." Lk 17:21. Resurrections in Jesus's day were practically a dime a dozen. Gods died and resurrected yearly in the spring, symbolizing the return of the sun and crops to the Northern Hemisphere. At the spring equinox, around March 21, day overcame night. The sun rose directly due east, east-er, eastre (the ancient word for spring) or Eostre (mother goddess of the Saxons), passing over the due east point, the origin of the word "Passover."[43] Competing for souls, early Christians taught that Jesus resurrected like other mystery gods, such as Dionysus. For the Gnostics, however, resurrection meant spiritual rebirth only. Though highly contested, the Nicene Creed unified the church in the West. The East balked at the doctrine of the Holy Trinity that, as in Egypt, India, or Nigeria, implied three gods.

In the Middle East, even though the Byzantine emperor Heraclius defeated the Sassanid rulers of Iran in 622 CE, both superpowers were exhausted and they withdrew from their Arab territories. Scholars called *revisionists* or *rejectionists* say that the Arabs, now left on their own, founded the Umayyad dynasty led by Mu'awiyah, the first caliph, not the fifth, as is usually reported, who united their populations. Known as Nabeteans then, some Arabs worshipped various gods, including one called Allah, while others belonged to one of three denominations in the East that believed in one God and considered Jesus a human prophet. Mu'awiyah's successor, Caliph 'Abd al-Malik, looked for a common creed that all could follow in Arabic, and a prophet to unite the different tribes. The Caliph was a Christian following the Syriac tradition that considered Jesus the messenger of God. In 691, he built the Dome of the Rock at Jerusalem, and inside the dome had the inscription: *muhammadun rasūl allāh*. Translated it read: "The messenger of God is to be praised," indicating that Jesus was God's (*Allāh's*) messenger (*rasūl*), who was to be praised (*Muhammadun*). This was the first time and place the name of the prophet Muhammad was ever introduced. 'Abd al-Malik's other inscriptions left little doubt who he was talking about: "God bless your messenger and servant, Jesus son of Mary," and "For the Messiah Jesus, son of Mary, is the messenger of God."[44] Jews, too, believed Jesus was the Messenger of Allah.[45] (In Medina, a vacant grave next to Muhammad's tomb waits for Jesus's return, as predicted by the prophet). Clearly, Islam and Christianity share a founding ideology, and their theological concepts are similar except in language and worship. So, if we all recognized each other as extended family, wouldn't that be something! Anyway, as DNA analysis reveals, we are all related. Acknowledging this fact should help to dispel hostilities among us and promote peace and harmony with our neighbors, whether it's the family next door, or a nation beyond our borders, and, most importantly, among the various creeds with the same patriarch, Abraham.

Debates about Jesus's nature raged on in the West. What did it mean that Jesus was of the same substance as the Father? What substance was God His Father? But with Constantine having assumed Titus Flavius' role as the resurrected Messiah, who dared to ask? Further still, who cared? Christianity had now hit the big time as the official religion of the Roman Empire. Those once persecuted were finally in position to return the favor.

A FLAT EARTH

Is the Earth flat? Of course! It had to be to have four corners, or quarters. Is 11:12. Many people say so, too. The organization, International Flat Earth Society, headquartered in Lancaster, California, promotes the idea. It doesn't matter that masts of a sailing ship appear before the vessel comes into view, or that the Earth casts a circular shadow on the moon. Those trusting their eyes see a flat horizon, and a round Earth is nonsense. But go east; fly supersonic. Lo and behold, eventually you'll find yourself back to where you started.

When the Bible was canonized, God was not consulted. God wouldn't have allowed allusions to a flat Earth. Texts like the Dead Sea

Scrolls, only discovered in 1946, weren't available for consideration in the inevitable contentious debates. In the scrolls, the book of Samuel was longer and better. Trusting in the historical accuracy of Genesis also means believing that we aren't related to other species, despite sharing genes with them, 98.8 percent with chimps. Failing to make sense of the creation story, Calvin considered it an example of baby talk, to simplify complex and mysterious processes for the layperson. This way, everybody could have faith in God.[46]

Even though mostly legendary, based on cosmic events, the Bible's popularity continues unabated. Found today in hotel rooms and libraries "'round the world, "the greatest story ever told," has sold more than five billion copies and is required reading for many. In spite of scientific innovations, its stories inspire and explain the world to believers in ways they understand, giving them comfort in times of stress and need. Like other religious texts to their followers – the Sutras for Buddhists, the Vedas for Hindus, the Quran for Muslims, the Tanakh for Jews, and the Odus for the Yoruba, to name a few – the Bible's readers derive both religious and spiritual benefits from its pages. It means, since the desire to invoke spirituality – the impetus behind religious practice – is universal, nondenominational and nonbelievers may likewise benefit through other related practices.

Our difficulty finding God suggests that He may not exist in physical or tangible form at all. In absence of a physical deity, whether it's God or Jesus, humans have tended to ascribe supernatural powers to the sun as a god, for instance, Aten or Amen-Ra, the Egyptians' gods, or Mithra, Sol Invictus, the Romans' sun god. Likely, our desire to define and find a source for these powers is but a component of our attempt to understand our spirituality. Psychoanalyst Sigmund Freud believed God was an illusion, a device of the subconscious. The reason for believing in Him was the human yearning for a father figure. Freud declared God was nothing more than "an exalted father figure," and that religion belonged to the infancy of the human race, finding it no longer relevant in the twentieth century.[47] Karen Armstrong took the argument one step further, saying that since it was an inner feeling that God existed, the mind must have hatched the idea of God in the first place. She called God a metaphor for our spirituality, echoing St. Augustine, who considered God a spiritual presence deep within the self. Indeed, if God was inside us, wouldn't that make belief in Him universal? Wouldn't belief, then, be a trait inherited from our parents?

HENRY KAKEMBO, M.D.

CHAPTER 10

OLD TIME RELIGION

Why should I be unable to regard the bodies of others as "I"?
It is not difficult to see that my body is also that of others.

~Shantidev[1]

While visiting Minnesota in the dead of winter, you and a group of friends, oblivious to the ominous greying skies, go cross-country skiing. When the time comes to head back to camp – *wham* – Snowy Armageddon descends. Gusty winds pelt snow in your faces with a vengeance. Blizzard-like conditions, poor visibility, and falling darkness separate you from the party. You trudge in circles in the snow for what seems like an eternity, freezing, hungry, and scared. Stop moving for even a minute and you'll find yourself hugging Enoch in Seventh Heaven.

Then, you hear voices behind you. Wheeling around, you see shadowy figures holding hands, moving in a circle, but they vanish when you approach them. "I'm hallucinating," you realize, and panic. Even though you're an irregular churchgoer, you pray to God for help. Like a drowning man clutching at a straw, prayers are your last hope for survival.

A chill runs through your body, as a bright light appears at the end of a tunnel. You abandon your body and float toward the light. There at the end of the tunnel, you find your deceased grandmother and other relatives. They welcome you with great joy and reminisce about what a Dennis-the-Menace you were as a child. Your whole life reels past, and then you sense a powerful presence and you're ecstatic. "Hey, this must be Heaven," you guess, then, "paradise must be real!" The presence dissolves and merges with you and everything else around you. Your soul joins those of others as you melt into the universe.

Grandma's piercing voice jolts you out of the trance, urging you to return to your body before it's too late. As you drift back, you see a search party approaching your body. "There must be two of me," you muse, recalling quantum superposition, where a subatomic particle occupies two places at once, "a levitating soul and a lifeless body." The search party, giving you little hope of survival, somberly hoists you into an SUV, where the warmth jolts you out of the Near-Death

Experience (NDE). Overjoyed, the group says a little prayer, thanking the Lord for pulling you back from the edge of the precipice.

WESTERN BELIEFS

When in dire straits, people automatically tend to entreat a higher power for help. In the West, this power could be a god, a spirit, or the sun. We sense it watching us all the time. Of the various higher powers, the God of Abraham, patriarch of the Jews, Christians, and Muslims, gets the call much of the time. In 2018, about half of the world population, or 3.5 billion people, belonged to one of these denominations.

On reading the Bible, we find that God is self-evident. People such as Abraham and Moses saw or talked to Him. Nothing beats eyewitness accounts, unreliable as they may be. Anyway, the idea of God, and hence an eternal rest in paradise, eases our fear of death, and renders life tolerable. The belief offers hope that our minds and thoughts will not cease to exist just because we die and that death here on Earth is not the end of us. But even with hope of such a rosy afterlife, I have yet to meet anybody anxiously waiting to go there.

FAR EASTERN DEVOTION

If only half of the world population believes in the Western God, what about the rest of the people on the planet, mainly those in the Far East? According to purists, not faith but philosophy guides those in Tibet and parts of India. Buddhism – considered something of a hybrid between philosophy and religion – which originated in India, and is still popular in the central, eastern, and southern regions of the country,, is more widely spread throughout neighboring countries and in Japan today. As in Hinduism, life is also suffering in Buddhism. For enlightenment, or Nirvana, the achievement that ends the cycle of suffering, Buddhists insist on proper conduct.

In India, eighty percent of the population practices Hinduism, which may invoke none, one, or numerous creators springing from a godhead. Instead of protecting and giving people comfort as in the West, though, Eastern religions stress proper conduct. The enlightened ones, those freed of misdeeds, transcend the painful, endless cycle of death and rebirth and associated suffering in life, never to come back to this cruel world. They attain peace and bliss, salvation known as moksha or Nirvana, by uniting with the universal world soul. A primary Hindu goal is to raise a son. Sons make offerings to ancestors hanging out with the unknowable godhead from which we all originated. In one branch of Hinduism, Jainism, creators don't even exist. Jains avoid violence and hurting life to free their souls of subtle karma, infra-atomic particles of matter that cause bondage. To attain salvation *(moksha)*, they free their souls of karma through practicing the right faith, right knowledge, and right conduct, called the three jewels.[2]

Before the government suppressed it, Confucianism set the moral standards in China. Unlike religion, Confucianism put emphasis on ethical conduct with social harmony as the goal. Confucius, a fourth century BCE Chinese philosopher, taught that to achieve social harmony one had to observe the whole of human duty, which comprises five virtues: kindness, uprightness, courtesy, wisdom, and faithfulness. Unlike some people that we know, Confucius practiced what he preached, saying that rulers could be great only if they themselves led exemplary lives. By following his own teaching as magistrate and then minister of Lu State, he was so successful in reducing crime that a neighboring state ruler, feeling threatened by Confucius's influence, conspired to have him fired.

So far, by religion promoting good conduct with the promise of a happy afterlife as the reward, and torture in Hell as the punishment for sinners, it seems to be encouraging people to look out for one another.

AFRICAN DEITIES

Africa used to have many religious beliefs and gods, mostly all local because of small groups living in isolated areas with poor communications. Even when Christianity reached Ethiopia in the fourth century CE, it didn't penetrate south into sub-Sahara until the Portuguese introduced it in the fifteenth century, and the Dutch in the seventeenth century. Despite Muslim conquests in the North, Islam only penetrated East Africa during the slave trade that peaked in the nineteenth century. Christianity reached the interior when missionaries crusading against slavery, and Europeans interested in colonizing the continent, arrived. Christianity found difficulty spreading in areas where Islam had already taken root. And in Uganda, King Mwanga II, favoring Islam because it allowed one to have more than one wife, martyred Christians between 1885 and 1886.[3]

In Africa, worship was usually centered on a local god in the sky, a spirit inhabiting a prominent landmark, or a demigod directing human affairs. As on other continents where a culture may worship several gods, each deity would be responsible for a specific function such as health, the weather, fertility, or war, as in Uganda, discussed further below.

In Nigeria, for instance, the gender-neutral Yoruba god, Olodumare, starting with an empty universe, appointed night and day and arranged the seasons as in Genesis. Then, after asking the sea to manifest on the planet, Olodumare sent soil to spread on the sea and make earth before turning attention elsewhere in the universe to create more worlds.

Olodumare formed a "trinity" with his intermediaries Olofi and Olorun. Each one was responsible for specific functions. As the ruler of heaven and the true creator, Olorun was responsible for directing the creation of the Earth from above, and for creating humans. Then Olofi made the planet's foundation and filled it with earth. Olofi stayed close to the people as Olodumare didn't intervene in human affairs. A large number of other intermediaries or energies, up to seventeen hundreds of them, known as Orisas, consisted of superior powers found in prominent natural objects such as the sky, sea, rivers, and volcanos.

The Yoruba religious scriptures, the Odus, say that humans should learn from their mistakes and pay the price for redemption. For their righteousness, they are rewarded here and in the afterlife. Olodumare was transcendent, a state of existence, and has never been seen. Similar to the biblical God and to Egyptian Amun, it was hidden and couldn't be found by searching. But as its intermediary Olorun, Olodumare allowed itself to be seen in its "highness," the bright mid-day sun. Since people were growing too old and weak, Olodumare created death.

In Uganda, the Baganda believed a Supreme Being, Katonda (which means creator), made heaven, earth, and humans. Katonda did not routinely intervene in people's lives. He was assisted by a host of other gods, mainly spirits of men who had exceptional attributes in life. The spirits were ubiquitous. In their several forms as ghosts (mizimu) and so on, each had a specific function such as responsibility for the sky, lightning, lakes, and disease. Despite embracing a foreign religion like Christianity or Islam, people could also pay respect to a local god for full spiritual sustenance or for a cure.

Unlike in the West, and to some extent, the East, neither Baganda gods nor religious doctrine prescribed or enforced morals. The community did that. People observed morals because they made sense. Murder was simply wrong, but not because gods forbade it. In fact, as in ancient Greece, gods were mischievous, much like people. Also, the Baganda didn't believe in an afterlife beyond spirits ceaselessly swirling about, scaring the daylights out of the living, who mistreated them.

Clearly, it appears that people have always believed in higher powers and that all religions are branches of the same spiritual tree. But what need do these beliefs satisfy other than moral guidance or promise of a reward—an afterlife in paradise—for having lived a righteous life? Such lofty ideas are not universal. As we find, the talent for spiritual experience is interwoven with human biology and so is etched in our genes, indicating that it provides an evolutionary advantage.[4] This could account for spirituality's staying power and its issue, religion, for instance, through the ages.

Underscoring the evolutionary nature of religious rituals is the similarity between human rituals and those of animals. For instance, according to Andrew Newberg, "virtually all species of animals perform some variation of a mating ritual." He notes that religious rituals turn spiritual stories (myths) into spiritual experience in every culture, and that humans have intuitively found ways (rituals) to bring their myths to life and pass them on to progeny. Such myths, then, are driven by biological compulsion.[5] This renders spirituality, on which religion, creation myths, and various rituals are based, universal.

Religion has been around since ancient times, as rock art and monolith arrangements attest. We find religious texts recorded on pottery and on rocks. Africa has the oldest and largest collection of such records inscribed on pottery shards called *ostracon,* and on rocks such as those found near the Kharga Oasis west of Nubia. Although dating rock art is not easy, it can be done through *Relative Dating*, such as inferring an approximate date by comparison with something else of known age, writing style, or subject matter. These methods date the writing to be at least from 5,000 BCE. Records found in the Western Sahara are dated to 3,000 BCE.[6] Sumerians recorded their gods' amorous exploits on Cuneiform tablets four thousand years ago, and the Torah kept the Israelites united in spirit after the Romans dispersed them in 135 C.E. Spirituality, then, endures because of its evolutionary advantages.

But, how can a mere belief provide a survival benefit? It seems that besides explaining nature, belief in a higher power could have helped people cope with the tragedy of death in the West, ended the cycle of death and rebirth in the East, and helped to cure disease, embolden warriors, and bring rain in Africa. Before we delve into the evolutionary benefits of religion, though, what do people give as their reasons for belief in God?

HOW WE BELIEVE

Today, because science has explained much of what mystified our ancestors, is religion going away? One-morning in 1882, German philosopher Friedrich Nietzsche proclaimed, "I mean to tell you. We have killed him—you and I!" when he told a parable of a ranting madman running into a marketplace crying, looking for God. When the man was asked where he thought God was, he replied, "We are all his murderers!" Eventually, Nietzsche also went mad.[7]

Apparently, the killing of God began when Copernicus and Galileo contradicted scriptures by saying the Earth moved, thus displacing us from the center of the universe to a remote corner. To add insult to injury, British geologist Charles Lyell discounted God's role in shaping the world landscape in *Principles of Geology* (1830). Tinkering with earlier estimates of the Earth's age, Lyell rolled it back much further than the six thousand years Bible chronology allowed.[8] The final blow came when Darwin suggested we were not specially created by supreme architects and that we had instead evolved by chance from lowly life forms. Instead of the six days of creation discussed in Genesis, scientists have established that life evolved over billions of years.

All of this evidence notwithstanding, belief in God today is as strong as ever, particularly in the Americas, South Korea, the Islamic world, and Africa.

A Gallup poll in the January 10, 1996 *Wall Street Journal* reported that 96 percent of Americans believed in God, 90 percent in heaven, 79 percent in miracles, 73 percent in hell, 72 percent in angels, and 65 percent in the Devil. Clearly, answers overlapped and some people could have answered affirmatively to all of them. According to Michael

Argyle, an afterlife was 100 percent real for those over ninety years old.[9]

In 1992, Finke and Stark reported that church membership rose steadily from 17 percent in 1776 to about 62 percent in the 1990s.[10] In *Understanding Church Growth and Decline, 1950-1978,* Roozen and Carroll reported membership remained constant at about 42 to 44 percent in Gallup polls.[11] And in a survey by political scientist George Bishop reported in 1999, believers in God and heaven were more likely to be female, older, African American, less well educated, American Southerners, and fundamentalist protestants.[12]

The opposite was true in Europe. According to Argyle, surveys in Britain since the 1940s showed a decline in belief in God from 79 to 68 percent. Church attendance had dropped from 24% in 1900 to 15% throughout the century. Thanks to hard-working missionaries, however, religion in developing countries saw an increase. Mega-churches seating up to 10,000 devotees sprang up in countries like South Korea, which lives under constant threat of its neighbor's missiles, a situation that increases magic-making in a society, as explained by Bronislaw Malinowski, below.

Citing a 1997 British journal *Nature* report in *How We Believe*, Michael Shermer, director of Skeptic Society and founding publisher of *Skeptic* magazine, wrote that about 40 percent of respondents from a random sampling of 1000 American scientists believed in God.[13] These results were unchanged compared to a similar study in 1916 by religion psychologist James Leuba.[14]

Outside the United States, belief in God was lower, as was belief in the literal translation of the Bible. Only 7 percent of respondents in Britain rejected evolution, compared to 45 percent in America. That number was even lower on the European continent. A Pew Research Center Study published in 2015 reported that in America, belief in God, daily prayers, and regular church attendance declined modestly by 3 percent while spirituality (probably those not affiliated with a religion) had grown by 7 percent from 2007 to 2014.

According to argyle, the fastest growth in religion came from radical sects splintering off from old Protestant churches. Their leaders often were dissatisfied church members. Perhaps as the main churches became more middle class, he explained, they tended to cater to everyone's interests, losing the authoritarianism and fervor to stress values that initially were their appeal. In contrast, the splinter sect leaders tended to be strict, demanding obedience. Manic or schizotypal personalities, traits of mental disorder, were common among the leaders, Argyle wrote. These schizotypal personalities (often relatives of people diagnosed with schizophrenia) showed marked religiosity, a capacity for religious experiences, and creativity.[15] According to neuroendocrinologist Robert Sapolsky, such individuals tended to be reclusive, choosing private occupations like film projectionist, and had flat emotions. Their personalities were marked by disconnection from the rest of the world, metamagical thinking, an intense interest in science fiction and fantasy, and a literal translation of the Bible. Wisemen, or shamans, witchdoctors, and medicine men could be counted among them. Although denying being religious, they interpreted the Bible very literally, and once they started talking on the subject of God, they would not stop.[16]

Because of their personalities, sect leaders are often considered "wacky." Followers tend to be young, often disadvantaged, seeking to improve their lot. Hare Krishna and the Jesus Movement have been around quite a while, but most splinter sects don't last and may even end violently. In 1978 in Jonestown, Guyana, after killing a congressman, 912 members of Jim Jones's cult voluntarily swallowed cyanide and died. Following an FBI siege at Waco, Texas, in 1993, David Koresh's Branch Davidians burned down their compound with seventy-six followers inside. And, when the end of the world didn't come at the dawn of the Millennium in 2000, leaders of a Uganda cult to restore the Ten Commandments ended the lives of 900 members.

Cults springing up after the Second World War often adopted religious ideas from the Far East, such as Zen and Transcendental Meditation, based on Buddhism and Hinduism, respectively. Members sought self-improvement through spiritual growth and physical development. They strove to improve their health and live longer through biofeedback. Also, according to Argyle, churchgoing, a type of meditation, improved health and promoted longer lives for members. In 1954, clinical pharmacologist Louis Lasagna and his associates reported that hospital patients who attended church responded better to placebo therapy.[17] Besides church encouragement of healthy lifestyles and promotion of social support, the main factors shown to repeatedly play a role serve to strengthen the immune system, as explained in the next chapter.

Many studies since 2000 have shown that highly religious people are more likely to live longer.[18] Also, church members' fear of death decreases, and attendees are somewhat happier. Healthier lifestyles aside, such benefits could come from the psychological relaxation brought by religious practices. Argyle suggests church social support and the optimism belief creates lead to such happiness because religious participation can lend believers a sense of purpose and meaning in life. Such benefits are not limited to the religious only, however. Mystics and people who practice yoga and related activities have them, too.

Belief in God seems involuntary and is universal, further evidence that it's inborn. Following his studies (see below), Andrew Newberg concluded that humans had a genetically inherited talent for entering unitary states (trances) that many of us interpreted as the presence of a higher power.[19] If so, for beliefs to endure, they had to confer survival benefits, and therefore people of faith are more likely to be selected by evolution. Otherwise, if they were by choice, what were people's reasons for them?

WHY DO YOU BELIEVE IN GOD?

In 1927, to explain why we believed in God, Sigmund Freud said that religion had its origins in the infant's relations with its parents.[20] Indeed, Argyle noted that religious behavior and beliefs of children resembled those of their parents. Students who were the most likely to leave their church, an act known in religious circles as apostasy, "…came from families where religious emphasis, especially from the mother, was weak."[21] So, nonbelievers might not have had a strong relationship with their parents, and could have separated early from them. As psychologist Channa Ullman reported in 1982, more converts to other religions had absent fathers and poor relations with parents than did children from supportive families.[22]

But in Freud's case, there was no such lack of family support to account for his dismissive attitude toward religion. Freud was one of seven children by the third wife of Jacob Freud. To escape anti-Semitic riots, the family fled from Freiberg, now in the Czech Republic, to Leipzig, Germany, when he was three years old. Freud's mother adored him, and his father thought he was destined for fame. Since his parents were not practicing Jews, Freud did not have a strict religious upbringing, probably the reason for his irreligiosity. Even his nanny was Roman Catholic.

Freud believed God was a projected father figure based on early experience of the real father and was needed as a source of protection. Freud argued that the god image was derived not only from childhood experience, but also from *inherited memory images* of the primal father, in concordance with Jungian thinking, discussed below. So, during stress, as after natural or social disasters, internal parts of the person were projected as parent-figures and seen as external (such as God in the sky). Freud called religion a neurosis, an illusion that would fade away as we moved into the rational age of science. He had no timeframe for this to happen. He believed it was futile to abolish religion, but thought people would outgrow God in their own good time. Sociologists and other scientists agreed for the same reasons, adding that industrialization and rational and scientific explanations of natural events would lead people to better understand the world and rely less on superstitions and myths. But surveys suggest that the opposite is happening, that belief is increasing. Why?

Reasons why people believe in God and why they think others do are conflicting. While we justify our personal beliefs, we think those of others are fated. Psychologists call it "biases in attribution," wherein we attribute the causes of our own behavior to a certain situation while attributing the causes of others' behavior to their incompetence (disposition). In situational attribution, we blame an outside cause such as a slippery pavement for falling. In dispositional attribution, we blame the other person for being clumsy.

So, what reasons do people give for believing in God, and what do you think their real reasons are?

In 1995, Shermer surveyed members of his organization and found 30 percent of this highly educated and well-informed group believed in God, about the same number as that of general scientists in the 1940s. As to why, the majority, 29 percent, gave intellectual reasons (situational attribution). Explanations included good design, beauty, perfection, and complexity of the universe. A smaller number, 21 percent, thought belief in God was comforting, relieving, consoling, and their beliefs gave life meaning and purpose. When asked why other people believed in God, the argument that it was comforting rose to the top. The question of design came in last. Results from a general survey of 1000 Americans were similar.[23]

But what about the nonbelievers, the skeptics? Skeptics gave lack of evidence as the most important reason they did not believe in God. They found no proof in design like the others did. They also said, in descending order, that there was no need, it was absurd, God was unknowable, and science provided all the necessary answers.

Clearly, people's reasons for belief in God vary widely, their answers depend on whether it's "they" (disposition) or "we," (attribution). The argument most frequently given for "they" is comfort, and for "we," good design. True, prayers offer comfort, at the very least to the one doing the praying. Prayers came in handy when terrorists crashed planes into the World Trade Center on 9/11.

During the Middle Ages, because so much was a mystery, superstitions ran rampant. When a plague broke out, people thought the gods were angry. Besides fearing God, they also lived in dread of sorcerers, blood-sucking vampires, werewolves, and demonic curses. Uncertainty and insecurity ran high. To allay anxiety, people often fell back on the same kind of magical thinking their deep ancestors had used in fearful times.

But today, thanks to science, we know thunderstorms aren't retribution by irate gods. Demons aren't real and have nothing to do with epilepsy or madness, even though the Bible tells us Jesus cast them out of people and into pigs and then drowned them in a lake. Mt 8:28-34. Also, disasters such as epidemics, volcanoes, and hurricanes have been explained and may now be predicted and prepared for. Today, with our belongings and loved ones insured and the security alarm set, we can sleep peacefully. It's no wonder that belief in superstitions has waned, shifting the importance of religion from protection to comfort. Yet, even though we have not seen God reprimand anyone, the fear of being punished for our sins is real and ever present.

Despite scientific advances, tranquillizers like Valium were the most prescribed drugs in the world in the early 1990s. To cope with stress from an increasingly complex world, people hankered for them more than they did for antidepressants.[24] Stressors included job insecurity, bills, chronic illnesses, onerous regulations, terrorists, wanton shootings, cybercrime, imagined adversities, and on. Such anxiety may account for the growing reliance on God. Among those who are church-going, attendance appears to reduce depression by as much as 30 percent. Freud noted that, "devout believers are safe guarded in a high degree against the risk of certain neurotic illnesses." But he added that their acceptance of a universal neurosis spared them the task of constructing a personal one.[25] Numerous studies have also shown that people who practice any mainstream religion live longer. They have fewer strokes, less heart disease, lower blood pressure, and better immune function than the population at large.[26]

WHY THE BELIEF ENGINE?

While living in the Trobiland Islands off the coast of New Guinea, Polish born British anthropologist, Bronislaw Malinowski, noted that the farther out to sea the islanders fished, the more superstitious they became. By the time they reached the dangerous and treacherous waters of the deep sea, they were heavily into magic. In contrast, those who stayed in the calm waters of the lagoon had few rituals. Malinowski concluded the islanders' beliefs were a means of dealing with the anxiety uncertainty brought, and that magical thinking depended on environmental conditions, not inherent stupidities. He wrote in a 1925 essay:

> "We find magic wherever the elements of chance and accident, and the emotional play between hope and fear, have a wide and extensive range. We do not find magic wherever the pursuit is certain, reliable, and well under control of rational methods and technological processes. Further, we find magic where the element of danger is conspicuous."[27]

In prehistoric times, predators, natural disasters, and the like were common, and people were generally terrified. Food supply was erratic, and starvation was a constant threat. As proper medical treatment was non-existent, death often came early and could be agonizing. For example, 1.5 million years ago, the eight-year-old *Homo erectus,* Nariokotome Boy, succumbed to a tooth abscess, a condition treatable only today in the antibiotics era. Nevertheless, Erectus itched to travel, much as humans do, and often confronted new challenges in unfamiliar territory despite the risks. Being human, for courage and to calm his nerves, he likely put trust in a higher power, which his brain, like ours, could have sensed, especially since ritualistic behavior similar to ours has been observed even in lower animals like butterflies. It means what we call spirituality could be very old. Neanderthals, too, could have practiced spiritual rituals when burying their dead, implying they trusted in supernatural phenomena, and could sense other fantastic worlds that seemed more real than ours, as mystics describe them.[28] As half of Neanderthals died before their twelfth birthday, according to radiometric dating, being human, their levels of anxiety must have been off the chart.

As discussed above, our ancestors' conditions were like the Neanderthals', and starvation, mental stress, and extreme exhaustion would have triggered a trance and hallucinations. During such events, people would visit other worlds, merging with spirits and animal helpers there, taking on their powers. They would return spiritually renewed and pass on the good cheer to embolden hunters and warriors and, via the mind-body link, heal the sick. Eventually, they would have regularly practiced rituals like dancing and singing, behavior that triggered trances.

Such practices tended to boost morale. They would have made people feel good. Trance experiences reportedly appeared more real than the physical world we live in. Besides making life meaningful, the trips afforded a glimpse of the Almighty. Shamans recorded such spiritual journeys in rock art in South Africa, France, and other places. Although the inward excursions relieved stress, those without the talent to induce them wouldn't have benefited, and with nothing to fall back on to relieve the stress, they would die early without offspring, leaving the stage to our magic-making ancestors. According to Newberg, "…religion may also have been a crucial reason the human race has managed to survive."[29] "It's very likely that natural selection would favor a brain equipped with the neurological machinery that makes religious behavior more likely."[30]

As our genes are 99.9 percent similar, we donate blood and organs to one another, and we behave alike: we laugh, cry, marry, bury the dead, and practice religion. Like it or not (and many don't), we are predictable (Spinoza says we're predetermined). How predictable? We discuss this in the section on free will. So, when people began leaving Africa about 70,000 years ago, they carried genes that conferred the talent for spirituality, and the trait endured.

To show that human behavior was indeed heritable, Thomas Bouchard and team conducted the famous "Minnesota twins" study. They found about 50 percent of identical twins raised separately were more similar in their religious interests, or magical thinking, than were fraternal twins raised together. What they did not inherit was the specific type of belief and commitment, which they acquired from experience through parental coaching and the like.[31]

So, while people in the West worship God, those in the Orient "experience" a universal force by delving within themselves through meditation. In the West, where meditation is not as widely practiced, similar experiences tend to occur spontaneously, as when churchgoers fall down in religious ecstasy, an event referred to as being "slain in spirit." Whatever the means, people say the experiences are very real. To evoke them, or the symbols that represent them, such as God, would be the drive behind building colossal shrines like Antoni Gaudi's Sagrada Familia Cathedral in Barcelona, Spain.

One group of people whose religious views are rarely sought is children. How do preschoolers believe in God and why? Whereas small children often seem to consider God real, painting Him as an actual figure, usually the face of an old man with a beard, teens see him symbolically, as the sun for instance. As they apparently are born ready to believe, it seems that only the type of belief, such as church affiliation, comes from parenting and other outside influences.[32]

Since kids are born ready to believe, faith is not a "choice." Yet, people assume it is and give reasons why they believe in God. Shermer says, "For centuries, many a theologian, scholar, and scientist has attempted to explain why people need religion and need to believe in God. Their efforts have left a legacy of theories and libraries of books on the subject.[33] After asking the evolution-loaded question "what is man," Zoologist G. G. Simpson said, "…all attempts to answer that question before 1859 – before Darwin published his work, that is – are worthless and we will be better off ignoring them completely."[34] Evidently, it's futile to explain faith rationally. As religious rituals offer evolutionary advantages, could our genes hold the answer?

OUR COLLECTIVE UNCONSCIOUS

In 1875, Carl Gustav Jung was born in Switzerland into a family steeped in religion. His father was a Protestant minister, and eight of his uncles and his maternal grandfather were also in the clergy. Grandpa dabbled in the occult, too, and talked to his deceased first wife as his second wife, Jung's mother – herself an enthusiast of the otherworldly – looked on.

After studying medicine, Jung practiced in a mental hospital. His work in psychology brought him fame and a close friendship with Freud, even though he disagreed with Freud's narrow sexual explanation for neurotic disorders. Jung also considered psychoanalysis useless in schizophrenia. For his part, Freud could not stomach "sanctimonious" Jung and his later parapsychology work into spooks and the occult.[35] During a long transatlantic voyage to North America to lecture, Freud refused to let Jung psychoanalyze him. He feared losing his authority if, for interpretation, he bared his closeted skeletons. Eventually, their friendship dissolved. Jung co-founded the study of analytical psychology and immersed himself in spiritualism big time. An extraordinary mystic, Jung was adept at exploring his unconscious. He also acted as a medium, holding séances, in addition to delving into hypnosis, clairvoyance, and telepathy.[36] Jung also had a near-death experience (NDE), like the skier at the beginning of this chapter, after a heart attack at sixty-eight years old, thus adding another resource to his in-depth focus on the inner workings of the brain.

For over fifty years, Jung developed theories based on his understanding of fantasies and dreams from his childhood, history, and mythology. During travel to study diverse cultures in Mexico, India, and Kenya, he was struck by the recurrence of certain common themes in several world religions that paralleled similar themes in myths, alchemy, cultures, and many dreams shared with him by his patients. Patients' dreams, for instance, could not all have come from ordinary experience, historical records, or religion and myth. He saw similar motifs in works of art. A close parallel between ancient myths and psychotic fantasies led him to view and study human motivation in terms of a larger creative energy.

Jung distinguished between a *personal unconscious* and a *collective unconscious*. Whereas a personal unconscious consisted of repressed feelings and thoughts developed during an individual's life, a collective unconscious consisted of inherited feelings, thoughts, and memories we all share. The collective unconscious, therefore, consisted of instincts, which determined our actions and inborn modes of perception, models, or archetypes. Archetypes corresponded to such experiences as confronting death or choosing a mate. They showed as symbols in religion, myths, fairy tales, and fantasies.

Below the individual unconscious lay the collective unconscious, which consisted of primordial images derived from the early prehistory of the human race, rather like instincts. The images were archetypes, "…dispositions to experience and respond to the world in the ways those ancestors did." They could not be known directly but only through symbols reflecting culture and personal experience. Being abstract forms, they had the potential to evoke certain mythical ideas, dreams, delusions, and religious beliefs, when shaped by individual and cultural influences.[37] In Christianity, the abstract forms would be God and Jesus his son:

> God is an archetype, an unknowable part of the collective unconscious, but experienced through symbols. These have taken a variety of forms through the history, of religion, including Christ, Buddha, kings and queens, dragons and other animals, and God is known through numinous religious experiences.[38]

From his trances, Jung drew diagrams that reminded him of Eastern sacred magic circles, also known as mandalas. Mandala images often emerged in the dreams and paintings of his patients and in Central Australian Aboriginal art. He assumed the father, like the mandala, was a symbol, not of the individual's father as Freud believed, but of a more generalized entity, a source of magical power people symbolized in different ways, such as God.[39]

In his 1907 six-volume *Encyclopedia of Religion*, Mircea Eliade traced every religious theme, not just from the New Testament to the Old Testament, but to earlier and quite different faiths as well. He showed that throughout the ages, cultures across the world held similar mythologies. They included virgin births, fallen angels, resurrections, floods to redress human corruption, and so on. Just as Jesus spent forty days in solitude in a desert enduring the temptations of Satan, the young prince Buddha spent forty-nine days under a Bodhi tree enduring the temptations of the demon Mara. Also, Mexican god Quetzalcoatl, born of a virgin as well, was tempted and fasted for forty days. These similarities pointed to a shared humanity, a common source. Gerald

Massey traced the Christ myth to lands drained by the Nile; Egyptologist Flinders Petrie similarly traced the earliest Egyptian religion to Central Africa.[40]

A JOURNEY TO THE OTHERWORLDLY

As shown in the BBC video series, "The Story of Human Evolution," for a long time, people wondered what the ancient rock and cave art in South Africa, Australia, France, South America, and the forests of Borneo meant. Believing that stories behind such art could be a way into the lost world of our ancestors thousands of years ago, archeologists and psychologists sought to connect with it by looking into the minds of volunteers with hypnosis. Their results showed that trance experiences all over the world, be they in a laboratory in the United States or among the Khoisan in South Africa, were very similar.

Although the Khoisan no longer hunt in the old ways, they have otherwise changed very little; their customs closely resemble those of their forebears. They still use one of a whole family of obsolete languages and their way of life most closely reflects that of our ancestors 20,000 years ago. According to the gene tracking Genographic Project, their DNA confirms, because of isolation, that the Khoisan group, mtDNA L0, as mentioned earlier, is the oldest in the world and also the closest to our earliest ancestors 200,000 years ago. Could their experiences reveal the stories behind the rock art?

To induce a trance, a Khoisan shaman pretending he was an antelope performed the eland dance around a fire at night, singing and chanting for hours. He would go into a trance and summon the animal's power to take him into the lost world of the ancient people. According to a description of the ritual by archeologist Thomas Dawson, the spirit of the eland would appear to the shaman in the darkness and he would merge with it, taking on its secret power. Afterwards, the shaman relates traveling underground, going through tunnels, dying, and coming alive again. He also mentions flying, a sense of weightlessness, and floating. Until a few hundred years ago, though, the Khoisan recorded their experiences on huge rocks like the Drakensberg Rock Shelter in South Africa. They considered these rocks to be sacred homes for spirits. Since rock and cave paintings in Australia and Europe were similar, it seems reasonable to conclude that the paintings were records of similar trance and hallucination experiences.[41] The art depicted spiritual experiences, the foundation of religion. According to Wayne Teasdale, "…mystical spirituality is the source of the world's religions."[42]

When he worked at the American Department of Defense, clinical psychologist Etzel Cardena studied states of altered consciousness in twenty-eight individuals with no knowledge of one another but all of whom had a special talent for self-hypnosis. He mentioned that 5 percent of the population is suggestible: it responds readily to hypnotism. Cardena kept hypnotic suggestions to a bare minimum, hoping that by shutting out the modern world around the volunteers they would derive experiences solely from the workings of their minds.[43]

Results were as Cardena expected. All participants hallucinated the same way the African shamans did. One in particular, Janet Crossley, told how at the start of her hypnotic journey, she lost track of where "she ended and where she began." After a long time away, she felt she was going back home and, sinking deeper into her trance, she saw a pulsating white light. As its color increased, she felt she was going through something that resembled a tunnel of colors, funnels, spirals, and grids. She saw shapes and was unsure whether they were fish or not. "They move and flow into each other…they keep changing and mixing and making new colors," she said. "I feel like I am gonna be able to float up into them." The other shapes and forms there, she could not quite make out.

Janet found the experience very real. It was wonderful and good for her. If other people went there, they would feel the same way, too, she said.

Like Janet, once their awareness of the outside world was totally gone at a certain stage during hypnosis, the subjects experienced about the same four to five different geometric shapes of spirals, tunnels, funnels, and above all, grids. Grid patterns seemed to appear when the visual part of the brain lacked input. These patterns, apparently universal, are also seen by patients who experience eye diseases, migraines, and strokes. As subjects experience the grids, the right visual cortex lights up on a CAT scan, showing that the grid is a product of the brain and common to us all. The common feature of these hypnotic experiences strongly suggests that there is a hereditary component, and likewise suggests that the patterns appearing in ancient cave art depicting shamans' experiences in the spirit world were the result of a similarly hereditary phenomenon. Being a brain product, they reappeared in our dreams.

Grids and colored dots appearing in South African and French rock paintings point to a common origin. The process of producing the art was similar, suggesting that they were both made by the same people, the Khoisan. Similar patterns are found in modern art, paintings, and rituals. At Nyero in eastern Uganda, rock art consists mostly of circular, rectangular, sausage, and spiral geometric shapes and dots and lines. They also appear during a "trip" or "high" from LSD and psilocybin (extracted from mushrooms the Mexicans call "little gods").

These chemicals are structurally similar to the same feel-good brain chemical – serotonin – that alters consciousness. An excess of serotonin distorts sensory perception, interrupting the inhibitory circuit of the thalamus and producing a hallucinatory or mystic like state. Biology is key, then, to explaining mystical experiences or spiritual experiences. Trances may also be triggered by music, prayers, fasting, low oxygen as experienced by mountaineers, stress, extreme gravity (strong g-forces, as in supersonic flight), and rituals leading to sensory or sleep deprivation. Once the conditions are right, subjects could then have out-of-body or trance experiences that some call spiritual or religious experiences.

Of all the trance-inducing rituals, music seems to be the most effective.[44] Since trance experiences are similar, it suggests that humans (and early hominids, although they couldn't have made heads or tails of them) have been having them since as long as we have been around.

Many religious texts, such as Jesus's parables or the Gnostic Gospels, are best understood mystically. According to Karen Armstrong, the Quran was also best read as a mystical work. No doubt, we are spiritual like our deep ancestors. Our minds are similar, our brains are "programmed" to respond in a trance the same way the shamans' did.[45] The idea stirred up controversy when it was brought to the public's attention in a voluminous book, *Sociobiology,* by Edward O. Wilson. Wilson suggested that human (basic) behavior, such as instincts, and that of animals like ants was inherited. According to biologist Stanley Rice, then, most aspects of human thoughts, feelings, and behavior may have been influenced by evolution.[46] Such traits include romantic love—mating comes naturally—music and dance, perception of status or rank, emotions, deception, phobias, and fear of strangers. The potential for such behaviors was present from birth and nurturing helped to express it.

Like schizotypal personalities, some people with right temporal lobe epilepsy become passionately religious during a seizure. In a Nova 2001 video, *The Secrets of the Mind*, neurologist V.S. Ramachandran presented such a patient, John. Although typically irreligious, during an epileptic episode, John ran down the street shouting he was God, the savior, and spoke about making the world right. He would be so elated his joy knew no bounds and tears would run down his face. He said he wouldn't trade the experience for anything. During the five-minute event, he felt like he floated over a world so beautiful that words could not describe it. It was so real he preferred to stay there.

Evidently, John's experiences derived from his right-brain. Studies of people with left brain damage who thus must depend on the right brain, report similar experiences supporting this finding. In his book, *The Brain That Changes Itself,* Norman Doidge writes about Michelle Mack from Falls Church, Virginia, who was born without a left-brain, and who fondly and plainly believed in heaven.[47] And Jill Bolte Taylor – who tells her story in *My Stroke of Insight* – reported that after a stroke damaged her left-brain, she developed mystical or religious experiences, and felt at one with the universe.[48]

In 2001, after radiologist Andrew Newberg injected a radioactive tracer into Buddhist monks during peak meditation and nuns during prayer, he scanned their brains with a SPECT (single photon emission computed tomography) camera. The dye, because of increased blood flow, concentrated in the active parts of the brain and not in the inactive areas. On the scan, the right parietal lobe lit up in what Newberg termed the "Orientation Association Area." This area created the sense of where the body was in space. As the left side was inactive and remained dark, it couldn't create the sensation of a physically delimited body and couldn't find a boundary between the self and the world. The brain then perceived the self as endless and intertwined with the universe and everything in it, which causes the feeling of levitation and floating. The nuns' prayers reportedly produced "…a tangible sense of closeness of God and mingling with Him." Participants in the study said they felt a merging of self with the Spirit, God, or the universe. These would be called mystical experiences in regular folk. Such experiences, Newberg concluded, were "biologically, observably, and scientifically real."[49]

Newberg goes on to say that mystical experiences happen to us all the time. For example, if you have ever lost yourself in music, been swept away by a rousing speech, fallen in love, or been wonder-struck by the beauty of nature, and felt your ego slip away, and for a moment you realize that you're a part of something larger, that's, in essence, the mystical reality.[50]

OUR SACRED HERITAGE

By conferring survival benefits, genes promoting spirituality, or a mystical experience, endured. Rituals found to invoke such experiences, like repetitive rhythmic sounds, dance, or worship took hold. They offered a communal means to induce such experiences going back at least to prehistory, when, according to anthropological evidence, religions in their crudest forms began.[51] Eventually, as records on pottery shards and rocks in the Sahara and on Egyptian temple walls, monuments, and obelisks show, rituals were organized and systemized. This led to institutional or formal religion as the number of followers grew and worship centers or shrines were established. Such sites would have started small and rudimentary, such as a circle of monoliths at Nabta Playa in the Sahara Desert almost 20,000 years old shows. By the tenth century BCE, such centers of worship had grown into monumental structures like Göbekli Tepe, called the first world temple, in southern Turkey, 10,000 years old, and religious complexes, such as the Karnak Temple in Egypt, 4,000 years old.

Humans tend to believe in supernatural agencies as if under a spell or in a trance. Trance experiences are very real, or *realer than real*, and could be spontaneous. They vary in stages ranging from the very mild, for most people, to the very deep, for a few people like seasoned meditators or gurus, or those who may have them spontaneously.[52] Such experiences are the foundations of religion, which Richard Dawkins discredits in *The God Delusion*. To him, people worship a void, or nothing. Oddly, Buddhists agree, calling the elusive reality *tathata*, or "nothing," as it's indescribable. Since religion is worldwide, unlike a delusion, perceived only by the individual, this "nothing" must be based on an illusion, experienced by all. So, programmed to walk in our ancestors' footsteps, we cannot help but follow.

When starving or fatigued from a grueling hunt, our ancestors would fall into a trance, unite with animal spirits, and fundamental truths would be revealed to them. For their conversions, St. Augustine and other early Christian thinkers received similar numinous experiences revealing to them the existence of a higher power. They abandoned their way of life and dedicated themselves to serving the hidden, unknowable powers. Like Teasdale, Newberg wrote, "...mysticism... is the source of the essential wisdom and truth upon which all religions were founded."[53] And, contrary to common belief, mystics are not crazy. Their experiences are rooted in biology.

Today, government, not religion, enforces most conduct, penalizing, for example, violators of the sixth commandment, "You shall not murder." Ex 20:13, Dt 5:17. Prison is a stronger deterrent than the threat of living in Hell is, and religion – relegated to the role of a guard dog sans teeth – comes in third behind the court of public opinion. Unlike countries where religious tenet is law, the United States keeps church and state constitutionally separate, with notable exceptions (for instance, Congress has a House Chaplain, and numerous Presidents and other national dignitaries have made appearances or been memorialized at the Washington National Cathedral). Communist countries like Russia and China banned religion altogether with questionable results. Banning religion generally tends to simply drive it underground, not just because government can't provide spiritual comfort or meaning, but because faith is in our DNA. So, today even though people have more physical and financial security than in centuries past, growing psychological stress and depression could be contributing to a religious revival in some parts of the world. Were it not a conduit to spirituality that engendered social benefits, and, most importantly, health benefits, religion would be obsolete today.

"I know. I don't need to believe. I know," Jung said in a BBC television interview when asked whether he believed in God. God was a psychic reality, he explained. Experienced subjectively by each individual, Jung's God was similar to that of the Gnostics. Unlike Freud, who did not believe in God, Jung had been indoctrinated not only in religion but also in paranormal phenomena by age three. Like others in his family, he was adept at exploring his own unconscious, where he met Biblical characters like Philemon, with whom he held philosophical conversations, even though he was talking to himself.[54]

In contrast to Jung, but much like Freud, French philosopher Jean-Paul Sartre lacked a strong family religious background. Born a Protestant to irreligious parents in a Catholic country—even his nanny was Catholic—Sartre grew up an atheist. In his-best known novel *Nausea*, the central character, Antoine Roquentin, constantly feels sick when he discovers that there is no reason for anything to exist at all, thinking, "Everything that exists is born for no reason, carries on living through weakness, and dies by accident."[55] As if to say that we're born by chance (we didn't have to come) and survived driven by hunger-pangs (lived on borrowed time) until we were called (not soon enough). Not only nurturing but also genes seem to have played a role in Freud and Sartre's denial of God.

On his eightieth birthday in 1936, Freud declared religion to be the neurosis of mankind. I believe he was partially correct, as our early ancestors, facing death daily, had to be extremely stressed. As Newberg above suggests, those prone to mystical experiences found relief in trances, experiences that evoked spirits or supernatural powers. They calmed down, staved off the ravages of stress, and lived longer. The next chapter explains how a longer life can be achieved this way.

Unwittingly, we drift off into the otherworldly daily. When at church, a concert, listening to skin-tingling music, or mating, even atheists wander into the otherworldly, where there's serenity and peace, and one is transiently devoid of thoughts. As mentioned earlier, some trance triggers include art, breathtaking sceneries, or yoga. Anecdotal evidence of such occurrences abounds. For example, following eighty-three hours of a grueling bicycle tour without sleep, agnostic Shermer reported an out-of-body experience, and a milder one while under Michael Persinger's *God helmet*. As part of an experiment, Persinger, a Laurentian Ontario University neurologist, used an electromagnetic hat to dissociate Shermer's brain hemispheres, effectively separating the corpus callosum (the highway joining the hemispheres). Shermer's rational left-brain, however, kept pulling him back in. He was resisting the out-of-body experience.[56] In Chapter 12, I explain what takes place in the brain when it's buzzed with electromagnetic waves from the hat.

Some argue the reason for the brain's capacity for transcendental experiences is to give humans a glimpse of the maker. Perhaps. But Andrew Newberg suggests the brain mechanism enabling spirituality may have arisen from the same neural pathways that evolved for love and mating and procreation, and spirituality may be a byproduct. The strong survival advantages of religious belief and the inherited ability to experience spiritual union is the real source of religion's staying power. "It anchors belief in something deeper and more potent than intellect and reason; it makes God a reality that can't be undone by ideas, and that never grows obsolete."[57] When Nietzsche proclaimed God dead, he meant the personal creator God of the Bible.[58] He didn't know that denying God, a metaphor for our spirituality, was futile, since spirituality was innate, and as such would never die. On the other hand, Gods, as metaphors, change all the time and do even die. People may choose not to believe in gods or discard them, but they can't get rid of spirituality.

For the brain to develop free of the talent for spirituality, then, its structure would have to change. I do not see this happening anytime soon. A new species of Homo may have to evolve in a stress-free world first before this happens. According to hominid's average life span, barring an act of nature, it would take a couple million years or more for this to happen. As religion promotes good health and longevity, what's the mechanism behind it?

CHAPTER 11

FOREVER YOUNG

The fountain of youth has eluded humans for millennia. Maybe we could just live longer instead. We find spirituality fundamentally beneficial to us in that people who practice religion, which it engenders, tend to live longer. As mentioned previously, religion encourages healthy lifestyles, such as turning us away from vices, fostering community cohesion, and offering social support. Above all, though, religion awakens the spiritual state that has a strong positive effect on health and survival. Other practices, besides religion, that may induce trances and similarly promote good health include meditation, yoga, and ritualistic behavior as was noted.

We wish to live long active lives, even though we have no clue about our purpose. We tend to forget, though, that we are here, not because we wanted to be, but by chance, and that we simply happened and really don't know what to do with ourselves. We're urged to behave ourselves, so when we're finished here, if we complied with societal norms and God's commandments, we shall go to Paradise instead of Hell. But why hold onto life if a better one awaits us elsewhere? Fear of the unknown? Leaving family and loved ones behind? Apart from those who find meaning through meditation and mystical rituals, perhaps the longer we live, the greater our chances are of discovering our purpose. Our health, however, usually begins to fail before we do, and we leave it to future generations to pick up the mantel of the quest.

But whether we ultimately realize our purpose or not, live longer or not, planet Earth keeps turning, and some people see no need for God or religion. They rely on themselves, or –like the members of Shermer's Skeptics Society – science, to survive. Other things may also supersede religion, for instance, hassling for a living, winning competitions, studying for a degree, or giving a rousing performance. The young and restless wedded to music, along with those doing well or those who are always busy, belong here, too. So, how can such outsiders lead a healthy life and live longer? Many do.

Those skeptical of a designer other than evolution may not put much stock in purpose, meaning, or, according to Sartre, a reason to be born. Many of us doubt, even look askance at them, though, maybe even look down on them. Our deep ancestors in Africa, living with uncertainty in a dangerous world, wouldn't have had such qualms about belief. According to Bronislaw Malinowski as discussed in the last chapter, magic making increased as danger and uncertainty increased. So, people would likely have been heavily into magic and, for protection, ritualized the means to invoke all kinds of spirits and benevolent powers, praising them and sacrificing to them, and offering them gifts. Such rituals would calm their nerves and lower stress, as explained below.

Today, we have many ways of coping with stress and associated depression. For instance, since stressors have become increasingly psychological over time, psychotherapy addresses individual problems such as the death of a loved one, pacifying the individual, or convincing the mourner their loss is not the end of the world. But also, doldrums could briefly, if not permanently, go away upon one encountering a magnificent work of art, beautiful music, or a lost friend or love. Giving flowers is well known to uplift mood.

The need for action may be dire.[1] A Salish Native American consultant to the Canadian Department of Health and Welfare decried the death of his people from accidents, suicide, and violence, but also the death of the soul, because of a "continuous pain." A sense of helplessness, existential frustration, and dehumanization characterized the Native American society. The youth felt they didn't belong and were merely existing. They wondered, "Where am I going, what is my purpose?" The despair could account for the alarming increase in self-destructive behavior, such as suicide, among the youths that the consultant mentioned.

At 33 percent, suicide rates in the United States are especially high among Native American men, second only to elderly white men at 55 percent. In 2010, suicides averaged 105 deaths per day, trumping homicides by 100 percent. In comparison, in 2014, there were twenty suicides a day by veterans. By 2017, though, middle-aged white males, particularly in western Montana, had surpassed them. According to the CDC, among people who have risk factors for suicide, it remains difficult to predict who will act on their suicidal thoughts, even among those suffering from severe depression, or hopelessness, or both. Only some of the Salish youths above, among those who lacked a sense of purpose and meaning putting them at risk, carried out their thoughts.

THE MONSTER STRESS

Accidents, disease, and people all kill people, but stress? Actually, stress ranks as the number one killer in the United States, damaging every organ in your body. According to the American Medical Society, stress is a factor in 75 percent of all medical disorders. At Johns Hopkins Hospital in Baltimore, Maryland, psychiatrist and neurologist Adam Kaplin said depression could even be a greater risk factor for heart disease than everything else, including diabetes and smoking. In 2009, medicine Nobel Prize laureate Elizabeth Blackburn reported that stress, or generalized stress disorder, shortened chromosome telomeres, which could be associated with aging and longevity. In 2018, more than three million people in the United States were diagnosed with stress, or anxiety.

In a study of seventeen thousand British civil servants, those at the lower end of the pay scale, such as janitors, were four times more likely to suffer a heart attack than those higher on the salary scale, such as the permanent secretary. Reverse the order and the fallen boss's risk of a heart attack skyrockets. Results from an American study of a million Bell Telephone Company employees were similar to those reported in *Lancet* in 1991.[2] And, in 2006, two months after Kenneth Ley, chief executive of energy giant Enron, was convicted of fraud and tax evasion and faced prison, he died of a massive heart attack, likely precipitated by the stress from his devastating fall from grace. As noted to in other cases mentioned herein, his is not an isolated incident.

In 2012, two months after eighty-five-year-old Penn State University football coach legend Joe Paterno was fired, he succumbed to previously unknown lung cancer, stress most likely being the aggravating factor. No wonder, in his book, *Genome,* Matt Ridley writes that job position was a better predictor of a stroke than was your cholesterol, blood pressure, or smoking habits.[3]

When monkeys at the bottom of the totem pole in a zoo were picked on by those higher up the hierarchy, their arteries clogged with cholesterol. Their self-esteem as well as their serotonin levels ran low, and the stress hormone, cortisol, ran high. The more the low-ranking monkeys were dominated, the more stress they had and the sicker they became.

Since humans are physiologically very similar to monkeys, it's reasonable to conclude that when higher-ranking or top executives fall from grace, their serotonin levels plummet, too. Contrary to common belief, it was not how hard you worked, but the degree of stress that determined your fate.[4] Normally calm and good at reconciliation, recruiting allies, and strategizing, fallen bosses become impulsive and aggressive, just like the monkeys.

Paradoxically, Dave Abbott at the University of Wisconsin found that marmosets at the top of their social hierarchy, having to defend their position, developed more stress-associated diseases than those at the bottom who were more laidback. Resting glucocorticoid (steroids) levels of stress hormones, mainly cortisol, were higher in subordinate marmosets only if they were harassed and lacked social support.[5] For the marmosets, then, stress was more related to the meaning of their relative position in the hierarchy than to rank.

A person's socio-economic status, usually rank and income, is a good predictor of longevity. This is true within rich countries, such as the United Sates, which has one of the widest income gaps, and it's true in developing countries like Uganda. In both, life expectancy between the haves and have-nots mirrors the gap between the "classes." According to Sapolsky, how long you live has less to do with access to healthcare than it has to do with feeling poor.[6]

What's this fiend, stress? Why is it so dangerous? The explanation lies in the mind-body link. Psychological conditions like stress release chemicals that affect the body in various ways. Such chemicals include glucocorticoid hormones mentioned above, steroids such as cortisol from the adrenal glands sitting atop your kidneys. Cortisol affects virtually every system in the body. Ordinarily, we encounter cortisol in hydrocortisone cream used to relieve allergic skin rashes and itching. Cortisol regulates the immune system short term, promoting healing and survival. But if elevated long term, cortisol clobbers the system, leaving you defenseless to disease, shortening cell life and longevity by about eleven years in humans.

Cortisol levels mirror stress: the two are virtually the same. Either may cause the death of white blood cells, mainly lymphocytes, the infantry that wards off harmful invaders. So, as in conditions like AIDS in which the lymphocyte count is low, resistance to infections and skin cancer is impaired. And, as psychologist Elissa Epel and co-workers have shown, stress accelerates aging.[7]

Initially, under cortisol influence, white blood cells and others throughout the body release inflammatory chemicals called *cytokines*.[8] Cytokines help fight inflammation short-term. But by stimulating the hypothalamus, a structure in your brain, they may release cortisol long-term and cause the body to attack itself. This may lead to autoimmune diseases, such as rheumatoid arthritis, lupus, multiple sclerosis, Sjögren's syndrome, and more.

Regulatory lymphocyte cells from the thymus gland located behind your breastbone aid cortisol and suppress autoimmunity as well. Like cortisol, these cells limit chronic inflammatory diseases, shutting down immune responses after an invader is eliminated. Long-term, high cortisol levels could also lead to diseases such as type II diabetes, depression (when combined with low brain serotonin levels), obesity, and high blood pressure. They may also lead to cholesterol abnormalities, causing the fat to stick to coronary arteries, narrowing them and giving you a heart attack.[9]

Not surprisingly, daily ravages by elevated cortisol shorten life. For example, fear in rats, particularly fear of the unknown, keeps cortisol levels high and shortens their lifespans by up to 20 percent. High cortisol levels may also contribute to dementia,[10] migraine,[11] and cancer progression, for instance breast cancer.[12]

In addition to job position and social rank, other situations such as refugee status or care for a relative with dementia could lead to mental strain and stress. Any change, good or bad, may lead to stress, increasing your chances for illness. Such situations contributing to ill health also include chronic anxiety, an impending exam, loss of a loved one or job, financial ruin, a wedding, a prison sentence, divorce, and retirement. The list goes on. It follows, then, that the lower the stress, the longer we can put off our Great Equalizer.

TAMING THE BEAST

Insecurity or death of a loved one may cause stress and depression. Many of us have heard stories of an elderly widow who dies six months after the loss of her spouse. Grief drops immunity, leaving the mourner vulnerable to sickness when exposed to disease. Also, caregiving, which often falls to the spouse of the terminally ill, is a significant stressor. The longer one remains stressed, the more vulnerable one can become.

Immunity can also tank in other stressful or life-changing situations. For example, after certain elderly Native Americans decide their time is up, they quietly pass away, or on retiring, the retiree quickly takes ill and dies, often of a heart attack. Incidents like these underscore the powerful effect attitude and behavior can have on biology and the course of disease. To demonstrate that humans can mentally control their own bodies through their attitude, for instance, by lowering cortisol, since we are similar to rats physiologically, psychologist Robert Ader at the University of Rochester School of Medicine fed saccharine to rats and then injected them with a stomach-upsetting drug. Their immune system dropped. It dropped again on expectation, when he gave them the sweetener alone, showing how the mind may influence the body.[13]

We can't control every environmental factor that impacts our health, but we can control stress to a certain extent through lifestyle, diet, and behavioral changes. According to physician Michael Roizen, founder of *RealAge*, lessening stress lowers your real age.[14] Roizen argues that lifestyle changes are more effective and safer than taking medication to reduce stress.

Such lifestyle changes include diet, group support, exercise, yoga, and meditation, all helping to reset your body clock to a younger age. A diet containing Omega-3 fatty acids, nuts, vegetables or the yellow pigment curcumin in turmeric, could fight stress and its associated inflammation. Likewise, exercise helps to reduce inflammation in fat cells and the liver, and releases morphine-like chemicals – endorphins – in the brain, thereby uplifting mood. The more you run, the more endorphins are released. No wonder some people crave running marathons.

According to Castillo-Richmond, yoga and meditation alone could ease stress and lower heart attack risk by 11 percent and reduce stroke risk by up to 15 percent.[15] Using these techniques, Dean Ornish's patients at San Francisco University in California lowered cholesterol, reversing heart disease by 50 percent. Meditation alone decreased chest pain and angina and increased exercise endurance.[16]

As indicated earlier, "prayers," a kind of meditation, appear to work for some as an antidote to depression, and may have played such a role for our ancestors going back millennia. How does that work? Such activities increase brain chemicals such as serotonin, whose removal (mop up) inhibitors are used to treat depression, and dopamine, the pleasure chemical, while they decrease havoc-wrecking body cortisol levels. So, prayers or meditation by participants increased focus, eased stress, led to ecstasy (or mild cases of it), and decreased depression. For illness or injury, it seems likely then that prayers, trance inducing rituals, and mysticism boosted immunity and helped our ancestors overcome infections. They survived and were able to have more offspring. No wonder the top candidates for evidence of positive selection in our genome are frequently involved in immunity when the human genome is examined.[17] Before antibiotics, 72 percent of deaths in our ancestors could have been due to infections, and life expectancy was only about 30 years, compared to 72 years globally today. For the chimps, even though we share 98.9 percent of our genes with them, their life expectancy has remained the same at 13 years.[18]

Giving up, then, may lead to stress hormones running amok, leaving us vulnerable. To prevent it, we could take up a purposeful activity such as volunteering for events that give you satisfaction, attending church, or joining a local social group to fill the empty days, particularly after retirement. Social contact is a great stress buster and can prove more effective than medical treatment. According to Sapolsky, in addition to social support, having a sense of control over one's life, as well as outlets for frustration and an optimistic outlook, are all good at keeping stress to a minimum.

Research has repeatedly shown that one may not need a specific therapy to get better. A placebo could do. According to psychologist Tor Wagner, studies show that placebos trigger expectation of relief in the brain, causing it to release dopamine. Such placebos can help to alleviate symptoms caused by Parkinson's disease. Apparently, sometimes thoughts and emotions alone, even something as simple as smiling, which triggers activity in the happiness centers, may help us to feel better.[19] Whether it's through placebos or other interventions, cortisol levels abate as stress comes down, invigorating the immune system. Such a mind-body effect, like the placebo effect, could be behind the favorable results of many alternative therapies such as acupuncture, imagery, aromatherapy, hypnosis, and naturopathy. To learn how thought, belief, emotion, and behavior affect the immune system, Congress created a division of National Center for Complementary and Alternative Medicine at the National Institutes of

Health in 1992. The center focused on alternative therapies, as many scientists were already doing.

AN OUNCE OF PREVENTION

Instead of wrestling with the beast to tame it, we could pacify it. We clobber our immune system with negative, sad thoughts or unwarranted guilty feelings, or obsess over imaginary threats. A lack of outlets for socializing, entertainment, exercise, and other activities to let off steam can allow such feelings to fester, weakening our defenses and pre-disposing us to the ravages of stress. This can be prevented in various ways, including eliminating feelings of guilt, setting goals, nourishing the spirit, and eliminating sources of stress.

NIP GUILTY FEELINGS IN THE BUD

Guilt causes stress, elevating your cortisol levels. It could be unfounded, brought on by a feeling that you're wrong or are an annoyance to others. But what is right or wrong? Unlike the Ten Commandments, moral codes are not carved in stone. Generally, a behavior promoting survival is moral and good. To judge conduct, then, weigh its evolutionary benefits. Our sense of right and wrong is ancient, related to our remote cousins' social tendencies that also belong to primates and other animals. Morality, then, is innate. Brain imaging shows different parts of the brain, some very old, lighting up during moral judgment.[20] The Golden Rule condenses all this moral stuff into one sentence. Summarizing the Torah, Jesus, as others had also done, said, "So in everything, do to others what you would have them do to you."[21] Ultimately, good – acceptable – conduct promotes one's chances of survival, as it's usually reciprocated, if not in kind, then at least by the good, stress-easing feeling it brings.

SET GOALS

A major concern in life is to avoid pain and suffering, both physical and mental. This should be our number one goal. For example, in Hinduism, life is suffering, so the goal is to escape it and not come back here after death (reincarnation) to suffer again. For this goal, the Hindu adhere to tenets spelled out in their religious texts, the Vedas. Pain and suffering elevate levels of stress hormones, which shorten longevity, as explained above. Safety is such a big issue that it leads to many product recalls, and courts award huge damages – more than for healthcare costs – for injuries from accidents negligence, and so on.

If you want to live a happy life, Einstein said, tie it to a goal, not to people or things.[22] Setting goals reins in the havoc-wreaking stress hormones. Ordinarily, people rely on a higher power, hoping that if they pray hard enough, the power will intervene and improve their lives. It's not guaranteed that prayers would produce tangible results, and as a hedge, we also need to bolster our mental health, which prayers actually can do, as described previously.

Maintaining sound mental health is more challenging than maintaining physical wellbeing. To nurture stable mental health and live longer, begin by curbing your stress demons; set your own goals to deflect negative thoughts and beat back the stress demons and prevent doom and gloom. A goal could be anything that brings a sense of purpose to your life: remodel your home, enroll in a course, spoil the grandkids, go to church, or practice tai chi, yoga, or meditation. Meditation (excellent therapy for PTSD) promotes mindfulness, a presence in the HERE and NOW by improving your focus and helping you to see and achieve your full potential. Being keenly aware of your surroundings can help to give you a heightened appreciation for nature, thereby uplifting your mood. Grecian Monks on Mount Athos pray 24/7, a ritual akin to continuous meditation. A hobby such as painting, gardening, or learning to play an instrument may be so consuming it keeps you rooted in the present, even as in fact the time is flying. Keeping company and socializing helps to combat loneliness, especially in retirement when days can all start to look the same and people one used to know aren't around anymore.

Having something to look forward to or live for, such as following a home team, playing the weekly lottery or Bingo, or embroiling yourself in politics (trashing opponents' ideas, originally yours before you "evolved") could uplift you. You're doing fine if you have what you really need in life. More won't make you any happier. According to Einstein, "try not to become a man of success. Rather become a man of value."[23] Ultimately, all these measures will ease stress and help to promote a healthy lifestyle and to live longer.

NOURISHING THE SOUL

When feeling down or running on empty, to calm your stress hormones, keep your immune system on an even keel, pep up your soul, and maintain mental balance, engage in art, music, or exercise. Art, like music, speaks to our spiritual legacy. It recorded our ancestors' trance experiences, and today it can help to repair wounded souls. Even as babies we are attracted to and soothed, comforted, or intrigued by colored pictures. Because of art's hold on us, we swoon at a Renoir, Rembrandt, or Picasso.

Before taking out your frustration on others or on a pet, if you don't have a bag to punch or a pillow to kick, listen to Beethoven's *Moonlight* piano sonata or the elegant melody of Nadama's *Inner Joy*. Pop music fans may like Khalid's *Free Spirit,* and smooth jazz enthusiasts may find Kenny G's *Forever in Love* skin tingling. Music can go around and around in your head, displacing unpleasant thoughts. In addition to endorphins that ease pain and elevate mood, music releases antibodies (immunoglobulins) that strengthen your resistance to disease. Learning to play a musical instrument may work even better. It's no wonder that music conductors continue to work well into old age. Music improves IQ to boot,[24] and, can motivate children to stay in school, helping them learn and score better on exams. Best of all, music is spiritual. It whisks us to the gates of Heaven for a glimpse of the Most High – through a trance, of course.

Other activities that prevent or fight everyday depression include physical exercise – a walk around the neighborhood, a bike ride, or a swim. Whereas these could be carried out at a moment's notice, prayers and other rituals may need scheduling or planning ahead of time and so may not be as readily accessible.

Confiding in someone else, even a stranger, can also release tension. In fact, many relaxing activities, be they a yawn, a smile, or exercise, may help to lower your cortisol levels, fight disease, and live longer. An old adage still holds true: laughter is the best medicine. Also, to comfort and renew, give (or get) flowers, or a perfume. Both may calm the demons and even kindle love in the bargain.

STRESS BUSTERS

For many people, the lack of a sense of purpose in life is a stressor and the search for it or one's calling may seem unending. Meanwhile, finding ways to effectively ease the stress of uncertainty may pose a challenge. Sometimes, people turn to physical exercise, or immerse themselves in exploration of spirituality through meditation, prayers or yoga, that make them feel better and abate their anxiety and consequently their stress hormone levels.

Humans have long found solace from the harshness of everyday life by seeking understanding of the bigger questions, and that search has led many to look for the guidance of a higher power through religious practice. The greatest philosophers throughout history have tackled these questions and come up with a variety of answers, including several already discussed in Chapter 1. Even some seemingly a-religious scientists have taken their turns at trying to figure out how to explain the unexplainable. Einstein, for example, had multifarious thoughts on the subject without a clear answer. He said, though, that the most beautiful experience we can have is the mysterious, and this endangered religion. A knowledge of this mystery, the perception of the profoundest reason and most radiant beauty constituted true religiosity. Thus, being aware of such a mystical experience meant he was "deeply a religious man."[25] Einstein was in good company, that of fellow physicists such as Edwin Schrödinger, Robert Oppenheimer, and Niels Bohr, who likewise were baffled by nature. Other prominent scientists included Carl Jung, mentioned earlier, the neurophysiologist John Lilly, astronomer Carl Sagan, and biochemist Erwin Chargaff.[26] Einstein is also quoted as saying, "Man can find meaning in life only through devoting himself to society.[27] Christians, however, found purpose in following the teachings of Jesus and crusading to save sinners from damnation and an everlasting inferno.

Each culture in its time period has added a special spin to this idea. As in Hinduism, above, to escape from karma, the cycle of death and rebirth, or reincarnation, one lives morally. Only then can one realize purpose and meaning, and achieve "enlightenment or Moksha." Alternatively, one could practice Tantra Yoga (yoga means to yoke or unite), and recreate sex between gods. In Tantrism, sex is divine, as it may evoke a fleeting spiritual experience, where it unites the individual soul to Brahman, the ultimate reality underlying all phenomena. This brings about enlightenment through a profound experience of sensual love in which "each is both." As this physical and sensuous side of human nature has always been associated with the female, Hindu goddesses abound (each house has a god).[28] Others, such as Saint Mother Teresa, in the hope that a happy hereafter will make sacrifices on Earth tolerable, motivated by a desire for salvation, dedicate themselves to the relief of suffering. Those who believe in theologian Martin Luther's theory, however, say that self-deprivation and ascetic conduct is irrelevant, because our destiny is fixed, or predestined from birth. Efforts at overcoming despair may also help open the door to discovering personal meaning. But one should know one's limitations to avoid disappointment. Reinhold Neibuhr's Serenity Prayer, adopted by Alcoholics Anonymous, offers the following approach:

> *God, give us grace to accept with serenity the things that cannot be changed, courage to change the things that should be changed, and the wisdom to distinguish the one from the other.*[29]

For novel means of managing substance abuse and hence stress, it dawned on Yohann Hari that the opposite of addiction wasn't sobriety. It was "connection." Such connection included love, compassion, and support. These could give life meaning and help to prevent or manage drug misuse.[30] According to psychologist Martin Seligman, inherited behavior is very difficult to change. For instance, depression and alcoholism, often inherited, are treatable but not curable. Anxiety and panic attacks, which are not inherited, are curable. Sexual behavior varies and may be divided into five layers. The first layer, the core, concerns *sexual or gender identity* (what you feel your gender is). Here, chromosomes are appropriate for the gender, but may vary in their effects leading to some people being unable to fit in the sex they are assigned at birth. The next layer, *sexual orientation* (the gender you love or you're attracted to) could be attributed to an atypical hormone secretion in the womb. So, these two layers are not something individuals choose, but rather, what they are, and could be long term or may shift over time. The other three layers might wane. They include *sexual preference* (what body part or situation turns you on), *sexual expression* (which gender role you play), and *sexual performance* (whether you function "normally," e.g., erectile dysfunction, or ED). ED is amenable to treatment with medication or surgery.[31]

We might also find meaning through psychotherapy, or *logotherapy* (*logos* denoting "meaning," and *therapie*, "healing") founded by Holocaust survivor Viktor Frankl. Using a four-step approach, a therapist would first advise you to focus away from yourself and your ego and from what you feel you need, want, or must have (dereflection). Next, you would be guided to turn the negative into positive and to discover new dimensions in something old (attitude modulation). These two steps would reduce your neurotic symptoms and prepare you for openness for change or healing (healing therapy). Symptoms may still be present or even return without the last step, which includes the search for meanings, commitments, tasks, and goals. This step helps the person to "take hold of life."[32] Frankl believed people could find meaning in their suffering and taught that one may not be able to end suffering, but may still be able to find lessons or purpose in it. So, instead of asking "why" in a tragedy, ask what you can do.[33]

Although none of these methods for stress reduction will make or keep you younger, they can help you to live a healthier, more satisfying, and, possibly, a longer active life. Some methods may require significant effort, while others may not be appealing. But finding one or two, or a few that work for you, such as prayers, music, and love (of any kind), may help to ease your stress.

HENRY KAKEMBO, M.D.

CHAPTER 12

WHAT - I'M NOT WHAT I SAY?

A great many people think they are thinking when
they are merely rearranging their prejudices.

~Unknown[1]

Although we sense God watching over us from above, we have yet to find Him there. Perhaps it's about time we looked elsewhere. What about searching within ourselves? Jesus, when asked, said that the kingdom of God was within us. Lk 17:20-21. But where exactly is *in us*? In the heart, the seat of love? But all the heart does is pump blood night and day. Is it in the brain, then, since the brain calls all the shots? We still don't understand fully how the brain works, how it generates its information, however.

Psycho analysis ("Psycho" from ancient Egyptian *Su* for she and *Khe* for soul, Greek *Psu-Khe* or Psyche)[2] aims to go deep within the brain, much as religion has long done via the confessional. For Catholics, such a baring of the "soul" opens the door to "salvation." But before we go there ourselves, a word of caution: our command central is not only stranger than outer space, it's also weirder than fuzzy muzzy quantum mechanics.

The brain takes us to fantasyland daily. There we stay 24/7, 365 days a year, absolving us of our ignorance and the dereliction of our duties. As Plato put it over two millennia ago, our perceptions are merely shadows of what is out there. Plato believed the real world was hidden from us and that we experienced representative "forms" of that reality. In the 1600s, when Isaac Newton demystified the nature of light by saying that it consisted of colorless electromagnetic waves, thus implying that things were not necessarily as they appeared, people scratched their heads, and justifiably so. What are we to make of the knowledge that the brain hoodwinks us about everything, including God, and that there's no way verify its information? It means that what we see doesn't exist as it appears to us.

Mystics say they experience true or absolute reality while in a trance. A trance enlightens them to the true world, akin to salvation. Most of us, though, lacking the talent to evoke such experiences, are left only with prying the brain open for answers.

The mind was a mystery for ages. In the 400s BCE, Hippocrates and Plato insisted that the mind was located in the head. But Aristotle, arguing we could not understand the mind without studying the body, located it in the heart, just like the Egyptians and the Bible had done. Is 46:8.[3] The idioms *learning by heart* and *call to heart* (to remember) point to this organ as the seat of the mind. The brain, then, appeared to be a functionless piece of grey matter good for receiving sensations only. During their mummification process, the Egyptians discarded it and the pharaohs went into the afterlife without one.

Among the first to identify the brain as the seat of intelligence were fourth century BCE biologists Herophilus and Erasistratus in Alexandria. Then, in the second century CE, Galen, a Greco-Roman physician, identified the brain's parts. Galen considered the cerebrospinal fluid the fluid of the mind, or psychic fluid, and the ventricles, small spaces at the center of the brain where the fluid collected, the seat of the mind.

In some quarters, though, the church for instance, thoughts and behavior were supernatural. In the eighteenth century, when French physician Julien Offray de La Mettrie (1709-1751) published that the soul, like thoughts, was only a function of the activity of the brain, and didn't exist separately, he drew sharp criticism from the clergy and physicians. Not yet finished, he also said that man was a complex arrangement of matter like other animals (only a superior one). That did it. Besides relieving him of his duties in the French Royal Guard in Paris, the authorities booted him to Prussia, where he died three years later.

In the nineteenth century, a Spanish histologist, Santiago Ramón y Cajal, revealed that the brain contained a huge number of individual nerve cells, or neurons. Each neuron extended a few to several thousand branches toward its neighbors and, without touching, they passed messages to one another. For this groundbreaking work, Cajal and another histologist, Camillo Golgi, won the 1906 Nobel Prize for medicine.

MAKING A MIND

The nervous system evolved about half a billion years ago, appearing in primitive sea creatures like the sea squirt. It was then only a collection of about 300 neurons crackling with electricity. As the system developed, embryo cells lined up in a tube that later became the brain. The sea creatures, starting with fish, evolved and developed over time, the tube's functions improved, and some fish eventually could survive on land as amphibians. Further developments led to the appearance of reptiles to birds to mammals, each new species adapting and gradually adding new brain parts on top of the old.

Humans have the same basic reptilian brain dinosaurs had, jam-packed with the skills required to survive on land. Included in the various brain parts is the cerebrum, the mastermind of the nervous system. The cerebrum runs higher order mental processes, the ability to attend to, identify, and act on complex messages. It houses the "I" capable of conceptualizing God. By filtering and processing input from outside, said David Suzuki, it "allowed memory storage, for learning, for planning."[4] This organization has remained the same in humans for 200,000 years.

CHEMICALS MADE ME DO IT

To understand how the brain works and imagines God, a knowledge of biology is key. The brain is awash with chemicals, or transmitters, for that purpose. Like most genes, instead of each of the sixty or so different chemicals in the brain controlling a specific function, they regulate each other or other genes that may turn a function on or off. For instance, the pleasure chemical dopamine may regulate the mood stabilizer chemical serotonin in this way.

Besides serotonin, dopamine also regulates behavior. Too much of it in the limbic system (our emotional, mood, and memory center) and not enough in the cortex could lead to an over-suspicious personality, paranoia, and impaired social functioning. When dopamine floods the brain, it produces feelings of joy and happiness, feelings that may be different from the high that comes from exercise such as jogging, feelings produced by morphine-like brain chemicals called endorphins and enkephalins.

On the other hand, serotonin plays an important role in creating spiritual experiences, interpreted as religious experiences when people sense supernatural powers in them. It also regulates mood, appetite, and learning. A shortage of serotonin may lead to depression, increased aggression, and criminal behavior that can land you in jail. The culprit, according to Sapolsky, could be one's socioeconomic status.[5] As noted in the last chapter, serotonin levels are higher in top-ranking people in a hierarchy, such as the CEO or secretary of state, than in subordinates. Medications like Prozac, which raise brain serotonin levels by blocking its removal may help to alleviate the adverse effects – such as aggressiveness – of a shortage of the chemical.

A few transmitters play their roles directly. For instance, we would be chronically stressed if not for benzodiazepines, a group to which many tranquillizers such as Diazepam belong. Such chemicals act through specific receptors populating areas where they function. For example, serotonin receptors pack the thalamus (the gateway of all sensory information) and the frontal cortex.

The limbic system, located below the corpus callosum – the highway connecting our brain hemispheres – comprises several structures including the amygdala and hypothalamus. The hypothalamus regulates body temperature, metabolism, thirst, hunger, satiety, and pain, as well as our primitive instincts such as anxiety, fear, and pleasure. The amygdala helps to regulate emotion and seems to play a role in emotional memories. Without the amygdala and a learned response, our fear of strangers would go unchecked.

The limbic system also contains a structure, the nucleus accumbens, that when stimulated in humans, moderates cravings. It also generates pleasure and orgasm during sex, similar to taking cocaine or amphetamine, as well as the experiences meditators have when sensing God or the Universal Force. The nucleus accumbens is likely the structure that American psychologist James Olds wired with electrodes in experiments with a rat as his test subject. The animal stimulated it nonstop "to the exclusion of everything else except sleep."[6] Such stimulation has been shown to release dopamine within the nucleus accumbens.[7] Together with other structures in the pleasure circuit, this area is responsible for addiction to drugs, alcohol, tobacco and sex. The amygdala plays a key role in processing emotions and, after the judging cortex weighs in, storing them. *See Illustration 12.1*

Certain drugs and low cholesterol diets can lower serotonin levels and lead to aggressive behavior. For example, in one notable study, monkeys whose cholesterol was reduced with diet became violent and ill-tempered. In another study, the seven- year MrFit heart disease prevention trial, researchers found that of 351,000 participants, those with the lowest cholesterol died at twice the rate of participants with average values. The main causes of death were accidents, murders, and suicides. Other studies have supported such adverse outcomes to low serotonin levels.[8]

Since medications elevating brain serotonin levels may control aggressive behavior, the implications are intriguing. Should violent offenders be required to take them? For how long? For life? Or, should the inmates just get a slap on the wrist since their criminal behavior is rooted in biology? Incidentally, many inmates already take such medications for one mental disorder or another, mainly bipolar disorder. Some argue that healthy people shouldn't be made to take medications if they don't want to. Also, some scientists believe it would be an unthinkable invasion of one's mind to force drugs on certain people. Were it acceptable, however, a sure bet would be a trance-inducing drug like psilocybin from mushrooms. Psilocybin produces profound and mostly positive experiences, resulting in a long-lasting sense of well-being and satisfaction as described next. As with other hallucinogens, psilocybin is structurally similar to feel-good, trance-inducing serotonin. An excess of serotonin produced when the inhibitory circuit of the thalamus is interrupted distorts sensory perception, leading to a hallucinatory or mystic (spiritual) state.

In the early 1960s, when psychologist Timothy Leary gave psilocybin to inmates in the Concord Prison Experiment, 64 percent of them stayed out of prison when released, instead of the usual 25 percent. But Leary became too enamored by the pill, and together with his unorthodox research methods, he unnerved his Harvard colleagues, and the university fired him. Still, he showed psilocybin could positively change criminal behavior. A follow up study indicated that addition of psilocybin-assisted psychotherapy, within a comprehensive treatment plan that included post-release non-drug group support programs, was necessary for proper evaluation. Despite efforts by the experimental team to provide the post-release services, they were not made sufficiently available to the subjects.[9]

Also, according to Dean Hamer, a single dose of psilocybin given to theology students in a double-blind study led to increased tolerance of others, appreciation of nature, and involvement in social and political causes compared to the control. Participants felt more at one with the universe and the people in it. The experience felt very real. The experimental group scored four times higher on the mystical scale than the control group, and again when tested twenty-five years later.[10] The downside of psilocybin use was fear and anxiety in some people. Surely, some – if not many – inmates wouldn't mind trying a single dose at release to keep them out of trouble for life. But as of 2022, psilocybin was legal only in Oregon, Colorado, and Washington DC, plus several cities in the United States. Today, it's being studied for the treatment of smoking, drug addiction and mental illnesses at various major medical centers, including Johns Hopkins, University of California, and New York University.

AND LOVE IS…

For Christians, the love of Jesus is their ticket to Heaven. But such love, which many people think about when the word is mentioned, may trigger mystical experiences and rapture. In fact, the worship of Jesus is essentially designed to induce a spiritual experience, just as the worship of gods in other religious practices, including in Voodoo, will do. During romance, blame it on the Bossa Nova or not, when love is in the air, dopamine surges, setting the basal ganglia afire. To heighten the feeling of bliss, noradrenaline crashes the party, releasing adrenaline and speeding up the heart, thus causing the palms and cheeks to sweat. All this takes place in the pleasure center unbeknownst to the thoughtful cortex. No wonder love is blind.

By and by, the surge of chemicals dries up, the cortex weighs in, and we begin having doubts. Dopamine plays a role here, too. It releases oxytocin, otherwise known as the "cuddling hormone" or "chemical of attachment," that encourages bonding. There is a pattern to romantic love in humans. According to Helen Fisher of Rutgers University, beginning as lust, it turns into romantic love, then long-term attachment.[11]

But doesn't all this chemical stuff turn you off of love, as a cadaver would turn a pathologist off a steak? No way, Jose. Romantic love is a drive, as hunger is a drive. Understanding its mechanism won't change how it feels—and the steak to the pathologist will still taste delicious. By understanding the mechanism of love, however, and since it is intertwined with the spiritual experience, demystifying it could help us to figure out its purpose and the nature of God.

ME, A ROBOT?

Whether we are aware or not, our bodily functions chug away nonstop. When it's cold or hot, the body turns switches on and off, and we develop goose bumps or we sweat to moderate our temperature, in that order. Heart rate and breathing are similarly controlled.

Such regulation is achieved through the two branches of the autonomic nervous system: the quiet (parasympathetic) and the active (sympathetic) branches. The parasympathetic branch maintains the body at rest through the chemical acetylcholine. The active branch controls most of the internal organs and fight-or-flight response through adrenaline and noradrenaline. The last two chemicals also regulate our sexual organs, and they play roles in the rapture or trance experiences that may be interpreted as religious experiences.

Thoughts, especially sad thoughts, release adrenaline from the hypothalamus and could lead to depression, compromising the immune system and causing ill health. Biofeedback may reverse this and other stress-related conditions such as high blood pressure. Relaxation techniques like meditation activate the quiet system, decreasing heart rate and blood vessel tone, or resistance, to lower high blood pressure. Biofeedback may also relieve stress by lowering adrenaline levels, temperature, and pain. The good news is, unlike a knee-jerk reflex, we are not automatons through and through. We may have as much as ten percent control of our bodies, enough to allow us to take part in our destiny.

AND THERE WAS LIGHT

Our senses inform us about the world. Of the main five senses: sight, hearing, smell, taste, and touch, none is more intriguing or more studied than sight. Before Isaac Newton came along, people believed the eyes emitted a vital energy that made the surrounding area visible, and that a person could burn a hole in the heart of someone they hated by staring at them. Newton put an end to the vision and light nonsense by showing that our eyes detect the light around us, instead of our eyes beaming out lasers of energy.

Newton broke down light into three primary colors: red, yellow, and blue. Light travels as electromagnetic waves – and as particles or quanta. The brain assigns color to the various electromagnetic wave lengths: red to long waves, yellow to medium waves, and blue to short waves, called the primary colors. All other colors, called secondary, are a combination of these three in different proportions. For instance, red with yellow produces orange, yellow with blue, green, and blue with red, violet. White is made up of a combination of all the three primary colors. The color of an object corresponds to the kind of wave it reflects. A red apple bounces off long waves and a black shirt none, as it absorbs all waves. On the other hand, a white shirt bounces off all the waves. Seeing color is not a must. Whales and squirrels see in black and white only. Photoreceptors for color, the cones, located in the center of the eye, are outnumbered by the rods around them. Rods are responsible for black and white vision in poor light and at night.

So, since electromagnetic waves don't have color themselves, we're colorless. For us to survive in an ever-changing world, the brain assigns colors to the waves. It means that we can't tell what we and the world really look like. Although mystics report experiencing the real world in a trance, they can't describe it and say, instead, that it's ineffable.

"I AM WHO I AM"

Who hasn't heard of God's response when Moses asked for His name at the burning bush? Karen Armstrong interprets the answer in Ex 3:14, "I am who I am," (in Hebrew, *Ehyeh asher ehyeh*), as evasive. It's an idiom expressing a deliberate vagueness, as if to say, "Mind your own business."[12] But considering the information provided so far, do you know who you are? Perhaps you're not so sure; you're wondering what's really real, or true. Normally, you would have no problem identifying what belongs to you, such as your hand or nose or dress. But what about the "I" that perceives them as yours, conceives of supernatural powers, and obliges you to worship them? Where does the human sense of self reside in the brain?

With 100 billion neurons and at least 100 trillion connections, the brain is the most complex anything in the universe. Its 2 percent ratio to body weight is the highest in animals. Relative brain size within species, though, does not necessarily correlate with intelligence. After studying Albert Einstein's brain, pathologist Thomas Harvey, said it was average in size, but it lacked Broca's area, a part of the brain partially responsible for speech. This deficiency was apparently compensated for by a wider inferior parietal lobe. This lobe enabled mathematical thought, visuospatial cognition, and imagery. Einstein's brain also had an abundance of support cells, called neuroglia, crucial in the development of nerves (neurons) and their covers (myelin sheaths) that speed up electrical signals.[13]

To identify the functions of various parts of the brain, scientists study deficiencies in the behavior of stroke and head injury victims. In 1941, a British neurologist, W.R. Brain, reported a remarkable change in the behavior of three patients who had each suffered a stroke in their right parietal lobe. They were getting lost in their homes. They neglected their left sides, turning only to the right and opening doors only on the right side. When asked to copy a drawing of a house, they drew only the right side. They also bisected a line to the right of the center.[14] *See Illustration 12.2*

Scientists also study the brain with technology such as functional magnetic resonance imaging (fMRI) and nuclear positron emission topography (PET). The brain has no pain nerves, so it can be stimulated directly during surgery while the patient is wide awake to provide feedback. To verify their results, scientists probe similar regions in nonhuman primates and other animals. Such methods are invaluable in our quest to find God. *See Illustration 12.3.*

MIND, BODY, AND SOUL

Not only is the location of the "I" in the body a mystery, but so is that of the location of the mind and soul. What are the last two anyway? Most people consider the mind, body, and soul to be separate. Experts – scholars, philosophers, theologians, and scientists – do not see eye to eye on this. The differing opinions on the answers to these questions can be bewildering to the non-scholar.

Take consciousness, for example: its definition, like that of life, is imprecise. Some say consciousness consists of brain processes; others say that it's supernatural. Alternative medicine guru Deepak Chopra claims the universe is conscious, too, and that it thinks. Understanding consciousness, the mind, and soul, though, is essential, as it could lead us to God.

Without question, we cannot be alert sans a brain. But there is more to consciousness than being awake or aware. Medicine divides consciousness into four stages: wakefulness, drowsiness, stupor, and coma. Whereas you would respond to a strong stimulus in a stupor, no amount of poking will arouse you from a coma.

To keep your systems running, the brain burns lots of energy. About 1.6 pints of oxygen rich blood bathe it each minute. Starve it that long and it sputters. After another four to nine minutes, it goes on the blink, turning you into a vegetable. Why so much energy? Besides generating electrical signals, the energy helps restock the chemicals. No doubt, consciousness is linked to brain activity based in biology.

Even though we barely understand consciousness, we have no problem altering it for healing, spiritual awakening, or pleasure. We turn it off to block pain during surgery, and set it adrift to fight stress during meditation. Meditation may transport us to spiritual worlds yonder from which we return refreshed and bent on saving humanity from its excesses. To help people stop smoking, hypnotists manipulate consciousness. We also dull consciousness for pleasure with drugs such as alcohol (spirits), cocaine, marijuana, or heroin. Fentanyl only kills. Chemicals such as LSD, the peyote plant mescaline, and psilocybin also trigger hallucinations and spiritual wanderings. The anesthetic Ketamine, or direct electrical stimulation of the right angular gyrus in the brain's parietal lobe, can create a sense of floating, an out-of-body experience akin to a near death experience (NDE). Dopamine and related transmitters in the limbic system play a role in such experiences.

How is consciousness generated? According to chaos theory, consciousness is an emergent property of brain activity. It results from the hum of zillions of neurons self-organizing into a consistent pattern, a holistic effect, where the whole is greater than the sum of its parts. So, to cytologist Christian de Duve, a Nobel Laureate and author of *Life Evolving*, consciousness could be linked to some resonance among parallel circuits carrying different aspects of the same information. It would be similar to the phenomenon that causes sound to shatter a glass at a certain frequency, or to oscillations generated by marching soldiers to break a bridge.[15]

According to Johnjoe McFadden, since the energy of brain activity creates a significant electromagnetic field, enough to light a ten-watt bulb, consciousness could be a quantum phenomenon. Changes to the field would lead to conscious awareness. For instance, EEG studies show "that synchronous firing in different regions of the cortex correlates with awareness and attention."[16] Neurons firing in unison, then, similar to a bridge collapsing from soldiers marching in step on it, lead to awareness or consciousness. So, as all three theories suggest, it seems that groups of neurons respond alike to a stimulus for consciousness to occur.

MIND MOVES MATTER

A thought seems intangible. But how can something so ethereal move an arm or produce speech? In the early 2000s, in a lab at Wadsworth Center, in Albany, scientists Jonathan Wolpaw and Dennis McFarland decked a volunteer's head with an EEG cap. After some practice, the volunteer could move a cursor on a computer screen with his thoughts. Willing the cursor up or sideways produced changes in the brain's electromagnetic field and the EEG passed them to the computer. Another volunteer paralyzed by Lou Gehrig's disease communicated similarly. He selected words on a computer screen with his thoughts and the computer synthesized them into speech. Before Wolpaw and McFarland's work became available, this kind of meaningful brain-to-computer communication was effected via probes directly implanted into the brain. In both cases, thoughts altered brain electrical activity, leading to changes in the brain's magnetic field. The changes stimulated appropriate neurons, producing the desired activity. This activity showed that thoughts were electrical, physical, then.

Since they are physical, it means thoughts are tangible and can be measured, documented, or transferred to a computer to a certain extent. Unless we are Buddhist monks, though, we have little or no control over our thoughts, as sundry, random ideas pop up constantly. The mind can seem like an out-of-control chattering box.

If we can't control our thoughts, how do we make conscious choices about anything? Apparently, we don't, just as we don't choose to sleepwalk. In the seventeenth century, philosopher Baruch Spinoza rejected the idea that humans had any freedom of will. He said, "Men are mistaken in thinking themselves free, for their opinions are made up of consciousness of their own actions, and ignorance of the causes by which they are determined."[17] According to philosopher Richard Popkin, Spinoza denied free-will because he believed that it was inconsistent with the nature of God and with the laws to which human actions were subject.[18] Our behavior, then, seemed predetermined and predictable.

241

Neuroscientist Benjamin Libet of the University of California San Francisco used MRI studies to show that the decision to carry out an action occurred *after* that action had already begun. The studies revealed that the test subjects started voluntary actions unconsciously and only became aware of the decisions to act after the action had already started. The performance of a simple task was preceded by a specific electrical change in the brain (the *readiness potential)* that began 550 milliseconds before the act. The study participants became aware of the intention 350 to 400 milliseconds later, a tiny time window in which they could choose to intervene and alter the action.[19]

To intervene, to stop the action, we can select from a variety of stored memories acquired during life and those inherited from our deep ancestors, what Jung called our collective unconscious. For instance, when angry, some lash out despite themselves, showing "will is not free." With practice, over time one might learn to control such impulses. Nevertheless, in light of Libet's findings, ophthalmologist Thomas Czerner wrote, "If the readiness potential occurs before your 'intention,' you are deluding yourself with your notion of freewill."[20] Philosopher neuroscientist Sam Harris agreed, writing, "Free will *is* an illusion."[21] For instance, do you remember the many times you resolved to break a bad habit and couldn't? Something kept holding you back, ignoring your will. It's no wonder people believe in supernatural phenomena based on their faith alone, without evidence to support their belief.

If thinking, aptly called reflection, is the awareness of our (re)actions based on what's stored in our memory banks, as the epigraph at the top of this chapter implies, it seems, then, that as products of brain activity, consciousness, mind, and soul are interrelated, if not the same. Still, even though we would like to believe that the mind and soul, the "self," or our life essence, are located in a specific area of the brain, we have yet to find them.

At death, the "self," or soul, supposedly departs the body for the next life. But according to Plato, chairs, tables, and "animals" also have souls. Where do their souls go? Evidence for a soul eludes us, anyway, unless it's air. In many ancient languages, the Greek pneuma or psyche, the Sanskrit atman, the Hebrew ruach, and Latin anima and spiritus, the soul is breath, not a spirit or a supernatural entity. David Hume didn't believe there was a soul or a mind either. As a soul accounts for nothing anyway, better to forget it. Consciousness is daunting enough.

THE SPLIT PERSONALITY

It seems all that we know about God comes from the brain. To find out how the brain mines this information, scientists use various methods. For example, to control seizures in a female patient, N.G., her corpus callosum, the highway connecting the two brain hemispheres, was cut, disrupting communication between the two sides. Later, N.G. was studied by neuropsychologist Roger Sperry. In the study, Sperry showed a cup to N.G.'s right eye, whose visual information goes to the left brain, and N.G named it. But when he showed a spoon to her left eye, whose visual information goes to the right brain, she couldn't name it. Then he mixed the spoon with other objects under a screen, and N.G. – reaching down with her left hand – selected it, but she still couldn't name it and called it a pencil.[22] The right-brain, lacking speech, could not correct the error. This same phenomenon could also account for the inability of mystics to describe their trances, or spiritual experiences, which also originate from the right side of the brain.

The two hemispheres differ in other functions as well, called *lateralization* (the tendency of some brain functions to be specialized to one side of the brain or the other). While the dominant left side is analytical and rational, asking "why," and seeking explanations and generating theories, the right side excels in processing sight and sound stimuli, spatial (space) awareness, geometry, art, creativity, and discerning meaning. The right side is also holistic, a magic maker, and it could be superstitious and unfalteringly spiritual. It's also prone to anxiety and harboring negative emotions such as depression. As discussed earlier in Chapter 10, when the right side of the brain is freed of its controlling left, it gets spiritual.

In *Aquarian Conspiracy,* Marylyn Fergusson, an enthusiast of the New Age movement, suggested that we would become more in tune with the rest of reality by learning to rely more heavily on our right brains, thereby discovering our connection with one another and the cosmos. No wonder the right side thumbs its nose at its patronizing, know-it-all left side in the battle of the hemispheres. "Who do you think casts spells?" it taunts. "Watch, soon enough you'll find out who you prostrate yourself and build colossal temples to, and it's not to spirits."

In daily life, however, one side holds sway — the left. The right then consoles itself with "coloring" events, lending them emotional texture.

As their left side functions decline, some Alzheimer's patients' musical and artistic abilities improve before they finally head south, a testament to our two selves, the speech endowed controller left and the silent spiritual right. After years of studying split brains, Sperry concluded we had two minds, a left and right.[23] He was awarded the 1981 Nobel Prize in medicine for his innovative work. But evidently, in the end, the silent magical right prevails, since belief invariably trumps reason. As David Hume said, "Reason is the slave of the passions."[24]

Now that we have a good idea which side of the brain supports religious beliefs, we find that such beliefs arise from spiritual phenomena, which people experience as a presence they interpret as God (in the West), or the Universal Force (in the East). We should be getting close, then, to unveiling the nature of supernatural powers.

Since we now know the side of the brain that supports religious beliefs, and that such beliefs are based on phenomena in which people experience a presence they interpret as God (in the West), or the Universal Force (in the East), we should be getting close to unveiling the nature of our creator.

CHAPTER 13

THE EMPEROR'S NEW CLOTHES

*I thought I had a long way to go until I looked back and
saw that I had passed my destination years before.*

~Zen saying

Evolutionary scientist Stephen Gould considered life meaningless, as the conscious mind was simply a fluke.[1] For him, everything would ultimately pass in the end: Our fate was sealed. Once our life here was completed, there was no more. And when the day of reckoning comes? That's already here, as we judge ourselves. Even though Gould may have had a point (lacking evidence to the contrary), many people strongly disagree with his view. Our rationalizing minds cannot imagine life existing for no reason, or bad behavior going unpunished.

THE REAL MEANING OF LIFE

When mystics go into a trance, they experience a world more realistic than the one we live in, and they find meaning therein. But, just like self-created, or endowed meaning, theirs is not universal and so not absolute. According to Bertrand Russell, such experiences imparted no true insight about the universe. To realize true meaning, then, we should heed Jesus' saying, "Seek and you will find." Mt 7:7. As we'll see, true meaning has a lot to do with our survival.

The Bible teaches us to devote ourselves to God totally. When asked which of all the commandments was the most important, Jesus said, "Love the Lord your God with all your heart and with all your soul and with all your mind and with all your strength." Mk 12:30. Christians love Jesus, and hope that their love for him will lead to salvation. We all yearn for love. It whisks us to seventh Heaven.[2] Whether it's the love of a significant other, children, or the unconditional love of God, it uplifts the mood and relieves depression. It pulls us through the doldrums, makes the world go around, and makes life worth living. In 2000, economists David Blanchflower and Andrew Oswald calculated its effect on mental wellbeing to be equivalent to earning $100,000 a year. Love motivates us to start a family, and to find sources of income to support it. In his video, *A Brief History of Time,* physicist Stephen Hawking is said to have had no interest in completing a Ph.D. until he fell in love and married.

Marriage bells, although denoting accomplishment, are not a prerequisite for a happy life. As we saw in Chapter 2, philosopher Baruch Spinoza never married, yet he led a satisfactory life. He said for happiness he only had to love God, whom he equated with nature or the material universe. The Stoic Seneca said, "If you want to be loved, love." Love is so cherished that we celebrate it in music, poetry, art, and so on.

As mentioned earlier, the spiritual state leading to meditation rapture arose out of the brain pathways for romantic love and sexual feelings, although the triggers can be different. So, we find that rituals, such as Tantric Yoga, which recreates sex between gods, can lead to the experience of God. And, as the god Min's phallus was worshipped in ancient Egypt, the lingam, a symbol of the god Shiva the creator, is worshipped in India. In Kundalini, the infinite divine creative energy is said to contact the finite physical sex energy at the base of the spine. And, in *The da Vinci Code,* Dan Brown notes that, since love and spirituality are intertwined, sex offers the opportunity to achieve the ultimate meditation experience and liberation in the secret Hieros Gamos (Greek for *secret marriage*) ritual.

Electrical stimulation of the brain septum (lying in front and center) generates sexual thoughts and pleasure, similar to large quantities of chemicals (acetylcholine or adrenaline), which when instilled in the area can produce even greater sexual arousal, elevated mood, and euphoria. Women seem to respond more to the stimulation with repetitive orgasms. Since quantities of chemicals are large, they could spill over onto other mid- or right-brain structures such as the hypothalamus.[3] For similar effects, some turn to drugs such as cocaine, that block removal of the love chemical dopamine, thus flooding the structures.

To explain the finding above, that women seem more responsive to brain stimulation, Dean Hamer found that they generally scored 18 percent higher than men on the spirituality – Self-Transcendence – scale. It's also likely that women had a gene variant, Hamer calls it the God gene, that predisposes one to spirituality and love.[4] Women have more orgasms during right temporal lobe seizures, too.[5] Karen Armstrong, with such seizures, would have spiritual visions, but she understood them to be neurological.[6] Stroking, psychological stimulation, drugs or meditation, as well as love, all release dopamine, which produces the pleasurable effects. But dopamine may also lead one to crave the stimulus, and, for example, lead to addiction to it.

Other creatures exempted, love is everywhere, never far from the mind. But then, we don't commonly see copulation in public, apart from lowly creatures that don't know any better. Whether it is romantic, fraternal, or the unconditional love of God, Jesus, Allah, or Bondye, love certainly makes life worth living. It gives life meaning, the Ultimate or Absolute Meaning that humanity tirelessly seeks.

But we are wired for love for a different goal. The reason it's pleasurable, or sublime, is to ensure that we seek it. It is a means to an end, that of self-preservation. Without it, many of us would be indifferent to starting a family, or wouldn't have existed at all. This way, by passing on genes promoting reproduction to offspring, the blind selection process of evolution guarantees that our *selfish genes* live on in future generations. Love, then, plays second fiddle to our mission. It is but a means to an end.

THE PURPOSE OF LIFE

Some people say that the purpose of life is a life of purpose, and others say that it's to look for purpose. Neither of these circular arguments tells us anything new. This we know: nature, like a hard-nosed CEO, selects those who can withstand its erratic conditions. Those who do are chosen and multiply, and those who cannot aren't chosen and die out. Were conditions to reverse, the "champs," now at a disadvantage, would be winnowed. Since the *selfish gene* can't predict the future, the more people who carry it, the better the chances that some will survive adverse conditions. The survivors, then - like our ancestors who suited the prevailing environment - reproduce and preserve the gene lineage. The philosopher Kant would call this drive, the instinct to reproduce that belongs to all animals, a "categorical imperative." Our evolutionary goal, then, is to produce progeny. Thwarting it is behind the acrimonious debates about when human life begins.

Celebration of a wedding, or a newborn, shows what a joyous event taking a step toward maintaining a lineage can be. The two easily top the most important events in life. For the nuptials, we pull out all the stops. Romantic Love, passion, and lust each serve as a means to this end, lest we become indifferent to mating or raising children.

Accordingly, our genital brain map area rivals that of the lips, hands, and feet combined, or that of the chest, abdomen, and back combined.

We are subconsciously so consumed by fertility that we dot our landscapes with its symbols. One, a most famous 555-foot phallic symbol inspired by the Freemasons, touches the sky as it towers over Washington, D.C. The monument is a replica of Egyptian Hatshepsut's Kernan Temple obelisk, complete with reflecting pool. Like all obelisks, it takes after the Benben, the original Egyptian stone of creation whose pyramid-shaped aluminum capstone represents god Atum's "seed," or sperm, and the shaft represents his phallus. Engraved on the monument capstone's east side, the words *Laus Deo* –Latin for *Praise be to God* –greet the rising sun. The obelisk stands at the center of a circular plaza, a dot within a circle, or a *circumpunct*, the Egyptian sun god Ra's hieroglyph. Supposedly residing inside the monument is the god's spirit, and many people still pray to it, facing east three times daily, when possible.[7]

Obelisks are highly valued. In 2005, Italy returned a 1,700-year-old stone obelisk weighing 160 tons to Axum in Ethiopia. The Ethiopians deemed it stolen and insisted on its return. It was broken into three pieces for transport by air. On the plane landing, it was accorded a stately reception and traditional dancers took to the streets. BBC NEWS, Africa, announced the event as, "Obelisk arrives back in Ethiopia."

Rome still boasts of more obelisks than Egypt, their original home. Other male member symbols include the pyramid (as a bigger Benben stone), cone, candle, tower, and spire.

According to Manly Hall, many everyday objects like bananas, flags, hotdogs, maypoles, totems, and ties are also male sexual symbols. Tented Russian churches, too, look phallic, while church entrances and pentagrams hide sex symbols. Female symbols include the Holy Grail, Ark, and Garden of Eden, and lozenge-shaped church windows are yonic symbols. In fact, nothing would remain were all sex symbols stripped from Christian Churches. In Egypt and India, the cherished Lotus flower and its twin in the West, the rose, esoterically stand for the maternal creative mystery while the Easter lily is phallic.[8] Such symbols attest to our bondage to fertility and hence the evolutionary imperative to increase our gene participation in future generations.

To preserve our genes, then, nature imbued us with love and lust and associated shenanigans. We are so bound by this duty that we structure our lives around it. Fulfilling it gives us a sense of mission accomplished, worthy of celebration. God gave us our marching orders when He told Noah to be fruitful and fill the Earth. Gn 9:7. In an essay, "The World as I See It," Einstein wrote that us mortals didn't know what our purpose was. But at the same time, he hinted at what one would call a family, saying that "...one exists for other people...for those upon whose smiles and well-being our own happiness is wholly dependent...."[9] So, as physician writer Lawrence Rifkin and others opined, reproduction *is* our purpose. It's no surprise, then, that the great evolutionist George C. Williams, bewildered by this, wondered at such a system in "...which the ultimate purpose in life is to be better than your neighbor at getting genes into future generations."[10]

GOD

Many a preacher has said that to find salvation, people must love Jesus. Otherwise, according to Paul, "If anyone does not love the Lord, let that person be cursed!" 1 Cor 16: 22. To come to God, said Saint Augustine, one must come by love and not by sail. But it's as though we're groping in the dark, trying to find the hidden and unknowable object of our love. The nuns and Buddhist monks in Andrew Newburg's studies experienced only a presence, which, according to one's faith, was interpreted by the nuns as a "mingling" with God, and by the Buddhists as a universal force.[11] A Christian might call this reality Jesus; a Muslim, Allah; and ancient cultures, a powerful spirit of nature.[12] We use such symbols to represent the sensed presence during a transcendental (spiritual) experience, or mystic state.

In Kundalini, God would be the presence, or deity, awakened by opening the Third Eye as the self within becomes "the god you truly are."[13] Such self-realization was the same as god-realization in Hinduism. In yoga, this true nature of "self" was the same nature as God.

According to Freud, the trance experience was that of a projected father figure; the image derived not only from childhood experience, but also from inherited memory images. During stress, then, internal parts of the person were projected as parent-figures and seen as external.

249

Jung called such inherited images the collective unconscious and said that they determined our archetypes, which then showed as symbols in religion and myths in abstract forms. In Christianity, these symbols and myths took shape as God.

William James, born a few decades earlier than Jung, believed God existed in inner experience. He wrote, "It is as if there were in the human consciousness a sense of reality, a feeling of objective presence, a perception of what we may call 'something there'...."[14]

As discussed earlier, studies show that this "something there" can be evoked by dissociating the brain hemispheres. According to Roger Sperry, we have two minds and so two senses of self, with the left the dominant side. In a woman who used either side at will, the left corresponded to "I," and right to "it."[15] When neurologist Michael Persinger bedecked subjects with his magnetic (God) helmet, disrupting the connection between the two sides of the brain, the left deemed the right separate, "sensing" it as a numinous presence or God. The hat also triggered intense feelings of spiritual transcendence and a sense of mystical presence in a-religious people. They experienced a savior, spirit, or universal force. One person saw Christ in a strobe light while another experienced God visiting her.[16] A Khoisan shaman, then, as in the animal dance mentioned earlier, would likely have sensed an eland.

On dissociating the hemispheres, the speech-adept analytical left side senses the space-orienting rapturous right as the "other," or "it." In such a situation, it would seem that there were two selves or two of you. Unlike the left side, which interprets the world and expresses experiences in words, the speechless right cannot remedy the error. The left, by refusing to accept the reality of two selves, then perceives the right as a spiritual reality. To borrow an idea from Freud: you externalize the presence and project a parent, God, or a spirit onto it.

It's no small wonder that, in Eastern thought, such as Vedanta Hinduism (the earliest sacred literature of India) or Abhidharma Buddhist psychology (explains how the mind works and how it can be liberated), God is shown as self. Building on Immanuel Kant, German philosopher Georg Hegel said we should get to know God as our true essential self. Ludwig Feuerbach, Hegel's pupil, went further, saying that worshipping God was worshipping the idealized self. So, we cast all the qualities we most desire for ourselves onto this elusive being we see as perfect and who supposedly created the universe.

During early Christianity, fully initiated Gnostics believed they became one with Christ. They also believed that to prevent His disciples from backsliding, Jesus said to them, "For the Son of Man is within you.[17] This idea is not unique to Christianity, and has found expression in religions from around the world. For example, many believe that just as Abdul Hamid al-Ghazali found God within himself through a mystical experience, so can we. Further East, Brahman (the ultimate power underlying the universe) and atman (the human soul) are seen as one. And in Kundalini, by opening the Third Eye, the sensed presence or the self within becomes "the god you truly are."

Then God, the "sensed presence" or your "other self," is no further away than in your right-brain. As the Buddhist epigraph above bemoans, it was apparent long ago that God is within us. Indeed, quoting Psalm 82:6 at the Feast of Dedication in Jerusalem, Jesus asked unbelieving Jews, "Is it not written in your law, 'I have said *you are gods?*'" Jn 10:34.

NO, IT CAN'T BE TRUE!

When you look in a mirror, whom do you see? Yourself of course: the face of God. As discussed earlier, the brain tricks us by externalizing from the right hemisphere the sensed presence, which we then interpret as the supernatural power that created the world and watches over us. Like us, people in the legendary Prophet Enoch's time interpreted the presence likewise, but projected it onto the sun. Thus - as in the text from the *Slavonic Enoch,* a manuscript by Francis Anderson - while on the seventh heaven, Enoch saw God's face from afar on the tenth Heaven, and it was strong and very glorious and terrible. "Who is to give an account of the dimensions of the being of the face of the Lord (big), strong (sizzling hot), and very [very] terrible?" In the longer edition of the text from the *J* manuscript by Andrei Orlov, the face was like "iron made burning hot in a fire and it emits sparks and is incandescent." Yet it was "so marvelous and supremely awesome and supremely frightening." But then he added that God was incomprehensible and His face extremely strange and indescribable.[18] Despite armies of fiery angels paying their obeisance to the "Lord," withdrawing in "joy and merriment," Enoch was frightened out of his wits by what he saw. He shook like a leaf. Since in the Bible we were created in God's image, it's not clear who Enoch saw. We are not fiery like the sun.

The idea of a god within, the idea that we're gods, is not new. Buddha is quoted as saying, "You are each a God." Plato also thought humans were divinities, while the Hermetics argued, "Know ye not that ye are gods?" And the lesser-known Tsunyota Kohe't "Eagle Spirit Ministry" regarded the "sensed presence" as the "God" within. Kohe't called God your higher or idealized self. Indeed, the Bible traces our ancestry from God through Adam, Enoch, and Noah. Lk 3:38. Small wonder, then, that Gnostic initiates were taught to experience God as their true selves. Science, the most reliable discipline and the one closest to the truth, supports this idea in a way.

Up until modern times, many prominent figures called themselves gods and were revered. Pharaohs and Roman emperors built themselves temples, keeping mum that everyone else was also a god. People never wised up when Ramses the Great worshipped himself, and Ramses III was called the Great God. Besides Alexander the Great, the Ptolemys, the Caesars, Emperor Constantine, cult leaders, gurus, and others, elected officials included, also anointed themselves gods. In Uganda, people believed kings were divine. Criticizing them was blasphemy and could prove to be very unhealthy. Even the sick, such as those with manic disorders or those with right temple brain lobe epilepsy may call themselves gods. So, instead of experiencing a presence or savior like most of us do, Dr. Ramachandran's patient with right-brain seizures, John, was himself God, showing that the right temple lobe was the source of the illusion.

The apostle Paul taught that God was not external and needed no "middleman" for us to communicate with him. "We are his offspring," he said, quoting Athenian poets. Acts 17:28. Similar views have aired throughout history. Gnostics, believing Jesus, or Christ, will be found within each of us, considered the Jesus story a fable representing secret mystical teachings and equated Jesus with the human spirit. Zen teaches likewise: "If you wish to seek the Buddha, you ought to see into your own nature, for this is the Buddha nature itself." It follows, then, that when we worship and pamper God to evoke the spiritual state, in reality, we're pampering ourselves. Godfearing, then, would mean obeying societal norms and staying in the good graces of our fellow human beings, who would reprimand or sanction us for bad behavior. Since all gods are based in the same inner reality, competition and hostilities among religions are regrettable. The ancients who worshipped each other's gods, then, were correct to consider all of them equal.

Since what people perceive as an external all-present and all-knowing power is really themselves, denying the existence of God would be absurd. It is the equivalent of denying that they themselves exist, for belief in God is belief in oneself.

HIS MAJESTY SCAMMED

Gods – as conceived of in our human minds – are only but symbols of our spirituality. Because gods are products of our brains, they couldn't have created the world. The forces of nature—or the Universal Force that the East evokes—get credit for that, as Stephen Hawking also believed. So, we don't see God in the sky. Like the astounded small boy, who on seeing the emperor in his birthday suit, cried out, "But he isn't wearing anything at all!" it doesn't mean that we are incompetent or fools for failing to see God in the sky, despite what the sham tailors, who bamboozled His Highness, had said spectators would be if they couldn't see the emperor's new fine clothes.[19]

The feeling that God watches over us all the time comes from the perpetual "sensed" presence. Some yogis and secret societies call it the Third Eye, or the All-Seeing Eye. Others refer to it as our Guardian Angel. Apparently, the dominant talkative left brain – the "self" or "I" – senses the speechless right side floating outside the body, obligating it to praise and worship it and prostrate itself, practically enslaving it. When stimulated by art, music, beauty, love, mating, or profound relaxation, the right brain makes its presence known, literally.

For over two thousand years, physical union with a female has provided a path for the male to experience God in various cultures. Tantric Yoga and secret societies' rituals like Hieros Gamos were the means by which a man could become spiritually complete and achieve knowledge of the divine. "By communing with woman, man could achieve a climatic instant when his mind went totally blank and he could see God." Meditation gurus realized similar enlightenment and called it Nirvana, a never-ending spiritual orgasm.[20] So, according to Dan Brown, sexual congress was inexplicably and profoundly intertwined with our spirit, leading to oneness. As Andrew Newberg found, mystics described their ineffable experiences using the same expressive terms such as bliss, rapture, ecstasy, and exaltation. This was because the neurobiology of a spiritual experience arose from the same mechanism as the sexual response. At the beginning of his book, *"Why God Won't Go Away,"* Newberg wondered whether evolutionary factors were involved in such a development. Indeed, they were.[21] Love and our purpose evolved to spread the *selfish gene*.

Ages-old religious wrangles, hopefully, can now be put to rest. In much the same way that Einstein disposed of theories about luminiferous aether, such debates about which religion is the true one, or which people are the chosen ones, are just so much ado about nothing; they are based on erroneous assumptions and can do no more than rationalize an illusion. Because of the biological nature of the spiritual experience, all of us have the talent to invoke it, although to varying degrees. Those with a strong mystical talent could achieve the upper stages of an experience, and would be strong believers in supernatural powers, even with the knowledge that such experiences were illusory – just a brain construct. Others without such a strong inclination to trigger the phenomenon would have such a mild response that they might deny it and look askance at those who report it and adamantly believe in the existence of such powers.

Also, unless meant figuratively, miracles don't happen. One won't recover from a "terminal" illness on command. Beware of wonder peddlers. The blind will not see simply by rubbing dirt on their eyes. Jn 9:6-7. And once we die, that chapter of our life closes as well. Imagine you were cremated, or taken by hyenas. What would be left for the afterlife?

What are we to make, then, of predictions like the following, from Mk 13:25? During Jesus's second coming, "the stars will be falling from heaven and the heavenly bodies will be shaken from the sky." In the ensuing battle, Jesus will annihilate the forces of evil and banish death from the earth, and henceforth, no one will die - ever. Knowing that His *Return* refers to the coming of the next zodiac Age of Aquarius, which is what Jesus was talking about when He mentioned the man with a pitcher of water, the sign of Aquarius, it makes sense that—as with all risen saviors before him—the resurrection of Jesus was meant to symbolize the return of the sun and revival of food crops to the

Northern Hemisphere that would save people from the winter cold and food shortages.

Nevertheless, basing his teaching on the story of Jesus's Ascension, Paul rebuked those who doubted resurrections. He said if people could not be resurrected, and if Jesus was not resurrected, then his message "was empty indeed." 1 Cor 15:12-17. Paul would likely be disappointed to learn what we now know for certain: as mummies in the Cairo Museum and cremation show us, bodies of the deceased remain here on Earth.

WHAT IS CHRISTIANITY?

Since Jesus appears to be mythical, as discussed in Chapter 8, is our trust in Him as the son of God misplaced? Many teachings and deeds attributed to Him resemble Gnostic texts – the Gospel of Thomas, for instance. Others seem to have originated with Greek philosophers such as the Cynics (Citizens of the World, as they preferred to be called), Platonists, and Skeptics. The saying, "Such as you wish your neighbor to be to you, such also be to your neighbor" comes from Sextus Empiricus, a third century CE Greek physician skeptic.[22] Using the same metaphors and language, Paul borrows copiously from Plato. For example, he repeats Socrates's saying in Phaedrus that the heavenly ideals are perceived as though "through a glass dimly." 1 Cor 13:12. So, in *The Jesus Mysteries,* Clement of Alexandria bills the gospels as perfected Platonism.[23] It seems that Paul's church was modeled after Plato's ideal city-state.[24]

Similarities between Christianity and the philosophy of the likes of Heraclitus, Socrates, and other Greek philosophers were so striking that church apologist Justin Martyr acknowledged that such philosophers were Christians before Christ.[25] As mentioned earlier, the Jewish philosopher Philo, also a mathematician and follower of Pythagoras, who, together with the Flavians and Josephus crafted the New Testament, is credited with authoring John's Gospel; Philo's writing is characterized by its scholarly tone, lack of proverbs or parables, and its portrayals of Jesus as God.

Pythagoras's teachings reached Plato through Socrates, and ultimately reached Christianity through the Platonists, Cynics, and Stoics. For example, once when someone kicked Socrates, he did nothing. To the astonished onlooker, Socrates merely asked, "If an ass did kick me, still would it have been in order to take it to court?"[26] Following Socrates's lead, Jesus also stressed forgiveness, saying. "If someone strikes you on one cheek, offer the other one as well." Lk 6: 29; Mt 5:39. According to Joseph Atwell, this was supposed to pacify the Sicarii, and stop their rebellions against the Romans.[27] Regarding leadership, Jesus's observation that "many are called, few are chosen" (Mt 20:16 KJV) came directly from Plato, who – in referring to the Greek Eleusinian Mysteries in *Phaedra* – wrote, "Many carry the wand [are called], but few become Bacchoi [are chosen]." Borrowing yet again from Plato, Jesus rephrased the maxim: "It is impossible for an exceptionally good man to be exceptionally rich."[28] His more relatable version: "It is easier for a camel to go through the eye of a needle than for a rich man to enter the kingdom of God." Mk 10:23-24. This would surely have been welcomed by the less privileged folk who comprised the Christian base.

Like the Cynics, Jesus treated everyone the same, including outcasts and "gentiles." He wandered from place to place, preaching, embracing a life of poverty. He defied authority in the name of God. His followers abandoned their possessions and roamed the countryside, begging for food and shelter, surviving on donations and leading ascetic lives like Diogenes, the best-known cynic.[29] Diogenes lived in a doghouse (cynic means *dog-like* in Greek), trading comfort for a humble but happy life. When Alexander the Great asked him whether there was anything that he could do for him, Diogenes requested the emperor to stand out of his sunlight. The emperor replied that if he weren't Alexander, he would choose to be Diogenes.[30] As did other Cynics, Diogenes anticipated a better life in heaven. Following the Cynic's chosen lifestyle, Paul urged Christians to be humble and accept suffering and self-sacrifice. Phil 2: 1-11. Indeed, in the first century CE, the two were so similar that one could mistake a Christian preacher for a Cynic.

Most of Jesus's teachings and actions, however, reflected Egyptian influence, as was revealed after Jean Champollion deciphered the hieroglyphics in the nineteenth century. Jesus's theology, deeds, and wisdom, then, had been around for millennia. For example, the affirmation, "Amen," was the Egyptian chief god's name, and the idea of the Last Judgment came from the *Book of the Dead*. Gerald Massey wrote, "… the Canonical Gospels are only a later literalized réchauffé (old reworked or rehashed material) of the Egyptian writings...."[31] Gospel history had first existed as Egyptian mythology, and the confessionals, fasts, and penance required to enter the church were originally Egyptian, too. The Archangel of Rome was the imitation of Anubis, the god of mummification and the afterlife, and rites during the "Holy Week" resembled the festival of burning lamps.[32]

According to Flinders Petrie, the earliest Egyptian Osiris religions originated in Central Africa. There, animal worship was prevalent even before 10,000 BCE.[33] This could explain Jesus' remark in the Coptic Gospel of Thomas, "Blessed is the lion that the human being will devour so that the lion becomes human. And cursed is the human being that the lion devours; and the lion becomes man."[34]

Our ancestors experienced their spirits merging with animal spirits during a trance. Such encounters seemed so real that people ritually ate animal flesh and drank animal blood to unite with them and take on their power.

Likewise, to become one with the dead, people from ancient Central Africa and Egypt practiced cannibalism. In Egypt, a king considered too old and feeble to lead an army into battle was killed and people consumed his flesh to take on his qualities. Jesus's parallel teaching, "He who will drink from my mouth will become like me. I myself shall become he...,"[35] is the gist of the Holy Communion.

Ancient rock art depicting animal worship across the continents shows such rituals were universal, making religions everywhere related. The people practicing them came from small groups that migrated from Africa and spread throughout the world, carrying their beliefs with them. In Christianity, "paying homage to the saints" mirrors the African worship of dead ancestors.[36] Also, although we do not venerate animals openly as our ancestors did, we honor them in emblems and symbols, especially the lion and dragon (probably representing dinosaurs, whose fossils we see in museums today). People also worship spirits like the Holy Ghost, whose nature is elusive. What is it exactly? Correctly translated from old Greek, the Holy Spirit means "holy breath."

Chaney and Messick trace the beginning date of much of what is labeled *Christianity* to the beginning of humanity. "Jesus' ministry," they write, "had existed since time immemorial through Priest-King Zoster (Djoser) in Egypt, Lord Krishna in India and Plumed Serpent

(Quetzalcoatl) in Central America. It only began to be labelled 'Christianity' with the coming of the great master Jesus Christ."[37]

To build a strong army and maintain his power, Constantine courted several religions for recruits. Like Ptolemy I and Simon Magus, who had joined sundry religions into one, Constantine merged Christianity's sects into a single universal product, and made Christianity the religion of the Roman Empire with himself at the head as god and savior. In 391 CE, Emperor Theodosius I issued an edict to close down all pagan temples. An imperial decree demanded all books at odds with Christianity must be burned. Alexandrian-born Hypatia, a woman mathematician, philosopher, and astronomer, was torn limb from limb in 415 CE, and Archbishop Cyril, who incited the murder, was made a saint.

Despite Theodosius eradicating links between Christianity and earlier Osiris-Isis cults, strong similarities remained. Like Horus's and Jesus's immaculate conceptions, that of Krishna was also without "sin." Krishna, too, died on a cross like Jesus, and darkness fell at noon. Just as Mary nurses Jesus in her arms in Christian iconography, Nari nurses Krishna, and Isis nurses Horus. Madonna and child statuettes are central icons for both the Catholic and Orthodox churches.

THE FIRST POPE

Simon Magus, the magician, was a thorn in the side of the early Christian Church. Simon was born in Samaria, of Cushite ancestry, like the Canaanites and Phoenicians. He studied in Alexandria, Egypt, where he met the Jewish Philosopher Philo, considered by nineteenth century German theology professor and historian Bruno Bauer, among others, to be the true founder of Christianity.[38] Influenced by his Babylonian background, Simon blended Babylonian, Egyptian, Greek, and Christian beliefs into a universal religion called One Faith, and he was known as the Messiah to his followers. As Jesus rescued Mary Magdalene from prostitution, Simon liberated his companion, Helena, and designated her a female Messiah. He amazed the people of Samaria with his sorcery and they exclaimed, "This man is rightly called the Great Power of God." Acts 8:10.

The Bible singles out Simon for scorn, the reason we hardly find anything flattering written about him. Instead, we have diatribes. To Irenaeus, Simon Magus was the Anti-Christ, the father of all heresies.

Simon Peter severely rebuked him after he offered to buy the power to bestow salvation (called the first heresy and the source of the word "simony"). Acts 8:9-25. To show he was more powerful than Jesus was, Magus levitated and flew over Rome. An indignant Simon Peter called on Jesus to smite him in mid-air, whereupon the "phony" savior fell, breaking his leg in three places. Despite his hostility toward Magus, Peter prayed to God to let Magus live. But having been "sorely cut" during treatment, Magus succumbed.[39]

Such absurd stories show how anathema Simon Magus was to the Church, and they support the scholarly impression that such views canonized in the book of Acts were forged to promote an historical Jesus and demonize Magus. Justin Martyr living in the early second century CE had not even heard of the Acts. But by the close of that century, Irenaeus and Tertullian were lauding them for upholding the doctrine of a factual Jesus. Acts countered the "heretical" Gnostics and altered Paul's message so it agreed with church doctrine. Obviously, Acts was invented, since they misquote the Hebrew Bible, being unfamiliar with it. The Acts' account of Simon's death, however, was nonsense, as Simon survived until Nero's time. The attacks speak to the church fathers' dread of his "false" religion. Judging by the rants and denials online today, the vilification of Simon Magus continues.

Like Paul and the Gnostics, Simon Magus's Jesus was not a person but was, rather, a spirit. His universal religion became the Catholic Church ("catholic" means "universal"). Simon Magus combined Serapis Christus with the dying and resurrecting Mystery gods like Nimrod (Tammuz), Dionysus, and Mithra, to create a mixed religion. No wonder the Roman Catholic Church took exception. The record shows, however, that the Roman Catholic Church succeeded in demonizing Simon Magus, all but erasing his name from history. Nevertheless, it failed miserably to quash his "false" but wildly popular religion and it took off big time.

THE CHURCH

Not Simon Magus, but Simon also called Peter was invented, just like Jesus and the rest of his apostles. John's gospel hardly mentions any apostles. Only about nine Hebrew disciples can be named. Andrew and Phillip were Greek names, and Thomas was Aramaic. Their number matched that of the "apostles" of other gods like Mithra and

Dionysus, who surrounded themselves with twelve satellites or signs of the Zodiac respectively.[40]

For the mathematical Pythagoreans, twelve was the exact number of touching spheres encircling one of equal size. The idea, according to Freke and Gandy, originated in Egypt, where it symbolized Osiris "as the spiritual center of the turning wheel of change represented by the 12 signs." "They were the reapers in the plains of Neter-Kar," Massey wrote.[41] If the apostles were but symbols of the twelve Zodiac constellations the sun traveled about with, who then was the first Bishop – or Pope – of Rome?

PETER THE ROCK

In ancient religions such as Mithraism, the name Peter and its variants – Pator, or Patre –represented a sacred rock, gods, or their priests. Peters were the interpreters of oracles in Egypt. Later, the name came to signify father or parent, and in Greece, where he was the chief priest of the family, the father became known as patriarch. Gods were also called Peters, or divine interpreters, in ancient Rome. According to Ewart Gladstone in *Homer and the Homeric Age I*, the Roman supreme god Jupiter and the Greek god Zeus were the same. Ju-Peter was the Roman way of saying Zeus-Peter.[42]

In Egypt, Amon's fertility priests, called Peters, held idols, poles, or stakes known as Pietaurums, which originated in the Benben stone of creation, representing the male member. Often shown standing by this male masculinity symbol or stone pillar was Artemis, the Greek goddess of childbirth and fertility. Nevertheless, Artemis was strictly a virgin. Pillars, then, and such other phallic symbols and obelisks, were known as Petras, or the sacred Peters, still the name for the phallus in some quarters. There was hardly a temple without a Peter in ancient times. By and by, the name came to mean a large rock, then a small rock, or Petros.

Also, temples were known as Peters, including the legendary Tower of Babel, Bel or Ninus, and were credited to mythical Nimrod. Like those of the sun gods Osiris, Baal, and Mithra, Nimrod's shrines were called Peters in the East. By each land having a chief oracle or Peter temple, when Rome became the religious center, Saint Peter's Basilica became the world's chief temple. The gatekeeper's job then went to the Peter-god, Janus (or gates), who was *The Keeper of the*

Gates of Heaven and Earth" to the ancient Romans. Janus, a Peter, was represented with a key in one hand centuries before the legend of the apostle Peter surfaced. According to the church, Janus, as Peter, gave to it (the church) the keys to the gates of heaven and no one could enter into God's presence unless the church opened the gates. The cardinals were the hinges upon which the gate (the Pope) was able to turn. The square's key shape of St. Peter's Basilica is based on this legend.[43]

Simon Magus, also called Pater or Peter ("priest" or "god" in Chaldean), and often confused with Paul, established his mixed religion in Rome, strongly influencing the Church. Orthodox priests called Simon a pretender and declared his religion false for its worship of idols. Despite all of this, Simon Magus was popular, the darling of emperors Claudius I and Nero. Historical evidence strongly suggests that it was Simon Magus, then, who spent the twenty-five years in Rome, not Simon Peter. Ignoring fabricated scriptures, there is no evidence a disciple named Peter ever lived in Rome. If he had, church fathers Martyr and Irenaeus would have known about it. Christianity, then, was not founded by anyone named Peter. Peter was simply a commonly understood name for gods, priests, temples, and so on.

Simon Magus's universal religion infiltrated the young Christian movement with its idolatry and "paganism," and took it over to form what ultimately became the Catholic Church.[44] Simon Magus, the heretic Samaritan, then, became the first Bishop of Rome and was deified by the Roman Senate. Although his pensive black statue sits in St. Peter's Basilica, Magus's name isn't inscribed all over Rome as fictitious Peter's is. In his honor, Claudius Caesar erected a statue in the Tiber in the likeness of Jupiter and inscribed it, "The Holy God Simon." The statue is now gone.[45]

A golden coffer in Saint Peter's Basilica in which Peter's bones supposedly rest contains only palliums, sashes the Pope gives new cardinals. Peter's "actual" remains – nine pieces of bone – should be in a bronze display case on the side of the altar, inside a jeweled box that Pope Francis showed the public at an open-air mass in St. Peter's Square for the first time ever in 2013. These bones, however, were in fact recovered, among many others – some belonging to "small" animals – in 1939 from grottoes underneath the altar. The bones have never been identified. Part of Peter's skull has supposedly rested in Saint John Lateran's Cathedral for over 1,000 years. Twentieth century attempts to locate and identify Peter's authentic remains were unsuccessful. The land where the Vatican now sprawls once belonged to the head of the Mithraic church, *Pater Patrum* ("Father of the Fathers,") or simply *Papa* ("Father"). Today we call his successor "the Pope." Lacking convincing proof to the contrary, it is reasonable to conclude that – like the gods Janus and Mithra, who once reportedly held the keys to the gates of heaven – Saint Peter wasn't a historical figure.

Besides the Catholic Church, Simon Magus was also considered the founder of Gnosticism. Many writings which were likely his appear to have been wrongly credited to Paul.[46] Gnosticism, however, probably originated with a first century CE Samaritan, Dositheos, who was reputed to have been a teacher of Simon Magus. As Freke and Gandy submit, Gnostics could have been the first Christians.[47]

THE SAVIOR

Throughout the book, we have encountered not only Jesus Christ (the two names combined literally mean *anointed savior*), but many other saviors and Christs as well – deities such as Horus from Egypt, Krishna from India, and Quetzalcoatl from Mesoamerica, all worshipped by their respective cultures and predating the New Testament Jesus story. As argued in chapter 11, spirituality, or rituals – like religion – that may evoke it, decidedly prolongs life, making religion a human need that promotes good health.

Since evolution winnowed out people incapable of spiritual transcendence, this benefit of worshipping may have dawned on our ancestors, although they assumed that it came from their gods, who they then regarded as *saviors*. Whereas God in the Old Testament saved the Jews from catastrophes, in the New Testament Jesus saved Christians from their sins. His followers believed He died on the cross to pay for their sins and to save them from the devil and death so they might live with God for eternity. They also believed that those who trusted in Jesus will be forgiven Adam and Eve's cardinal sin, which they had shouldered since the couple disobeyed God's command in the Garden of Eden. They believed they'd be forgiven the sin by trusting in Christ.

Our forebears could not know of the physiological changes mystical transcendence produced in the brain that helped to promote good health. Nonetheless, they outlived those without the talent to evoke spiritual experiences.[48] By living longer, they passed their "God" genes to future generations, while those without the "God" genes – those who lacked a modicum of superstition, so that spirituality, their true *Savior*, couldn't redeem them – died out.

Where do we turn for guidance and spiritual sustenance when our confidence in a supreme being or protective power has been shaken? Do not despair. Alternatives are available. As always, to live well, we could learn from one another, with Richard Dawkins's *selfish gene* in the driver's seat. Or we could summon the "other," evoke self-realization, for spiritual nourishment through meditation and related practices. According to Andrew Newberg, "As long as our brains are arranged the way they are...spirituality will continue to shape the human experience," and God will not go away.[49] Indeed, spirituality helped our ancestors live longer and multiply by controlling stress, depression, infections, and other diseases.

Regardless of the method we adopt, learning and engaging in ethical conduct is up to each one of us, as we monitor ourselves. For correct conduct, morals are inborn, and, as we saw earlier, even babies have a sense for them.[50] David Hume believed that morality was more of a feeling than a judgment. If it is correct that we can instinctively tell good from evil, it's reasonable to think that we also would – barring any competing interests – and as cultures worldwide do, instinctively follow evolution's Golden Rule as Jesus recited it to His disciples. *See* Mt 7:12; Lk 6:31. Other rules and values are mere social conventions, customs that may vary by region. It seems, to me at least, that we go through life mainly by trial and error, learning from our mistakes.

Greek Sophists taught their pupils to make the best of any case, regardless of their own convictions.[51] In other words, truth was what you made it. Such a shifty attitude got the Sophists in trouble and earned them Plato's contempt. But, like Niccolo Machiavelli, Sophists were realists. Five or so centuries after the sophists, in *The Prince,* Machiavelli gave politicians and rulers tips on how to survive in power. His advice was that it was better to be feared than loved; better to be miserly than generous; better to be untrustworthy than honest. This is an approach some people embrace anew today, particularly in politics and the media. Machiavelli advocated for prioritizing cunning and fighting like a lion over compliance and compassion. Not surprisingly,

Machiavelli's ideas riled the church. Upstaged, the church censored him. Many leaders also openly trashed his book, although they devoured it in private. That's politics and human nature for you.

In the *Republic,* Plato disparaged democracy, or government by the people, pointing out that politicians would always avoid making difficult choices, kowtow to deep pocket donors, and pander for votes. Rarely is a politician financially independent and direct, even brash, like Donald Trump. Plato ranked democracy just a step away from tyranny. He preferred a handpicked philosopher king, an extremely intelligent and knowledgeable person – someone like himself – or a pharaoh (or king). Anyway, democracy can be a quasi-mirage. It implies freedom of choice but doesn't really offer it, and what little benefit it has, when it works, is characterized as fairness. In actual practice, it may be dysfunctional, sometimes devolving into a circus, as egos and grandstanding get in the way of attending to the people's business, with open voting potentially an invitation to intimidation and fear of retaliation. Plato could have added diplomacy to the mix, as it may delude participants with half-truths and outright lies.

Back to Hume for another moment. Hume said that true reality eluded us, that we couldn't know anything with certainty, and – since our convictions were intuitive, we could only deal in probabilities.[52] The brain, however, evolved so that certain events and practices produced beneficial transcendent experiences. For the solace, comfort, and good health that came from such experiences, we formalized practices and rituals that evoked them, organized them, and called it religion. Today, to induce such experiences, people may go to a concert, visit a park or an art museum, or attend church. For the latter, it could be any place of worship. Atheists (yes even Richard Dawkins is spiritual), agnostics or Hindus could go to a mosque and Jews to a church, or to a shamanic ritual. Those unaffiliated with any religion could visit a non-denominational church such as "life coach" Joel Osteen's Lakewood Church, with its simple universal message of hope and coaching to help people lead healthier, happier, more productive lives.[53] Meditation also helps. It could be practiced in a class or individually at home following simple instructions.[54] Belief in something numinous could embolden us to face challenges. The bigger the challenge, the greater the emboldening effect. We regularly see sports figures, politicians, combatants, and other victors attribute their successes to a higher power.

As explained earlier, nature selected those with the talent to evoke mystical experiences, as they tended to have better health and lived longer, thus improving their chances to procreate and pass along their spirituality-prone genes to future generations. In the final analysis, we find that *love* and *spirituality* are intertwined, a manipulative tactic by the *selfish gene* to pass on an independent mechanistic survival process into future generations.[55]

In Kabbalist numerology, since love equals one equals God, to be one with God, love. Pope Benedict XVI cut no corners, when he said, "Love is God." Love, then, as in Tantrism, is divine.[56] But, as already mentioned, love is a trick by nature to reinforce our desire for mating and so fulfill our evolutionary purpose to multiply, resulting in the survival of our species.

KARMA

Since Jesus's ascension, each generation has awaited His return. Despite His promise of the Second Coming during His generation, two thousand years and counting, we're still waiting. Mt 24:27-34. Forty years after the crucifixion, Titus Flavius did claim that he was the returned Messiah, as promised. The Second Coming, however, was not about the return of Jesus, but rather, as mentioned earlier, the coming of the Age of Aquarius, which astrologists anticipated would occur between 1885 and 2500. Titus had appropriated Jesus's story to convince the Sicarii that he was the Messiah they were waiting for.[57]

People also braced themselves for Armageddon, the final battle between the forces of good and evil. But the red dragon with seven heads, seven crowns, and ten horns still languishes in the abyss. As the story of the Apocalypse originated from Persian Mithraism, the promises of calamities as detailed in the Book of Revelation are much older than the story in the New Testament.[58] The prophets had no inkling that the real end of the world would come when, in its death throes, the sun swelled and engulfed our tiny planet in six billion years.

A true definition of life still eludes us. Without it, a definition of death is also elusory. We likely simply recycle through nature, a kind of reincarnation. Nonetheless, as the universe slowly loses steam and winds down over the next one hundred billion years, it will grow cold and dark and disintegrate, except for our hardy protons. Protons will tour the heavens for decillions of years more, extending our lifespans by orders of magnitude, virtually for eons, before decay and heat death set in, in roughly one googol year. That's a 1 followed by 100 zeroes.

Even though our minds will be long gone by then, according to karma – the never-ending chain of cause and effect – we steer our foreseeable destiny, our individual fate, by our actions. Since life is a fluke, our destiny cannot have been predetermined by a planner, but is dependent on the quirks of nature. According to the second law of thermodynamics, the universe is relentlessly decaying, increasing in disorder, called *entropy*. Eventually, as life runs out of energy, it will become disorganized, and disintegrate together with the universe. So, our final fate is predetermined, as it's tied to that of the Universe.

Meanwhile, we should endeavor to live well as we strive to achieve our full potential, circumstances permitting. We need to protect our environment so it takes care of us in turn. We share the planet with approximately 8.7 million species (could be as many as 2 trillion), of which we have identified only 1.3 million. Though, in an extinction unparalleled since the disappearance of the dinosaurs, we humans destroy up to 156 species a day. We cut down trees, pollute rivers, and pump carbon dioxide into the atmosphere, which contributes to rising global temperatures. Already, global warming is wreaking havoc on earth's poles, where glaciers are melting. Half of all species could have disappeared by 2050.

For us to live comfortably, good health and renewable energy supplies are crucial. Our natural resources of coal, gas, and oil – resources that took millions of years to form – are being depleted at an alarming rate. According to British Petroleum estimates in 2013, we could run out of known recoverable oil reserves in fifty-three years. But these reserves are those from conventional sources, so called "elephant fields," which are large reservoirs that are cheap and easy to operate. With modern technology, however, other sources, such as oil sands or fracking, have come to the fore. As they are expensive to operate, the higher price of their oil could stop us from using it, forcing us to turn to alternative energy sources such as wind power, solar power, or hydroelectric. Also, conservation and electric vehicles are bound to decrease our dependency on fossil fuels.[59]

But how do we avoid laying waste to Mother Earth, and preserve it for our grandchildren, when it is so difficult to curb population growth and human impact in many parts of the world? Obviously, natural catastrophes such as glaciation, earthquakes, extreme weather, and wildfires are beyond our control. A killer asteroid strike is also always at the back of our minds. We need a global defense plan against this potential catastrophic threat. Various agencies, such as NEAT (Near-earth Asteroid Tracking) by NASA, monitor such asteroids regularly, considering nudging them away from us were we in their crosshairs. The moon was part of Earth before an asteroid knocked a chunk out of it into space. Neither did the dinosaurs survive an impact despite their size. In September 2022, a spacecraft, DART (Double Asteroid Redirection Test) was launched and crashed into an asteroid moon to gauge if the impact could knock off course a rock headed our way. Even if worst comes to worst, however, someone usually survives such cataclysms and repopulates the planet. Still, our species can survive only for so long. If we do not succumb to a deep freeze, a broiling sun may eventually hasten our souls to the "next life."

No need to panic, however. According to *chaos theory*, weather, as the most typical chaotic system, displays irregularity. Yet, it remains predictable within a broad range (a strange-chaotic- attractor or the state to which a mathematical field of dynamical systems, such as climate, unpredictably eventually settles). Short-term weather predictions have about a million variables, with more developing all the time. Climate is so sensitive to minute agitations that eddies from Mount Ruwenzori in Africa may spawn hurricanes in the US. At a Washington conference in 1972, chaos theory pioneer, Edward Lorenz asked, "Does the flap of a butterfly's wings in Brazil set off a tornado in Texas?"[60] Lorenz merely asked the question. He gave no clear answer. That butterfly flapping its wings seems just as likely to prevent a tornado.

Short-term weather predictions have to account for about a million variables, and more are developing all the time. Not surprisingly, sometimes forecasters get their predictions wrong. For long-term forecasts, however, like the weather in the next century, they look for the shape of the whole climatic attractor. Long-term weather patterns tend to repeat, maybe once, twice, or a thousand times. Being cyclical, since hot periods alternate with glaciation over thousands or millions of years, a series of warm winters or hot summers do not necessarily mean long-term permanent change has set in, and they do not mean a development such as global warming will arise. But if we fail to control greenhouse gas emissions and the temperature rises, say, by 2 to 10 degrees over the next 200 to 300 years, the frigid zones today might then beckon, I reckon. People tend to follow the snow line as glaciers melt, mainly in the north where land masses are larger.

The next glaciation period should be due soon – in approximately two millennia – though global warming could delay it. Much of Earth will become uninhabitable then. New York City skyscrapers will be ground into the Atlantic Ocean, and the European population driven south as in ice ages past. As water becomes tied up in ice, rainfall will decrease, leading to drought and food shortages. Competition for food supplies could dwarf today's oil wars. To avoid the advancing ice, the tropics may become the prime destination. Before all that, though, a mindless nuclear war could wipe us all out, or a virus could annihilate us. Ebola, Nile Fever, or the Corona Virus (Covid-19) will be minor by comparison. Potential disaster lurks below the surface, as well. Earth's interior will run out of nuclear fuel in 2 billion years, dropping temperatures precipitously. By the time the sun burns itself out, our planet will already be lifeless like Mars.

Even though the world population is growing by 82 million a year, birth rates are still falling thanks to birth control. Growth should level off at around nine billion by 2050, despite 98 percent of births – mostly in developing countries – being unplanned. That's still a lot of people. Based on desirable carbon dioxide emissions, food supplies, and energy resources, this number far exceeds Earth's optimum population levels of 1.5 to 2 billion. To slow and eventually reverse this overload, couples should have a maximum of two children.[61] But, a large population doesn't necessarily spell disaster. In the twentieth century, the world human population quadrupled, but food shortages and turmoil did not arise as Robert Malthus predicted. With mechanized high-yield farming and equitable distribution of technology and produce, food supplies could keep pace.

So here we are. A bonafide theory of everything still eludes scientists; Stephen Hawking died in 2018 without making headway. Without a clue to the true nature of reality, figuring it out is very much like a dog chasing its tail. Even more puzzling is the big question of

"What was here before the universe?" Confounded by true reality, like Leibniz, scientists don't know where to start looking for answers. Mystical reality, like a dream, is only a brain construct. Most likely, in the beginning, space consisted of energy only. Currently, the energy is divided into dark energy, dark matter, and a small amount of normal matter made from fermion particles. Einstein, however, showed that normal matter could also convert to energy. The question, then, is where did the energy come from? This is like asking, "When did time begin?"

Many people are also still equally in the dark about the true nature of our beliefs and thoughts; how can we tell if they are real and not illusions or – as Marcus Aurelius intimates in the epigraph at the introduction of this book – mere opinions and perspectives? Many people believe in Heaven and an afterlife, but not nearly as many take seriously the saying attributed to Jesus that the Kingdom of God is within us. Nevertheless, our search here suggests this to be figuratively true.

Given all that we've learned, the face of god is not like what Enoch saw, the God of the Bible. People wouldn't crave spending an eternity with Enoch's God, whose face shines brighter than the sun, glows hot, and emits sparks.[62] Despite insight into the true nature of god, yourself, and so the whereabouts of the kingdom of God, you needn't equivocate in your faith. All of us benefit from awakening the sensed presence; it makes life meaningful and promotes good health and survival. If you evoke the presence through religion, spiritual practice, or by other means, you're spot-on.

For a meaningful, satisfactory life, then, besides aspiring to a life of value, cultivate living in the HERE and NOW, and – under the watchful eye of your "Other" – love well.

HENRY KAKEMBO, M.D.

APPENDIX

ILLUSTRATIONS

Images in this appendix are arranged in order of their appearance in the text and are numbered to correspond with the chapters where they are referenced.

Illustration 3.1

Lindsay-Shapely Ring galaxy. Furious star birth gives the ring a bright blue color and the regions around it a pink color.

Image Credit: Hubble Heritage Team (AURA/STScI"), J. Higdon (Cornell) ESA, NASA.

HENRY KAKEMBO, M.D.

Illustration 4.1

A British ten-pound note depicting Charles Darwin on the right, the HMS Beagle in the middle, and a hummingbird on the far left.

HENRY KAKEMBO, M.D.

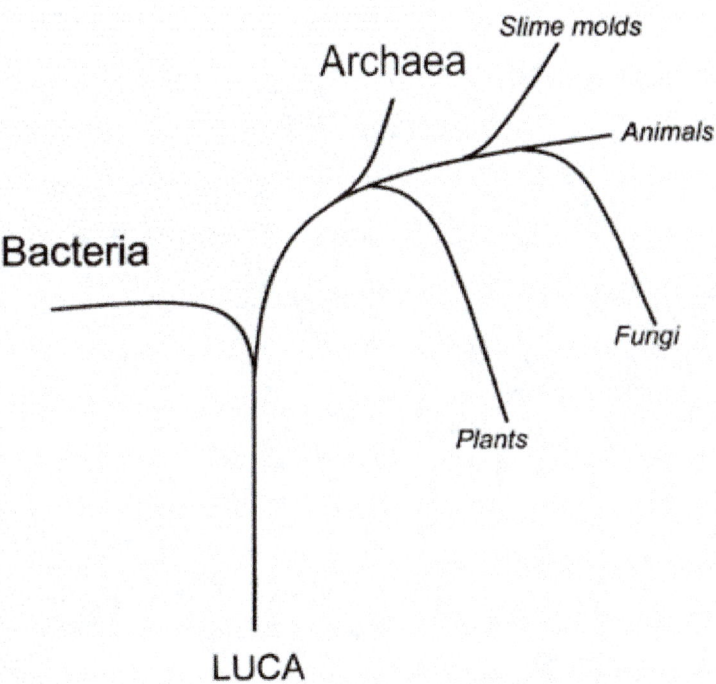

Illustration 5.1

The tree of life: According to Carl Woese, LUCA, the last universal common ancestor, gave rise to three domains—Bacteria, Archaea, and Eukarya. Eukarya branched into four kingdoms: Plantae, Protista or Slime Mold, Fungi, and Animalia, our group.

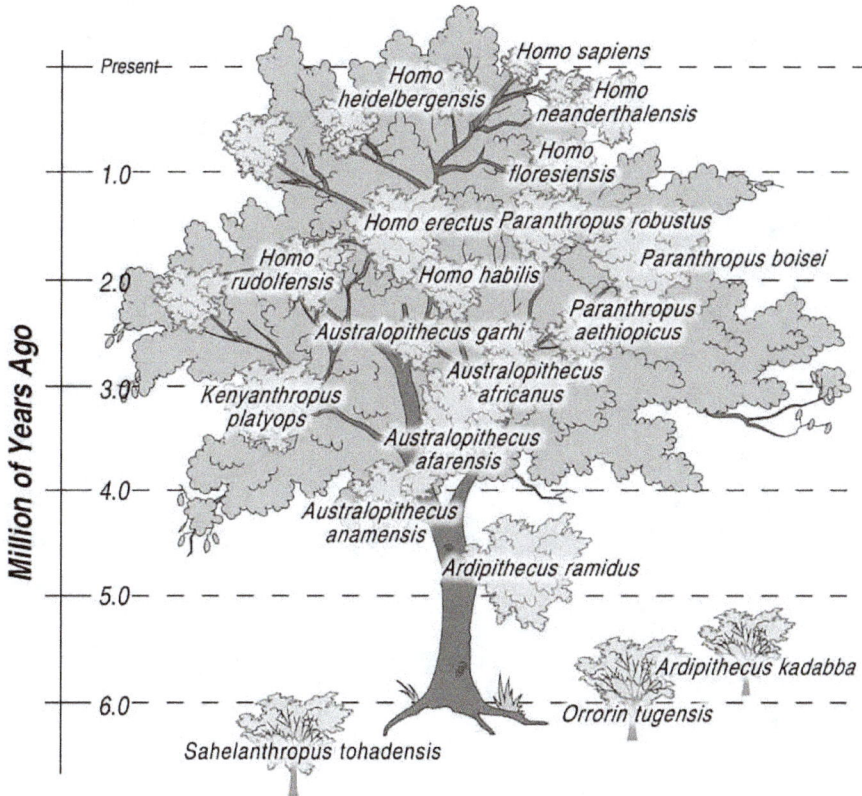

Million of Years Ago

Present

1.0

2.0

3.0

4.0

5.0

6.0

Homo sapiens

Homo heidelbergensis

Homo neanderthalensis

Homo floresiensis

Homo erectus   Paranthropus robustus

Homo rudolfensis

Paranthropus boisei

Homo habilis

Australopithecus garhi

Paranthropus aethiopicus

Kenyanthropus platyops

Australopithecus africanus

Australopithecus afarensis

Australopithecus anamensis

Ardipithecus ramidus

Ardipithecus kadabba

Orrorin tugensis

Sahelanthropus tohadensis

Illustration 6.1

Human evolution from about five million years ago. Progression to *Homo sapiens* is clearly nonlinear, as several species existed simultaneously. The diagram portrays a likely route. It is still debated whether *H. habilis* and *H. rudolfensis* belonged to the same species or to another genus such as *Australopithecus*.

279

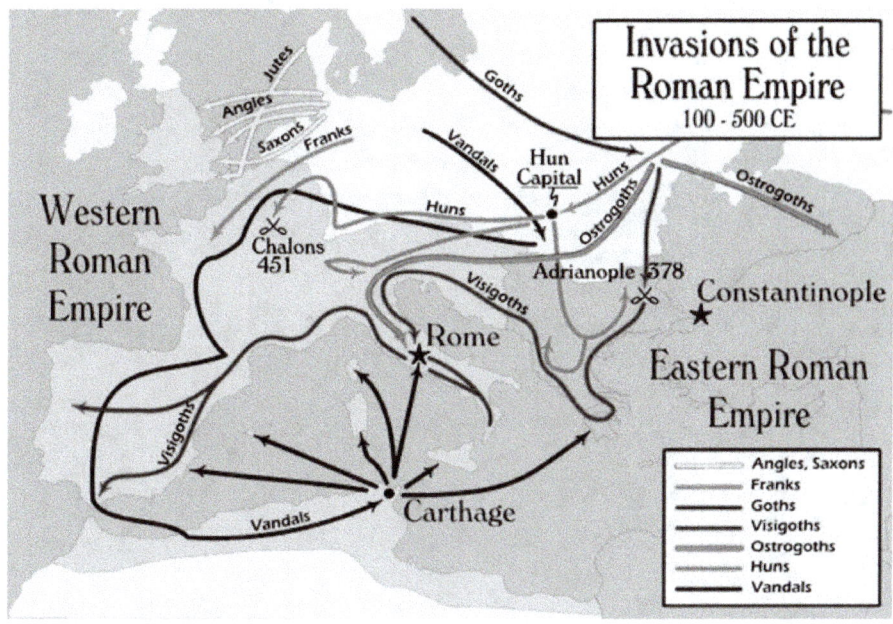

Illustration 7.1

European migrations (*Völkerwanderung*) between 400 CE and 700 CE.

Image credit: Hayden Chakra, "Invasions of the Roman Empire," in "The Great Migration Period - European Migrations" (About-History.com, June 25, 2018), accessed June 25, 2018, https://about-history.com/the-great-migration-period.

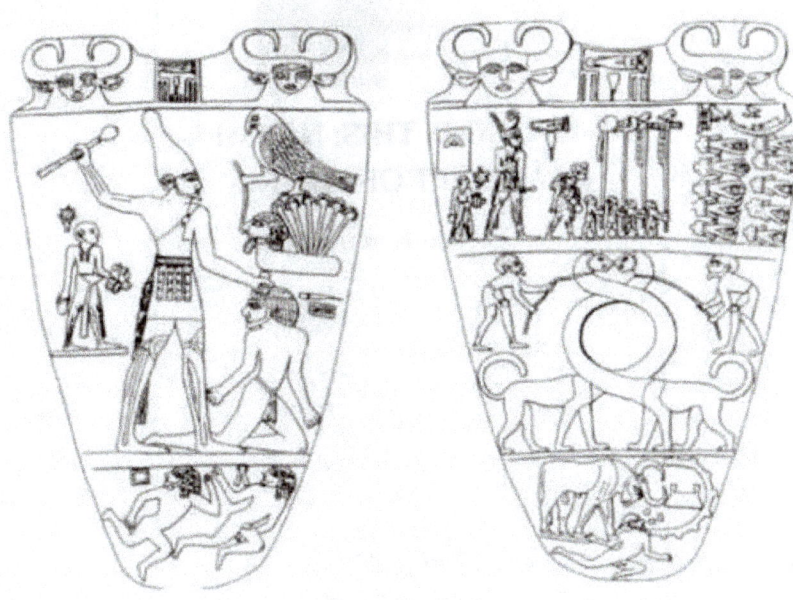

Illustration 7.2

An outline of the Narmer Palette housed in the Cairo Museum. It was discovered in 1897/98 by J. E. Quibell at Hierakonpolis, an ancient capital in the Egyptian south.[1]

Image credit: Histor Eidenai, "The Narmer Palette," published online at https://www.flickr.com/photos/51033794@N06/albums/72157624299551622, accessed June 15, 2021.

Henry Kakembo, M.D.

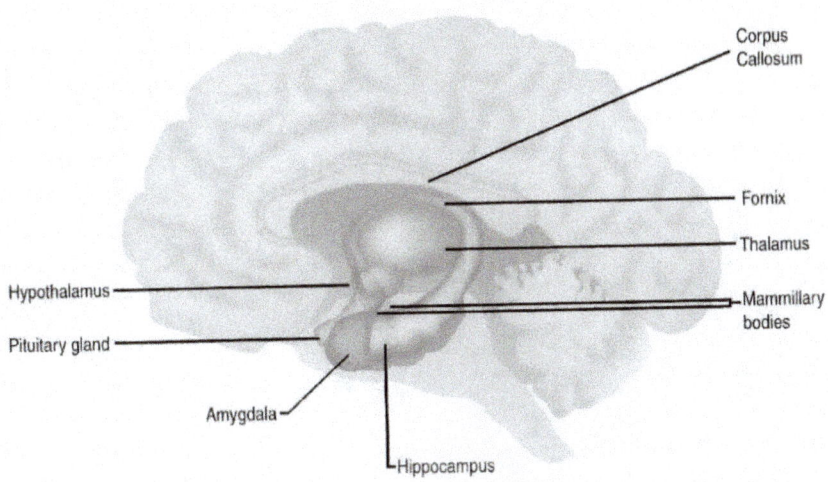

Corpus
Callosum

Fornix

Thalamus

Hypothalamus

Mammillary
bodies

Pituitary gland

Amygdala

Hippocampus

Illustration 12.1

The limbic system, a group of brain structures in the temporal lobe with a role in memory, emotion, motivation, and smell. The hippocampus and surrounding structures are crucial in long-term encoding and retrieving memories. The amygdala helps to regulate emotion and seems to play a role in emotional memories.

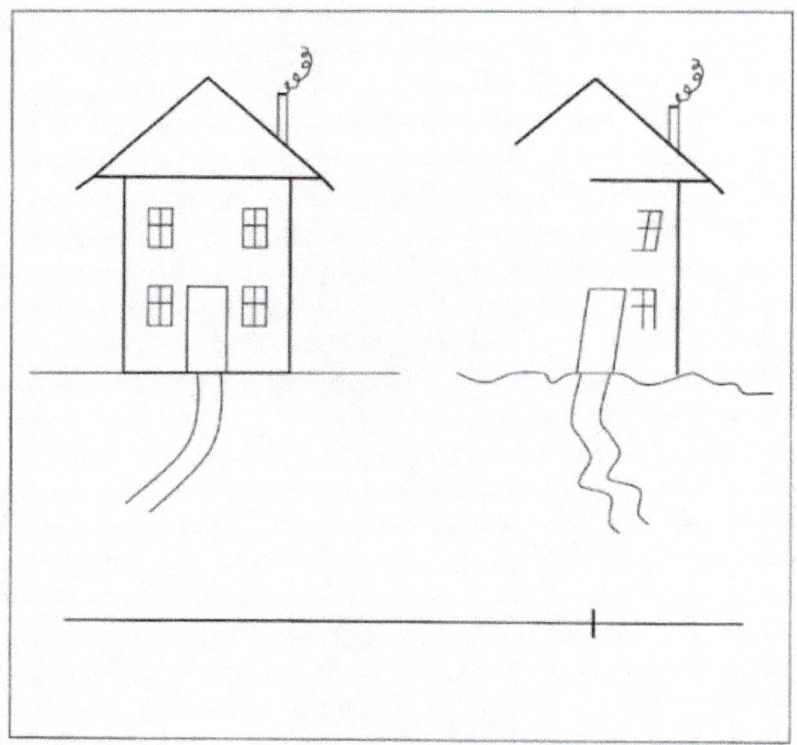

Illustration 12.2.

The Neglect Syndrome: A person who had damage to the right parietal cortex was asked to copy the figure on the left; his imitation is on the right. He also bisected a line on the right instead of the center (bottom of image).

Image credit: Michael Posner and Marcus Raichle (eds.), Images of Mind (New York: W.H. Freeman, 1994); E. Bisiach and C. Luzzatti, "Unilateral neglect of representational space," Cortex, 14, 129-133.

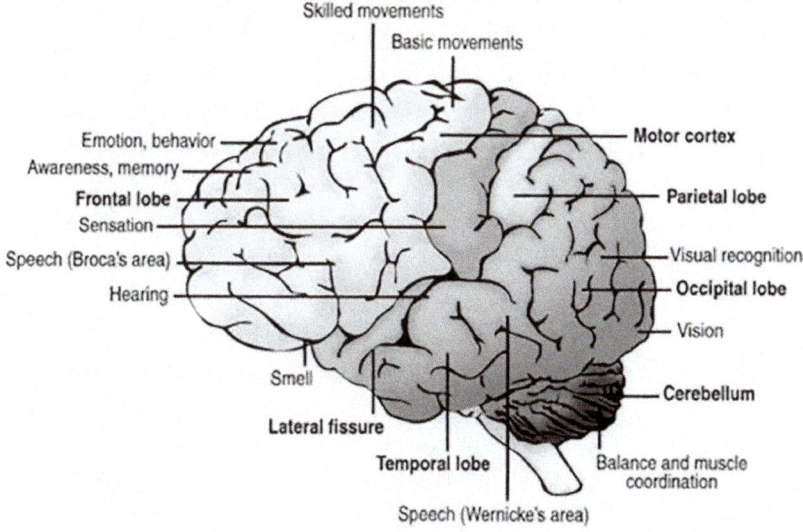

Illustration 12.3.

Some of the functional areas of the human brain distributed in the cortex and cerebellum. In general, these areas are found on both sides of the cortex, each controlling the opposite side of the body.

Image Credit: Encarta Encyclopedia 2000.

REFERENCES

ENDNOTES – INTRODUCTION

[1] Stephen Hawking, *A Brief History of Time* (New York: Bantam. 1998), 191.

[2] Bill Murphy, "China, officially atheist, could have more Christians than the U.S. by 2030." *Houston Chronicle,* February 24, 2018. https://www.houstonchronicle.com/news/houston-texas/houston/article/China-officially-atheist-could-have-more-12633079.php

[3] Henry Havelock Ellis in Tom Morris, *Philosophy for Dummies: A reference for the Rest of Us* (Foster City, CA: IDG Books Worldwide, 1999), 288.

[4] Henry Miller in *Brainy Quotes,* accessed January 06/2020, https://BrainyQuote.com/authors/henry-Miller-quotes

[5] Eric Fromm in Tom Morris, *Philosophy for Dummies: A reference for the Rest of Us* (Foster City, CA: IDG Books Worldwide, 1999), 288.

[6] Daniel Hill argues in *Philosophy Now* (35).

[7] Fritjof Capra, *The Tao of Physics: An Exploration of the Parallels Between Modern Physics and Eastern Mysticism, 4th Ed.* (Boston: Shambala, 2000), 99, 139.

[8] Ibid, 138. According to David Bohm, "…inseparable quantum interconnectedness of the whole universe is the fundamental reality, and… relatively independently behaving parts are merely particular and contingent forms within this whole." (D. Bohm & B. Hiley, 'On the Intuitive Understanding of Nonlocality as Implied by Quantum Theory,' *Foundations of Physics,* vol. 5 (1975), pp. 96, 102.

[9] Joseph B. Lumpkin, *The Books of Enoch: The Angels, The Watchers and The Nephilim,* 2nd Ed., Slavonic manuscript, Book 2 Enoch (Blountsville, AL: Fifth Estate, 2011), 177-186; In Seventh Heaven or "What Enoch did next: the seven heavens of Slavonic Enoch," accessed January 25, 2019, https://paradiseandperdition.weebly.com

ENDNOTES – CHAPTER 1 – THE QUEST

[1] Andrew Newberg, D'Aquilli and Vince Rause, *Why God Won't Go Away: Probing the Biology of Religious Experience* (New York: Ballantine Books, 2001), 79-81.

[2] Ibid, 80-81.

[3] R.S. Cavan. et al., *Personal Adjustment in Old Age* (Chicago: Science Research Associates, 1949) in Michael Argyle, *Psychology and Religion: An Introduction* (New York: Routledge, 2000), 28.

[4] Bryan Magee, *The Story of Philosophy: The Essential Guide to the History of Western Philosophy,* 1st Ed. (New York NY: DK Publishing, 1998), 36; Aristotle, *metaphysics,* 1.980a 22.

[5] Eugene d'Aquili and Andrew B. Newberg, *The Mystical Mind: Probing the Biology of Religious Experience* (Minneapolis: Fortress Press, 1999), 53.

[6] Joseph B. Lumpkin, *The Books of Enoch: The Angels, The Watchers and The Nephilim,* 2nd Ed., 2 Enoch (Blountsville, AL: Fifth Estate, 2011), 185-186.

[7] Karen Armstrong, *A History of God: The 4000-Year Quest of Judaism, Christianity and Islam* (New York: Ballantine Books, 1993), 389.

[8] Ibid, 364.

[9] Ibid, 186-191. *See also* Fritjof Capra, *The Tao of Physics: An Exploration of the Parallels Between Modern Physics and Eastern Mysticism,* 4th Ed. (Boston: Shambala, 2000), 24.

[10] Fritjof Capra, *The Tao of Physics: An Exploration of the Parallels Between Modern Physics and Eastern Mysticism,* 4th Ed. (Boston: Shambala, 2000), 24.

[11] Robert Kane, "The Quest for Meaning, Values, Ethics, and the Modern Experience," The Great Courses, DVD by The Teaching Company (Chantilly, VA: 1999).

[12] Timothy Freke, Peter Gandy, *The Jesus Mysteries: Was the "Original Jesus" a Pagan God?* (New York: The Three Rivers Press, 1999), 228.

[13] Paul C. Boyd, *The African Origin of Christianity: A Biblical and Historical Account,* vol. 1 (London: Karia Press, 1991), 64-65.

[14] Carl Sagan, *Cosmos* (New York: Ballantine Books, 1985), 2.

[15] John Pappademos, "The Newtonian Synthesis in Physical Science and its Roots in the Nile Valley," in *Nile Valley Civilizations: Proceedings of the Nile Valley*

Conference, September 26-30, 1984, ed. by Ivan V. Sertima (Atlanta, Georgia: Morehouse College Edition, 1985), 98.

[16] Richard Pogge, "An Introduction to Solar System Astronomy," *Astronomy* 16. Accessed August 04, 2020, http://www.astronomy.ohio-state.edu/~pogge/Ast161/unit3/response.html

[17] Stephen Hawking, *A Brief History of Time* (New York: Bantam. 1998), 191.

[18] Immanuel Kant in Bryan Magee, *The Story of Philosophy: The Essential Guide to the History of Western Philosophy,* 1st Ed. (New York NY: DK Publishing, 1998), 137.

[19] Karen Armstrong, *A History of God: The 4000-Year Quest of Judaism, Christianity and Islam* (New York: Ballantine Books, 1993), 378. Karl Popper criticized Logical Positivism in his book, *The Logic of Scientific Discovery* (1959).

[20] Robert H. Kane, "The Quest for Meaning: Values, Ethics, and the Modern Experience," Part I, section 3. *The Teaching Company* (1999).

[21] Michael Shermer, *Why Darwin Matters: The Case against Intelligent Design* (New York: Henry Holt and Company, 2006), 23-29; Herbert Thomas, *Human Origins: The Search for our Beginnings* (New York: Harry N. Abrams, 1995), 134-135; Daniel J. Fairbanks, *Relics of Eden: The Powerful Evidence of Evolution in Human DNA,* (Amherst, New York: Prometheus Books, 2007) 137-138.

[22] Stephen Hawking, *A Brief History of Time* (New York: Bantam. 1998), 191.

[23] Bryan Magee, *The Story of Philosophy,* 1st Ed. (New York NY: DK Publishing, (New York NY: DK Publishing, 1998), 225.

[24] Cheikh Anta Diop, *The African Origin of Civilization: Myth or Reality* (Chicago: Lawrence Hill Books, 1974), 250, 258-259.

[25] Michael Argyle, *Psychology and Religion: An Introduction* (New York: Routledge, 2000), 26.

[26] Fritjof Capra, *The Tao of Physics: An Exploration of the Parallels Between Modern Physics and Eastern Mysticism, 4th Ed.* (Boston: Shambala, 2000), 99.

[27] Ibid, 313. Hundreds of years ago, the Buddhist sage Nagarjuna put it thus: *"Things derive their being and nature by mutual dependence and are nothing in themselves."*

28 Ibid, 7.

29 Ibid, 67.

ENDNOTES – CHAPTER 2 – X-RATED ORIGINS

1 Tenzin Gyatso, the Fourteenth Dalai Lama of Tibet, *From Freedom in Exile* (New York: Harper Collins Publishers, 1990).

2 Karen Armstrong, *A History of God: The 4000-Year Quest of Judaism, Christianity and Islam* (1993), 5.

3 Enuma Elish - New World Encyclopedia, accessed August 19, 2020, www.newworldencyclopedia.org/entry/Enuma_Elish

4 John G. Jackson, *Christianity Before Christ* (Parsippany, NJ: American Atheist Press,
1985). 9.

5 Gerald Massey's Published Lectures (5), "The Historical Jesus and Mythical Christ," paragraph 63, accessed June 30, 2021, https://tringlocalhistory.org.uk/massey/dpr_05_hebrew_and_other.htm.

6 Jimmy Dunn, "The Wonderful Land of Punt," accessed February 10, 2016, http://www.touregypt.net/featurestories/punt.htm

7 Gerald Massey, *A Book of Beginnings* (Baltimore, Black Classic Press, 1995), 1:16.

8 DNA Tribes, a genetic ancestry analysis company once based in Arlington, Virginia, USA, cited the article, Jump up ^ Hawass et al. 2012, Revisiting the harem conspiracy and death of Rameses III: anthropological, forensic, radiological, and genetic study. BMJ2012;345doi: http://dx.doi.org/10.1136/bmj.e8268 Published 17 December 2012

9 DNA Tribes Digest, accessed January 3, 2015, http://www.dnatribes.com/dnatribes-digest-2013-02-01.pdf

10 Theophile Obenga, "The genetic relationship between Egyptian (ancient Egyptian and Coptic) and modern African languages" (1978), in UNESCO (Ed.), *The peopling of Ancient Egypt and the Deciphering of the Meroitic Script* (65-72). Paris: UNESCO. Obenga, T. (1993). Origine commune de l'Egyptien Ancien du Copte et des langues Négro-Africaines Modernes.

[11] Gerald Massey, *Ancient Egypt: The Light of the World,* vol. I (Medford, NY: Martino Fine Books, 2014), 423.

[12] Brian Flemming, Director, "The God Who Wasn't There" (Microcinema, 2005), back cover preview.

[13] Dan Brown, *The da Vinci Code,* special illustrated ed., 1st ed. (New York: Doubleday, 2003), 128.

[14] Charles S. Finch, "Africa and Palestine in Antiquity" in *African Presence in Early Asia: Incorporating Journal of African Civilizations,* April 1985, 7, no. 1, ed. by Runoko Rashidi and Ivan V. Sertima (New Brunswick and London, Transaction Publishers, 1988), 193-194. Abraham's name, (Ibrahima in Arabic), in Egyptian breaks down into *ib* for heart, desire and wisdom, *ra* for the Sun God, and *im* for fire or light, meaning "the desire or wisdom of Ra's light or fire." Isaac (Ysak in Hebrew), his son's name, breaks down into *Ys* for place, and *akh* for "offering by fire or burnt offering," linking it to Ra by his relation to fire. Changing Yaqub (Jacob) to Israel, where *ys* stands for place, *ra* for Sun God, and *ir* for creation, gives us "ys-ra-ir" or "the place of Ra's Creation."

[15] John Maddox, "Plagiarism Is Worse Than Mere Theft," Pythagoras was a systematic plagiarist. The tale is that Pythagoras had been urged by his elder, Thales, to travel to Egypt and attach himself to the priesthood, who were then the custodians of the Nile Valley's wisdom. Young Pythagoras did just that, returning to Greece in his early fifties brimful of notions about immortality of the soul, the square of the hypotenuse of a right triangle, and the symbolic significance of simple geometrical shapes that became the hallmark of the Pythagorean School. Newsome relates how upset the Egyptians were to see their credit stolen. *Nature* 376 (31 August 1995): 721, accessed February 18, 2016, http://www.albany.edu/~scifraud/data/sci_fraud_3655.html

[16] Cheikh A. Diop, "Africa's Contribution to World Civilization: The Exact Sciences," in *Nile Valley Civilizations: Proceedings of the Nile Valley Conference,* September 26-30, 1984, ed. by Ivan V. Sertima (Atlanta, Georgia: Morehouse College Edition, 1985), 72.

[17] John Pappademos, "The Newtonian Synthesis in Physical Science and its Roots in the Nile Valley," in *Nile Valley Civilizations: Proceedings of the Nile Valley Conference,* September 26-30, 1984, ed. by Ivan V. Sertima (Atlanta, Georgia: Morehouse College Edition, 1985), 94.

[18] Asa G. Hilliard III, "Kemetic Concepts in Education," citing Egyptologist Mohamoud Abdullah from a 1984 Seti 1 Luxor tour in *Nile Valley Civilizations: Proceedings of the Nile Valley Conference,* September 26-30, 1984, ed. by Ivan V. Sertima (Atlanta, Georgia: Morehouse College Edition, 1985), 157.

[19] Beatrice Lumpkin, "Mathematics and Engineering in the Nile Valley," in *Nile Valley Civilizations: Proceedings of the Nile Valley Conference,* September 26-30, 1984, ed. by Ivan V. Sertima (Atlanta, Georgia: Morehouse College Edition, 1985), 115-117.

[20] In Karen Armstrong, *A History of God: The 4000-Year Quest of Judaism, Christianity and Islam* (New York: Ballantine Books, 1993), 196.

[21] Fritjof Capra, *The Tao of Physics: An Exploration of the Parallels Between Modern Physics and Eastern Mysticism, 4th Ed.* (Boston: Shambala, 2000), 7.

[22] Antonio Damasio, *Looking for Spinoza: Joy, Sorrow and the Feeling Brain* (New York: Harcourt, 2003), 267-289.

[23] Bryan Magee, *The Story of Philosophy,* 1st Ed. (New York NY: DK Publishing, (New York NY: DK Publishing, 1998), 99.

[24] Ibid, 99

ENDNOTES – CHAPTER 3 – HEAVENLY EXTRAVAGANZA

[1] Spencer Wells, "The Journey of Man," *The Genographic Project,* DVD by National Geographic, 2003.

[2] Gerald Massey's Published Lectures (5), "The Historical Jesus and Mythical Christ," paragraph 63, accessed June 30, 2021, https://tringlocalhistory.org.uk/massey/dpr_05_hebrew_and_other.htm.

[3] Timothy Ferris, author and podcaster, said in *Creation of the Universe,* PBS video (2003), part II.

[4] John Gribbin, *Hyperspace: Our Final Frontier* (New York: DK Publishing, 2001), 227.

[5] John Gribbin in *Hyperspace* writes that, because its mass-energy is cancelled out by its negative gravitational energy, a universe destined to expand to infinity can appear out of nothing at all. *Ibid.*

[6] Timothy Ferris, author and podcaster, said in *Creation of the Universe,* PBS video (2003), part II.

[7] Fritjof Capra, *The Tao of Physics: An Exploration of the Parallels Between Modern Physics and Eastern Mysticism, 4th Ed.* (Boston: Shambala, 2000), 97.

[8] Stephen Hawking, *A Brief History of Time* (New York: Bantam. 1998), 120.

[9] Ibid, 7.

[10] CERN: Conseil Européen pour la Recherche Nucléaire (European Organization for Nuclear Research).

[11] Fritjof Capra, *The Tao of Physics: An Exploration of the Parallels Between Modern Physics and Eastern Mysticism, 4th Ed.* (Boston: Shambala, 2000), 330.

[12] Ibid, 331.

[13] Ibid, 95.

[14] Leah Anderson and Vlatko Vedral, "CQC Introductions: Quantum Entanglement," accessed July 16, 2020, http://cyber.sibsutis.ru/FIONOV/QC/CQC%20Intros%20Quantum%20Entanglement.htm

[15] Meinard Kuhlmann, "What is Real?" *Scientific American,* August 2013, 40.

[16] Here *Logos* means the Word of God, or principle of divine reason and creative order, identified in the Gospel of John with the second person of the Trinity incarnate in Jesus Christ.

[17] David Hume, accessed April 24, 2019, https://salirickandres.altervista.org/david-hume/

[18] Jim Bridenstine, "NASA Administrator Says Pluto Is Still a Planet, And Things Are Getting Heated," accessed August 11, 2020, https://www.sciencealert.com/nasa-administrator-jim-bridenstine-says-pluto-is-a-planet

[19] Paul C. Boyd, *The African Origin of Christianity: A Biblical and Historical Account,* vol. 1 (London: Karia Press, 1991), 31-32.

[20] Joseph B. Lumpkin, *The Books of Enoch: The Angels, The Watchers and The Nephilim,* 2nd Ed., Book 1 Enoch (Blountsville, AL: Fifth Estate, 2011), 14: 15-22; The Book of the Watchers, "A world of fire and ice," accessed January 25, 2019, https://paradiseandperdition.weebly.com/blog/a-world-of-fire-and-ice-heaven-according-to-enoch

[21] Alis J. Deason, Azadeh Fattahi, Carlos S. Frenk, et. al., "The Edge of the Galaxy," https://arxiv.org/pdf/2002.09497.pdf

[22] Peter Joseph, "The Greatest Story Ever Told" in *Zeitgiest, The Movie* (2007), part I, accessed January 11, 2015, http://zeitgeistmovie.com

[23] Ibid. This is why in early occult art Jesus is shown with his head on the cross, for He was the sun, the son of God, the Light of the World, the Risen Savior, who was to "come again," as the sun does every morning. He was the Glory of God who defended against the works of darkness, as He was "born again" every morning and could be seen "coming in the clouds," "up in Heaven," with His "Crown of Thorns," (*sun rays*). Charles S. Finch, "The Kamitic Genesis of Christianity," in *Nile Valley Civilizations. Proceedings of the Nile Valley Conference*, September 26-30, 1984, ed. by Ivan V. Sertima (Atlanta, Georgia: Morehouse College Edition, 1985), 183.

[24] Joseph B. Lumpkin, *The Books of Enoch: The Angels, The Watchers and The Nephilim,* 2nd Ed., Book 1 Enoch (Blountsville, AL: Fifth Estate, 2011), 14:15-22; The Book of the Watchers, "A world of fire and ice," accessed January 25, 2019, https://paradiseandperdition.weebly.com/blog/a-world-of-fire-and-ice-heaven-according-to-enoch

[25] Paul Strathern, *Hume in 90 Minutes,* (Chicago: Ivan R. Dee, 1999), 59.

ENDNOTES – CHAPTER 4 - LIFE

[1] Although this quote is often attributed to Darwin, Darwin might have not made or said it in exactly those words. Accessed June 28, 2021, https://quoteinvestigator.com/2014/05/04/adapt/

[2] Robert M. Hazen, *Genesis: The Scientific Quest for Life's Origin* (Washington, DC: Joseph Henry Press, 2005), 27.

[3] Neil deGrasse Tyson and Donald Goldsmith, *Origins: Fourteen Billion Years of Evolution,* (New York: W. W. Norton & co., 2004) 233-234.

[4] Carl Sagan, "Life," Encyclopedia Britannica, 14th ed (Chicago: Encyclopedia Britannica Incorporated, 1970), 1083–1083A.

[5] Richard Dawkins writes that we evolve by Natural Selection, and the gene – not the species - is the unit of selection, in *The Selfish Gene* (Oxford: Oxford University Press, 1976); Richard Morris, *The Evolutionists: The Struggle for Darwin's Soul,* (New York: W. H. Freemen and Company, 2001) 75.

[6] Karen Armstrong, *A History of God: The 4000-Year Quest of Judaism, Christianity and Islam* (New York: Ballantine Books, 1993), 124.

[7] David Galton, "Did Darwin read Mendel?" *QJM: An International Journal of Medicine*, Vol. 102, Issue 8, August 2009, Pages 587–589, accessed February 08, 2020, https://doi.org/10.1093/qjmed/hcp024

ENDNOTES - CHAPTER 5 – QUANTUM GENESIS

[1] Karen Armstrong, *A History of God: The 4000-Year Quest of Judaism, Christianity and Islam* (New York: Ballantine Books, 1993), 28.

[2] John M. Hayes, "The earliest memories of life on Earth," Nature 384 (1996): 21 – 22.
Summary: How old is life on Earth? Its chemical traces may have been recognized in rocks from Greenland that are more than 3.5 billion years old, at least 300 million years older than any previous evidence.; Stephen Moorbath, "Paleobiology: Dating Early life," Nature 434 (2005): 155.

[3] Douglas Fox, "Primordial Soup's On: Scientists Repeat Evolution's most Famous Experiment," *Scientific American* (175). Accessed October 1, 2020; https://www.scientificamerican.com/article/primordial-soup-urey-miller-evolution-experiment-repeated/

[4] John R. Gribbin, *The Origin of the Future* (Yale University Press, 2006), 229-230. For instance, certain molecules, *pyran rings*, easily form long chains. Each chain then attaches to its neighbor by latching on to carbon and oxygen. When it snaps into two, the strands grow anew. Certain molecules, *pyran rings*, easily form long chains. Each chain then attaches to its neighbor by latching on to carbon and oxygen and, when it snaps into two, the strands grow anew.

[5] Robert M. Hazen, *Genesis: The Scientific Quest for Life's Origin* (Washington, DC: Joseph Henry Press, 2005), 115-116.

[6] Johnjoe McFadden, *Quantum Evolution: The New Science of Life* (New York: W.W. Norton & Company, 2004), 95-98.

[7] Robert M. Hazen, *Genesis: The Scientific Quest for Life's Origin* (Washington, DC: Joseph Henry Press, 2005), 113, 205-211.

[8] To reduce their energy in water, oil droplets spontaneously form double layers with the water-soluble surfaces outside and inside and the water repelling side in the middle. Water on the inside forms a vesicle. Such a primitive structure could have trapped either an RNA molecule or a protein as its pores opened and closed with changes in temperature. Eventually, the lipid bilayer became a cell wall protecting RNA cells making them nigh unstoppable.

[9] David H. Lee, et. al., "A Self-replicating peptide," *Nature,* 382 (1996): 525-528; Johnjoe McFadden, *Quantum Evolution: The New Science of Life* (New York NY: W.W. Norton & Company, 2004), 97-101. In nature, the peptide would sequester in a tiny space, in a rock perhaps, "as a quantum superposition of all possible peptides," a phenomenon described in Chapter 3.

[10] Johnjoe McFadden, *Quantum Evolution: The New Science of Life* (New York NY: W.W. Norton & Company, 2004), 99-101; Britannica, T. Editors of Encyclopaedia. "Anthropic principle." *Encyclopedia Britannica*, August 14, 2013. Accessed July 10, 2021, https://www.britannica.com/science/anthropic-principle.
The principle is not a hypothesis or theory, as it's not falsifiable. It's like a rule or belief, a method.

[11] Johnjoe McFadden, *Quantum Evolution: The New Science of Life* (New York NY: W.W. Norton & Company, 2004), 99-101, 221-232. David Lee cut corners when engineering the self-replicating peptide, raising the question of how such a peptide could assemble naturally. So, in the beginning, the first amino acid would have existed as a quantum superposition of all the essential twenty amino acids and reacted with the other amino acids, adding them to itself to form a superposition of all the possible peptides inside the proto-cell. For life to form, the peptide would have had to spill out of its superposition state. RNA peptides with enzyme activity could have helped. RNA enzymes going in and out of the proto-cell would have added amino acids to the peptide, until a self-replicating enzyme reacted with the growing peptide chain and irreversibly amplified it. Information about the peptide would leak out of the cell, coupling the cell with the environment (the detector), which would have collapsed the peptide's superposition quantum state into a classical state and made the peptide real. From there, the self-replicator could lead to the emergence of life, in keeping with the anthropic multiverse principle.

[12] Ibid, 236.

[13] Ibid. McFadden, citing a John Cairns' 1988 paper, writes that to stave off starvation, E. coli bacteria increased their mutation rates to obtain energy from lactose, a source they had never used, 77-78, 263. Accounting for such adaptation, groups of RNA molecules, called riboswitches, regulated whether a protein was made or not according to need or supply, through feedback. Also known as ribozymes for their protein-like catalyzing powers, these switches hailed back to the ancient bacterial world before DNA became the mastermind of the cell. The adaptive mutations took a cell in a beneficial direction. Jeffery E. Barrick, Ronald R. Breaker, "The Power of Riboswitches," Scientific *American,* January (2007), 50-57.

[14] Johnjoe McFadden, *Quantum Evolution: The New Science of Life* (New York NY: W.W. Norton & Company, 2004), 239-240.

[15] In 1928, Alexander Fleming, working at St. Mary's College in London, accidentally discovered penicillin, saving many a soldier's life during the Second World War.

[16] Lynn Margulis, *What is Life?* (Berkley and Los Angeles: University of California Press, 1995), 136-137.

[17] Peter Ward, "Impact from the Deep*," Scientific American,* October (2006), 64-77.

[18] Kara Rogers, "Will There Be Another Ice Age?" Ask SciFri, accessed 08/04/2020, http://www.Sciencefriday.com/articles/will-there-be-another-ice-age/

[19] Bill Bryson, *A Short History of Nearly everything* (New York: Broadway Books, 2003), 356-360.

[20] Matt Ridley on the relationship between rank, stress hormones, and heart disease in both monkeys and people in "Chromosome 11: Personality," *Genome* (New York: Perennial, 2000), 165.

[21] Georg Halder, Patrick Callaerts, Walter J. Gehring, "Induction of Ectopic Eyes by Targeted Expression of the eyeless Gene in Drosophila," *Science* 267 (1995): 1788-1792. A Pessimistic Estimate of the Time Required for an Eye to Evolve. Author(s): Dan-E. Nilsson and Susanne Pelger Reviewed work(s): Source: Proceedings: Biological Sciences, Vol. 256, No. 1345 (Apr. 22, 1994), pp. 53-58. Accessed: 31/08/2012 11:57. Published by: The Royal Society Stable URL: https://www.jstor.org/stable/49593

[22] Herbert Thomas, *Human Origins: The Search for our Beginnings* (New York: Harry N. Abrams, 1995), 26.

[23] Rebecca Quiring, U Walldorf, U Kloter, W.J. Gehring, "Homology of the eyeless gene of Drosophila to the Small eye gene in mice and Aniridia in humans," Science 265 (1994): 785-789.

[24] Bill Nye, *Undeniable: Evolution and the Science of Creation* (New York: St. Martin's Press, 2014), 48-52.

[25] Stephen J Gould, *Ever Since Darwin: Reelections in Natural History* (New York: W.W. Norton & Company, 1977), 83.

[26] Dan E. Nilsson and Susanne Pelger, "A pessimistic Estimate of the Time Required for an Eye to Evolve," *Proceedings: Biological Sciences*, 256, No. 1345 (1994), 53-58.

[27] Sabine Wilkins, "The Evolution of Cichlids," accessed February 13, 2016, http://www.cichlid-forum.com/articles/evol_cich_pt1.php

[28] Dick Teresi, "Lynn Margulis," *Discover Magazine,* April 2011, 66.

ENDNOTES - CHAPTER 6 – OUR DEEP ANCESTORS

[1] Audrey Smedley, "The History of the Idea of Race…and Why it Matters," a paper presented at the conference, "Race, Human Variation and Disease: Consensus and Frontiers held March 14-17, 2007, in Warrenton, VA, accessed November 7, 2018, http://www.understandingrace.org/resources/pdf/diseases/smedley.pdf. (You may search by putting the title in a search bar.)

[2] Albert Lin, et al, "Crowdsourcing the Unknown: The Satellite Search for Genghis Khan," 2014. Accessed August 24, 2020, https://www.researchgate.net/publication/270290607_Crowdsourcing_the_Unknown_The_Satellite_Search_for_Genghis_Khan/citation/download

[3] Such disputes used to be deadly serious: During the nineteenth century Bone Wars between Edward Cope and Othniel Marsh, their assistants carried guns for protection. Othniel Charles Marsh, Edward Drinker Cope, "The Two Paleontologists Who Had a
Bone to Pick with Each Other," accessed June 28, 2021, https://interactive.wttw.com/prehistoric-road-trip/detours/the-two-paleontologists-who-had-a-bone-to-pick-with-each-other

[4] The decay of potassium to argon is so slow it helps date rocks at least 500,000 years old. The decay of uranium to thorium helps identify rocks from 40,000 to 500,000 years ago, and uranium to lead to the beginning of the Earth. The most frequently

used method was invented by Willard Libby in the 1940s. It measures the decay of carbon-14 in organic material less than 40,000 years old.

[5] Genographic DNA testing was discontinued in 2019. Results would be up until 2020. For help in finding your relatives and recent ancestors, many DNA ancestry testing services are available online.

[6] David G. Poznik, et al., "Sequencing Y Chromosomes Resolves Discrepancy in Time to Common Ancestor of Males Versus Females," *Science,* 341, no. 6145 (2 August 2013): 562-565.

[7] Weiner, J. S., Oakley, K. P., and Clark, W. E. Le Gros. Bull. Brit. Mus. (Nat. Hist.), Geol., vol. 2, p. 141, 1953. Dawson was the chief suspect for the fraud. Besides being a cheat and swindler, Dawson had faked most of his other forty-some fossils. He was likely assisted by Sir Arthur Smith Woodard, who presented the forgery, and Martin Hinton, a curator, both working at the Natural History Museum in London. Among many others, a 1990 New York Times News Service story implicated Sir Arthur Conan Doyle, Sherlock Holmes' creator and Dawson's neighbor. The paper also fingered the venerable Sir Arthur Keith, who, besides certifying the forgery, unveiled a memorial marking the site of Piltdown Man's discovery.

[8] "Cannibalism," accessed January 17, 2020, https://www.gwern.net/docs/rotten.com/library/death/cannibalism/index.html; Tim D. White, "Once were Cannibals," *Scientific American,* (August 2003): 86-93.

[9] Katherine S. Pollard, "What Makes Us Human?" *Scientific American,* (May 2009): 44-49. The gene ASPM, or Abnormal Spindle Microtubule Assembly, is linked to brain size in humans and leads to small brains (microcephaly) when reduced. HAR1, or human accelerated region 1, helps the cortex wrinkle, increasing its area exponentially.

[10] Green et al., "A Draft Sequence of the Neandertal Genome," *Science* 328, no. 5979 (May 7, 2010): 710 – 722; Tina Hesman Saey, "Neandertal genome yields evidence of interbreeding with humans," *Science News,* 177, no. 12 (June 5, 2010): 5. Jill Neimark, "Meet the New Human Family," *Discover Magazine,* May 2011, 48.

ENDNOTES – CHAPTER 7 – ODYSSEY

[1] Dawn of Man: *The Story of Human Evolution,* BBC Video (2000), vol. 2, part II, Exodus.

[2] Armitage, S.J. et al., "The Southern Route 'Out of Africa:' Evidence for an Early Expansion of Modern Humans into Arabia," *Science* 331 (January 2011): 453-456.

[3] Vincenzo Formicola and Brigitte M. Holt, "Tall guys and fat ladies: Upper Paleolithic burials and figurines in an historical perspective." *Journal of Anthropologic Sciences,*
vol. 93 (2015), 71-88; Accessed October 3, 2020.
http://Creativecommons.org/licenses/by-nc/4.0/.

[4] Wilford, N. John, "A Lost European Culture, Pulled from Obscurity," accessed November 9, 2018, http:// www.nytimes.com/2009/12/01/science/01arch.html

[5] Phoenicians, "Sea Peoples and the Phoenicians: A Critical Turning Point."
Accessed August 24, 2020, http://www.phoenician.org/sea_peoples.htm#_edn37

[6] Ibid.

[7] "Ancient Man and His First Civilizations: The Original Black Cultures of Eastern Europe and Asia," accessed August 06, 2016,
http://realhistoryww.com/world_history/ancient/Dobruja_Thrace_1.htm

[8] Sarah Tishkoff, Floyd Reed, Françoise Friedlaender, et al, "The Genetic Structure and History of African Americans." *Science* (May 2009): 50, accessed October 26, 2018, https://www.sciencemag.org.

[9] Spencer Wells, "The Journey of Man," *The Genographic Project,* DVD by National Geographic, 2003.

[10] Morten K.B. Bogh, Anne V. Schmedes, Peter A. Philipsen, et al, "Vitamin D Production after UVB Exposure Depends on Baseline Vitamin D and Total Cholesterol but Not on Skin Pigmentation." *Journal of Investigative Dermatology* (February 2010): 546-553.

[11] Yuval Noah Harari, *Sapiens: A Brief History of Humankind* (New York: Harper Perennial, 2018), 303.

[12] Interview with Audrey Smedley, "Race – The Power of an Illusion," accessed July 21, 2020, https://www.pbs.org/race/000_About/002_04-background-02-06.htm

[13] Fouad Zakharia, Analabha Basu, Devin Absher, et al. "Characterizing the admixed African Ancestry of African Americans," *Genome Biology* 2009, 10:R141 accessed October 26, 2018,
https://genomebiology.biomedcentral.com/articles/10.1186/gb-2009-10-12-r141

[14] *"The First Americans were Australian,"* BBC News Online (August 26, 1999), accessed March 07, 2020, https://www.historyfiles.co.uk/FeaturesAmericas/PrehistoricFirstHumans02.htm

[15] Glenn Hodges, "The First Face of the First Americans belongs to an Unlucky Teenage Girl who fell," National Geographic (January 1915): 127-137.

[16] Nikhil Swaminathan, "America in the Beginning," *Archeology*, Sept/Oct 2014, 22-29.

[17] Ian Tattersall, in a six-part television and video series, *Dawn of Man: The Story of Human Evolution,* BBC Video (2000), vol. 2, part II, Exodus.

[18] Clyde Winters, *Ancient Black Civilization: A short World History of Black People in Ancient Times* (Lexington, Kentucky: 2013), 8-9.

[19] Christopher Henshilwood, in a six-part television and video series, *Dawn of Man: The Story of Human Evolution*, BBC Video (2000), vol. 2, part II, Exodus; Kate Wong, "The Morning of the Modern Mind," *Scientific American,* June 2005, 86-95; Curtis C. Marean, "When the Sea Saved Humanity," *Scientific American,* August 2010, 54-61.

[20] Thomas Dawson, in a six-part television and video series *Dawn of Man: The Story of Human Evolution,* BBC Video (2000), vol. 2, part II, Exodus.

[21] Ibid.

[22] Christopher Henshilwood, *Dawn of Man* (2000). BBC Books, Dorling Kindersley Publishing, New York.

[23] Audrey Fletcher, "The Narmer Plate and Osiris the Lord of Precession," part 8, accessed February 06, 2020, http://ancientegypt.hypermart.net/narmerplate/index.htm

[24] Yuval Noah Harari, *Sapiens: A Brief History of Humankind* (New York: Harper Perennial, 2018), 218.

[25] Clyde Winters, "Ancient African Writing Systems and Knowledge," accessed September 11, 2020, http://bafsudralam.blogspot.com/2008/08/thinite-writing.html

[26] Dan Eden, "Ancient Human Metropolis Found in Africa." Condensed by Native Village, accessed August 21, 2020, http://mondovista.com/adamscalendar.html

[27] John G. Jackson, "Sitting at the Feet of a Forerunner: An April 1987 Meeting and Interview with John G. Jackson," in *African Presence in Early Asia: Incorporating Journal of African Civilizations*, April 1985, 7, no. 1, ed. by James E. Brunson and Runoko Rashidi, ed. by Ivan V. Sertima and Runoko Rashidi (New Brunswick and London, Transaction Publishers, 1988), 203; Cheikh Anta Diop, *The African Origin of Civilization: Myth or Reality* (Chicago: Lawrence Hill Books, 1974), 168-169.

[28] Runoko Rashidi, "More Light on Summer, Elam and India" in *African Presence in Early Asia: Incorporating Journal of African Civilizations*, April 1985, 7, no. 1, by Runoko Rashidi and Ivan V. Sertima (New Brunswick and London, Transaction Publishers, 1988), 15, 163-164, 169.

[29] Runoko Rashidi, "Africans in Early Asian Civilizations: A Historical Overview," *African Presence in Early Asia: Incorporating Journal of African Civilizations*, April 1985, 7, no. 1, ed. Ivan V. Sertima; Runoko Rashidi, "The Nile Valley Presence in Asian Antiquity," in *Nile Valley Civilizations: Proceedings of the Nile Valley Conference*, Atlanta, September 26-30, 1984, ed. by Ivan V. Sertima (Atlanta, Georgia: Morehouse College Edition, 1985), 209; Runoko Rashidi (New Brunswick and London, Transaction Publishers, 1988), 173.

[30] Drusilla D. Houston, *Wonderful Ethiopians of the Ancient Cushite Empire* (Baltimore: Black Classic Press, 1985), 20.

[31] Ibid, 38-39.

[32] W. J. Perry challenges the view that Sumerians came from India in *The Growth of Civilization,* 2nd ed. (Harmondsworth, Middlesex, England, Penguin Books, 1937), 60-61, quoted by J.G. Jackson in *Ethiopia and the Origin of Civilization* (Baltimore, Black Classic Press, 1985), 18.

[33] Cheikh Anta Diop, *The African Origin of Civilization: Myth or Reality* (Chicago: Lawrence Hill Books, 1974), 3-4.

[34] Diodorus Siculus, *Universal History 3*, trans. Abbé Terrason (Paris: 1759), 341.

[35] Gerald Massey, *A Book of Beginnings 1* (London: Williams and Norgate), Frontpiece.

[36] Andrew Curry, "Egypt's Lost Fleet," *Discover Magazine,* June 2011, 60.

[37] Ivan Sertima, *They came Before Columbus: The African Presence in Ancient America* (New York: Random House, 2003), 58-60, 62. A papyrus boat, *Seti II,* successfully made it across the Atlantic from Africa better than studier boats; Drusilla

D. Houston, *Wonderful Ethiopians of the Ancient Cushite Empire* (Baltimore: Black Classic Press, 1985), 134.

[38] Ivan Sertima, *They came Before Columbus: The African Presence in Ancient America* (New York: Random House, 2003), 13; Patrick Huyghe, *Columbus was Last: From 200,000 B.C. to1492, A Heretical History of Who Was First* (Black heritage. New York: Hyperion, 1992), 182-192.

[39] Ivan Sertima, *They came Before Columbus: The African Presence in Ancient America* (New York: Random House, 2003), 254.

[40] Ivan Van Sertima, "The Nile Valley Presence in America B.C.," in *Nile Valley Civilizations: Proceedings of the Nile Valley Conference*, Atlanta, September 26-30, 1984, ed. by Ivan V. Sertima (Atlanta, Georgia: Morehouse College Edition, 1985), 209; Runoko Rashidi (New Brunswick and London, Transaction Publishers, 1988), 242-244.

[41] Ibid, 229-233.

[42] D.J. Thompson, *Memphis Under the Ptolemies* (Princeton, 1988), 99.

[43] Ibid.

[44] Ivan Van Sertima, "The Nile Valley Presence in America B.C.," in *Nile Valley Civilizations: Proceedings of the Nile Valley Conference*, Atlanta, September 26-30, 1984, ed. by Ivan V. Sertima (Atlanta, Georgia: Morehouse College Edition, 1985), 209; Runoko Rashidi (New Brunswick and London, Transaction Publishers, 1988), 226-227.

[45] Ibid, 234.

[46] Ibid, 239.

[47] Ibid, 245-246.

[48] Clyde Winters. "The Untold History: 'Blacks' were the 1st Americans," pt.2, accessed February 1, 2015, https://www.youtube.com/watch?v=Sttg1A5Ncfs

[49] Michael D. Coe, *America's First Civilization* (American Heritage Publishing Co., Inc., 1968), 92.

[50] Cheikh A. Diop, "Africa's Contribution to World Civilization: The Exact Sciences," in *Nile Valley Civilizations: Proceedings of the Nile Valley Conference*,

Atlanta, September 26-30, 1984, ed. by Ivan V. Sertima (Atlanta, Georgia: Morehouse College Edition, 1985), 81.

[51] The Moscow Mathematical Papyrus, also named the Golenishchev Mathematical Papyrus after its first non-Egyptian owner, Egyptologist Vladimir Golenishchev, is an ancient Egyptian mathematical papyrus containing several problems in arithmetic, geometry, and algebra. For additional information, *see* https://en.wikipedia.org/wiki/Moscow_Mathematical_Papyrus, particularly the list of reference works at the end of the entry.

[52] The movie, "Great Pyramid K 2019," by director Fehmi Krasniqi, is available online at https://youtu.be/KMAtkjy_YK4

[53]The Rhind Mathematical Papyrus, also designated as papyrus British Museum 10057 and pBM 10058, is one of the best known examples of ancient Egyptian mathematics. It is named after Alexander Henry Rhind, a Scottish antiquarian, who purchased the papyrus in 1858 in Luxor, Egypt. For additional information, *see* https://en.wikipedia.org/wiki/Rhind_Mathematical_Papyrus, particularly the list of reference works at the end of the entry.

[54] Ibid, 82.

[55] Philo Judaeus cited in Timothy Freke, Peter Gandy, *The Jesus Mysteries: Was the "Original Jesus" a Pagan God?* (New York: The Three Rivers Press, 1999)1999), 185.

[56] Jared Diamond, *Guns, Germs, and Steel: The Fates of Human Societies* (New York: W. W. Norton & Company, 1999), 232, 400. Although Diamond says that writing developed independently in Sumeria and Mesoamerica, and that it might have diffused into other countries such as China and Egypt, in this chapter we learn that writing originated in Africa and many scripts such as the Egyptian hieroglyphs, Olmec, Minoan, Uruk, Meroe, and Harrapan languages were deciphered using Mande. The script on the Shang bones in China could also be deciphered using Mande signs.

[57] Clyde Winters, "Ancient Writing in Middle Africa," 2009, accessed September 11, 2020, http://www.oocities.org/ekwesi.geo/anwrite.htm

[58] Abibitumi Kasa, "The Decipherment of the Olmec Writing System," paper presented at the 1997 Central States Anthropological Society Meeting, accessed September 11, 2020, http://olmec98.net/olmecDecip.htm

[59] Clyde Winters, "Ancient Writing in Middle Africa," 2009, accessed September 11, 2020, http://www.oocities.org/ekwesi.geo/anwrite.htm

[60] Robert Bauval, *The Egypt Code* (London: Century, 2006), 7, 30.

[61] Thomas Brophy, *The Origin Map: Discovery of a Prehistoric Megalithic, Astrophysical Map and Sculpture of the Universe* (New York: Writers Club Press 2002), xvi.

[62] Audrey Fletcher, "The Narmer Plate and Osiris the Lord of Precession," part 9, accessed February 06, 2020, http://ancientegypt.hypermart.net/index.htm

[63] Robert Bauval, *The Egypt Code,* (London: Century, 2006), 101, 111-116.

[64] Robert Bauval, *The Egypt Code* (London: Century, 2006), 111.

[65] Ibid, 57-66.

[66] Thomas Brophy in "The Astronomers of Nabta Playa," reported by Mark H. Gaffney in *Atlantis Rising,* March/April 2006.

[67] Thomas Brophy, *The Origin Map: Discovery of a Prehistoric Megalithic, Astrophysical Map and Sculpture of the Universe* (New York: Writers Club Press 2002), xxiv.

[68] Robert Bauval and Thomas Brophy, *Black Genesis: The Prehistoric Origins of Ancient Egypt* (Rochester, Vermont: Bear & Company 2011), 106.

[69] Cheikh A. Diop, *The African Origin of Civilization: Myth or Reality* (Chicago: Lawrence Hill Books, 1974), 22.

[70] Charles C. Mann, "The Birth of Religion: The World's First Temple," *National Geographic,* 219, 2011, 34-59.

[71] Gerald Massey, *A Book of Beginnings II* (Baltimore, MD: Black Classic Press, 1995), 599.

[72] Herodotus, *The Histories* (Bungay, Suffolk, Great Britain: Penguin Classics, 1972), 167-168; Runoko Rashidi expands on the identity of the Colchians. "The Nile Valley Presence in Asian Antiquity," in *Nile Valley Civilizations: Proceedings of the Nile Valley Conference*, Atlanta, September 26-30, 1984, ed. by Ivan V. Sertima (Atlanta, Georgia: Morehouse College Edition, 1985), 213.

ENDNOTES – CHAPTER 8 – GODS GALORE

[1] Mt. Llaima is at Conguillío National Park, Chile.

[2] Karen Armstrong, *A History of God: The 4000-Year Quest of Judaism, Christianity, and Islam* (New York: Ballantine Books, 1993), 389. Psalm 28, John Bowker translation, *The Religious Imagination and the Sense of God* (Oxford: Clarendon Press, 1978), 73.

[3] Ibid, 14.

[4] Michael Argyle, *Psychology and Religion: An Introduction* (New York: Routledge, 2000), 56.

[5] Andrew Newberg, D'Aquili and Vince Rause, *Why God Won't Go Away: Probing the Biology of Religious Experience* (New York: Ballantine Books, 2001), 80, 122-123.

[6] Kevin Nelson, *The Spiritual Doorway in the Brain: A Neurologist's Search for the God Experience* (New York: Dutton, 2011), 224; Bertrand Russell, "Mysticism and Logic and other Essays." Accessed August 28. 2020; http://www.archive.org/stream/mysticism00russuoft

[7] Karen Armstrong, *A History of God: The 4000-Year Quest of Judaism, Christianity, and Islam* (1993), *xxi*.

[8] Richard Dawkins, *The God Delusion* (New York, NY: First Mariner Books, 2006), 134.

[9] David Hume, cited by Karen Armstrong, in *A History of God: The 4000-Year Quest of Judaism, Christianity and Islam* (1993), 341-342.

[10] Steven Weinberg and John Polikinghorne on "Mysteries of the Universe's design" in a debate of the universe's design by Alan Boyle in Washington (1999). Accessed August 8, 2018; https://www.nbcnews.com/id/wbna3077380#.W2tPdihKhEY

[11] Bryan Magee, *The Story of Philosophy,* 1st Ed. (New York NY: DK Publishing, (New York NY: DK Publishing, 1998), 99; G.W. Leibniz "Monadologie (1714)." Nicholas Rechner, trans., *The Monadology: An Edition for Students* (Pittsburg: University of Pittsburg Press, 1991), 135.

[12] Christopher Hitchens, "What can be asserted without evidence can be dismissed without evidence" in *God is not Great: How Religion Poisons Everything* (New York, NY: Twelve Books, 2007), 150.

[13] William James Pragmatism: Lecture 3: Some Metaphysical Problems Pragmatically
Considered (1908), 42, accessed January 13, 2019,
http://www.nashvillegreatbooks.com/2006/11/william-james-pragmatism-lecture-3.html

[14] Xenophanes, "Goodreads: Quotes," accessed August 11, 2018.
https://www.goodreads.com/author/quotes/853837.Xenophanes
The Ethiops (Africans) say that their gods are flat-nosed and black,
While the Thracians say that theirs have blue eyes and red hair.
Yet if cattle or horses or lions had hands and could draw,
And could sculpt like men, then the horses would draw their gods
Like horses, and cattle like cattle; and each they would shape
Bodies of gods in the likeness, each kind, of their own.

[15] Gerald Massey, *The Historical Jesus and the Mythical Christ: Separating Fact from Fiction* (Escondido, California: The Book Tree, 2000), 163-173, 223.

[16] The Israelites were still worshipping the golden calf, likely a reference to the bygone age of Taurus the Bull, instead of the ram, the sign of the new age, Aries. Even though the Jews still blow a ram's horn today, the true sign is that of Pisces the fish that dawned in 1 CE (Gerald Massey dates it to 255 BCE. So we could be in the age of Aquarius today).

[17] Karen Armstrong, *A History of God: The 4000-Year Quest of Judaism, Christianity, and Islam* (New York: Ballantine Books, 1993), 310.

[18] Ibid, 354.

[19] Ibid, 28.

[20] Faheem Judah-El, "Meaning of Karast." Krst (the K is aspirated like Ch) is an Egyptian word meaning burial, accessed March 17, 2020,
https://www.scribd.com/document/266864395/Karast

[21] Joseph Atwill, *Caesar's Messiah: The Roman Conspiracy to Invent Jesus* (Charleston, SC: CreateSpace, 2011), Introduction.

[22] Emperor Hadrian's correspondence to Servianus in 134 A.D. refers to Alexandrian worshippers of Serapis as calling themselves Bishops of Christ: "Egypt, which you

commended to me, my dearest Servianus, I have found to be wholly fickle and inconsistent, and continually wafted about by every breath of fame. The worshipers of Serapis (here) are called Christians, and those who are devoted to the god Serapis (I find) call themselves Bishops of Christ," Quoted by Giles, ii p. 86.

[23] Arthur Weigall, *The Paganism in Our Christianity,* (New York and London: G. P. Putman and sons, 1928), 52.

[24] Gerald Massey, *The Historical Jesus and the Mythical Christ: Separating Fact from Fiction* (Escondido, California: The Book Tree, 2000), 27.

[25] Aviram Oshri, "Where was Jesus Born?" *Archaeology Magazine,* Nov/Dec. 2005, accessed February 08, 2015, http://www.archaeology.org/

[26] René Salm, "The Myth of Nazareth: The Invented Town of Jesus. Does it Really Matter? *American Atheist,* March 2007, 13-14.

[27] Gerald Massey, *The Historical Jesus and the Mythical Christ: Separating Fact from Fiction* (Escondido, California: The Book Tree, 2000), 32.

[28] Vassilios Tzaferis, "Jewish Tombs at and Near Givat HaMivtar." *Israel Exploration Journal,* Vol. 20, 1970, 18-32; Emanuel Gualdi, Ursula T. Hohenstein, Nicolet Onisto, et. al, "A Multidisciplinary Study of Calcaneal Trauma in Roman Italy: a Possible Case of Crucifixion?" *Journal of Archeological and Anthropological Sciences,* Vol.11 (5) April 12, 2018.

[29] Gerald Massey, *The Historical Jesus and the Mythical Christ: Separating Fact from Fiction* (Escondido, California: The Book Tree, 2000).

[30] Ibid, 19-20.

[31] Horus (Sirius) was the son of Osiris (Orion) and Isis (Canis Major); Timothy Freke, Peter Gandy, *The Jesus Mysteries: Was the "Original Jesus" a Pagan God?* (New York: The Three Rivers Press, 1999), 160, 295.

[32] Iamblichus attributed this power to Pythagoras in *Life of Pythagoras,* trans. Thomas Taylor (1986); Massey, *The Historical Jesus* (2000), 74.

[33] In May, the starry band known as the Milky Way stretched across the heavens like the lake Jesus crossed in the New Testament story. Egyptians believed the Milky Way was the source of the Nile. Accessed February 11, 2015, http://home1.gte.net/deleyd/religion/solarmyth/christ2002.htm

34 Timothy Freke, Peter Gandy, *The Jesus Mysteries: Was the "Original Jesus" a Pagan God?* (New York: The Three Rivers Press, 1999), 1.

35 Ibid, 89.

36 Massey, *The Historical Jesus and the Mythical Christ: Separating Fact from Fiction* (Escondido, California: The Book Tree, 2000), 2-3.

37 From Martin Luther King, Jr. papers, "Called to Serve," 1, January 1929-June 1951.

38 Timothy Freke, Peter Gandy, *The Jesus Mysteries: Was the "Original Jesus" a Pagan God?* (New York: The Three Rivers Press, 1999), 123-124.

39 Dan Brown, *The da Vinci Code,* special illustrated ed.,1st ed. (New York: Doubleday, 2003), 241.

ENDNOTES – CHAPTER 9 – WHO WROTE THE BIBLE?

1 Michael D Magee, "When Was Exodus Written?" accessed January 27, 2020, https://www.academia.edu/23772970/When_Was_the_Bible_Written_In_the_Persi an_Era; "Judaism, Moses and the Exodus," accessed February 22, 2020, https://www.academia.edu/23250467/Moses_and_the_Exodus

Magee mostly agrees with the late date of the Bible posited by Russell E. Gmirkin in his *Berossus and Genesis, Manetho and Exodus: Hellenistic Histories and the Date of the Pentateuch* (Library of Hebrew Bible/Old Testament Studies, 433; Copenhagen International series, 15; New York/London: T & T Clark, 2006) Pp. xii + 332.

2 Israel Finkelstein and Neil A. Silberman, *The Bible Unearthed: Archeology's New Vision of Ancient Israel and the Origin of its Texts* (New York: The Free Press, 2001), 24.

3 Charles S. Finch, "Africa and Palestine in Antiquity" in *African Presence in Early Asia: Incorporating Journal of African Civilizations*, April 1985, 7, no. 1, ed. by Runoko Rashidi and Ivan V. Sertima (New Brunswick and London, Transaction Publishers, 1988), 193.

4 Nephilim were the angels' children (angels were all males, no females) by earthly women.

5 David A. Leeming, *The Handy Mythology Answer Book* (Canton, MI: Visible Ink Press, 2015), 67. *See also* Daniel Doggett, "Greek and Roman

Mythology," *Mythology, Myths, Legends, & Fantasies,* ed. Janet Parker, Julie Stanton (Willoughby, Australia: Global Book Publishing, 2003) 32-35.

[6] Charles S. Finch, "Africa and Palestine in Antiquity" in *African Presence in Early Asia: Incorporating Journal of African Civilizations*, April 1985, 7, no. 1, ed. by Runoko Rashidi and Ivan V. Sertima (New Brunswick and London, Transaction Publishers, 1988), 193.

[7] Tom Vail, *Grand Canyon: A Different View* (Green Forest, Arkansas: Master Books, 2003).

[8] John G. Jackson, "Krishna and Buddha of India: Black Gods of Asia," in *African Presence in Early Asia: Incorporating Journal of African Civilizations*, April 1985, 7, no. 1, ed. by Runoko Rashidi and Ivan V. Sertima (New Brunswick and London, Transaction Publishers, 1988), 106.

[9] Michael D Magee, "When Was Exodus Written?" accessed January 27, 2020, https://www.academia.edu/23772970/When_Was_the_Bible_Written_In_the_Persi an_Era

[10] Ibid.

[11] Ibid.

[12] Donald B. Redford *Egypt, Canaan, and Israel in Ancient Times* (Princeton, NJ: Princeton University Press, 1992), 429.

[13] Jack P. Lewis, "Jamnia Revisited," in *The Canon Debates.* Ed. L. M. McDonald & J. A. Sanders (Peabody/Massachusetts: Hendrickson Publishers, 2002), 146-162.

[14] G. Massey, *The Historical Jesus and the Mythical Christ: Separating Fact from Fiction* (Escondido, California: The Book Tree, 2000), 32.

[15] Gerald Massey's Published Lectures (1), "The Historical Jesus and the Mythical Christ," accessed June 29, 2021, https://minorvictorianwriters.org.uk/massey/dpr_01_historical_jesus.htm .

[16] Timothy Freke, Peter Gandy, *The Jesus Mysteries: Was the "Original Jesus" a Pagan God?* (New York: The Three Rivers Press, 1999), 159-168. Answers the question of whether Paul was a Gnostic. On page 295, out of the fourteen letters attributed to Paul, Gerd Lüdemann lists the seven thought to be authentic: Romans, Philippians, Galatians, Philemon, 1 Corinthians, 2 Corinthians and 1 Thessalonians.

[17] This, according to Joseph Atwill in *Caesar's Messiah: The Roman Conspiracy to Invent Jesus (Charleston, SC: CreateSpace, 2011), 24.*

[18] Michael D Magee, "When Was Exodus Written? Exodus a Late Addition to the Jewish Scriptures." Accessed January 27, 2020, https://www.academia.edu/23772970/When_Was_the_Bible_Written_In_the_Persi an_Era

[19] Joseph B. Lumpkin, *The Books of Enoch: The Angels, The Watchers, and The Nephilim,* 2nd Ed., Book 1 Enoch (Blountsville, AL: Fifth Estate, 2011), 14:8.

[20] Matthew 1:1-17, Luke 3:23-38.

[21] John W. Loftus, *Why I Became an Atheist: A former preacher rejects Christianity* (Amherst, New York, 2008), 137.

[22] Donald B. Redford, *Egypt, Canaan, and Israel in Ancient Times* (Princeton, NJ: Princeton University Press, 1992), 429.

[23] Ibid, 263-265, 408-429.

[24] Robert Graves, *The Greek Myths* (London: Penguin Books, 1992), 196; Gerald Massey, *A Book of Beginnings* (London: Williams and Norgate, 1881), 1: vii; Runoko Rashidi, "The Nile Valley Presence in Asian Antiquity," in *Nile Valley Civilizations: Proceedings of the Nile Valley Conference*, Atlanta, September 26-30, 1984, ed. by Ivan V. Sertima (Atlanta, Georgia: Morehouse College Edition, 1985), 207.

[25] Steve Olson, *Mapping Human History* (Boston, New York: Houghton Mifflin Co.), 109-110.

[26] Joseph B. Lumpkin, *The Books of Enoch: The Angels, The Watchers, and The Nephilim,* 2nd Ed., Book 2 Enoch (Blountsville, AL: Fifth Estate, 2011), 15.

[27] Ibid, 12-14, 212.

[28] Ibid, 12.
[29] Donald B. Redford, *Egypt, Canaan, and Israel in Ancient Times* (Princeton, NJ: Princeton University Press, 1992)*,* 391-394.

[30] Ze'ev Herzog, "Deconstructing the Walls of Jericho," accessed January 14, 2015, http://mideastfacts.org/facts/deconstructing-the-walls-of-jericho/#more-7

[31] Simcha Jacobovich, "Paganism, Osiris Mithraism: Constantine's Christianity," 2016, documentary directed by Simcha Jacobovich, accessed March 8, 2020, https://www.youtube.com/watch?v=cQWmL-aKmWM. *See also*:
Timothy Freke, Peter Gandy, *The Jesus Mysteries: Was the "Original Jesus" a Pagan God?* (New York: The Three Rivers Press, 1999).

[32] Ibid, 241.

[33] Simcha Jacobovich, "Paganism, Osiris, Mithraism Constantine's Christianity, 2016, documentary directed by Simcha Jacobovich, accessed March 8, 2020, https://www.youtube.com/watch?v=cQWmL-aKmWM

[34] Bruce Metzger, *The Canon of the New Testament* (Oxford University Press, 1997), 98.

[35] CHRISTIANITY, the true story, accessed August 07, 2019, https://www.youtube.com/watch?v=it_HPL-E1_U
See also: Joseph Atwill, *Caesar's Messiah: The Roman Conspiracy to Invent Jesus,* Flavian signature ed. (Charleston, SC: CreateSpace, 2011), 19.

[36] Joseph Atwill, *Caesar's Messiah: The Roman Conspiracy to Invent Jesus* (Charleston, SC: CreateSpace, 2011), 24.

[37] Joseph Atwill, *Caesar's Messiah: The Roman Conspiracy to Invent Jesus* (Charleston, SC: CreateSpace, 2011), 18; Michael Goulder, *Type and History in Acts,* (London: William Clowes and Sons,1963), 2-4.

[38] Coptic Gospel of Thomas, saying 4. Also, Bart D. Ehrman, *Lost Gospels: Books that did not Make it Into the New Testament (*New York: Oxford University Press, 2003), 20, Saying 2.

[39] Asa G. Hilliard III, "Kemetic Concepts in Education," citing Egyptologist Mohamoud Abdullah from a 1984 Seti 1 Luxor tour in *Nile Valley Civilizations: Proceedings of the Nile Valley Conference,* September 26-30, 1984, ed. by Ivan V. Sertima (Atlanta, Georgia: Morehouse College Edition, 1985), 159.

[40] Timothy Freke, Peter Gandy, *The Jesus Mysteries: Was the "Original Jesus" a Pagan God?* (New York: The Three Rivers Press, 1999), 241.

[41] Richard Carrier, *Proving History: Bayes' Theorem and the Quest for the Historical Jesus* (Amherst, NY: Prometheus Books, 2012). The ancients believed there were seven or nine levels in Heaven, and Enoch visited them. According to Richard

Carrier, the earliest Christians believed that Jesus stayed in one of the lower levels and came on Earth.

[42] But if Joseph was not His father, then how could Jesus be descended from the line of King David? Until the nineteenth century, a woman was considered a vessel for the man's "seed" with no contribution to a child's (gene) makeup. Anyway, in Matthew 1:1-17 (KJV), Jesus' genealogy diverges from that given in Luke 3:23-38 (KJV) from the outset.

[43] Solar Mythology and the Jesus Story, accessed January 6, 2015, http://members.cox.net/deleyd/religion/solarmyth/christ2002.htm; also, Charles S. Finch, "The Kamitic Genesis of Christianity," in *Nile Valley Civilizations: Proceedings of the Nile Valley Conference*, ed. Ivan V. Sertima (Atlanta, Georgia: Morehouse College Edition, 1985, September 26-30, 1984), 186.

[44] Quran, 4:157. *See also* Jay Smith: The Historical Origins of Islam, 2016. Accessed January 23, 2023, https://www.youtube.com/watch?v=jorBwia9yFw.

[45] Nicholas Wade, *The Faith Instinct: How Religion Evolved & Why it Endures* (New York: The Penguin Press, 2009), 183-184.

[46] Karen Armstrong, *A History of God: The 4000-Year Quest of Judaism, Christianity, and Islam* (New York: Ballantine Books, 1993), 289.

[47] Sigmund Freud, *The Future of an Illusion* (Standard Edition), Translated by James Strachey (NY: W. W. Norton & Co., 1989), 18-30]; Karen Armstrong, *A History of God: The 4000-Year Quest of Judaism, Christianity, and Islam* (New York: Ballantine Books, 1993), 357-358.

CHAPTER 10 – OLD TIME RELIGION

[1] Shantidev, *Guide to the Bodhisattva's Way of Life*, 8.112.

[2] Jainism, accessed September 14, 2020, Jaihttps://www.qcc.cuny.edu/socialSciences/ppecorino/PHIL_of_RELIGION_TEXT/CHAPTER_2_RELIGIONS/Jainism.htm

[3] Uganda Martyrs, "The Christian Martyrs of Uganda-Buganda." Accessed October 9,2020, https://www.buganda.com

[4] Andrew Newberg, Eugene D'Aquili and Vince Rause, *Why God Won't Go Away: Brain Science and the Biology of Belief* (New York: Ballantine Books, 2001), 82.

[5] Ibid, 171.

[6] Ta Neter Foundation, "Ancient African Writing," accessed March 8, 2020, http://taneter.org/writing.html; Clyde Winters, "Ancient Writing in Middle Africa," accessed March 8, 2020, http://olmec98.net/anwrite.htm.

[7] Friedrich Nietzsche, *The Gay Science* (New York: Random House, 1974), No. 125.

[8] Charles Lyell in Michael Argyle, *Psychology and Religion: An Introduction* (London and New York: Routledge, 2001), 2.

[9] Ruth. S. Cavan, et al., in "Personal Adjustment in Old Age," Chicago, Ill: *Science Research Associates*, (1949) quoted by Michael Argyle in *Psychology and Religion: An Introduction* (London and New York: Routledge, 2001): 28.

[10] Roger Finke and Rodney Stark, *The Churching of America 1776-1990* (New Brunswick, NJ: Rutgers University Press, 1992).

[11] Roozen D. A. and Carroll, J. W. (1979). Recent trends in church membership and participation: an introduction. In D.R. Hoge and D. A. Roozen 9eds), *Understanding Church Growth and Decline* (New York: Pilgrim Press, 1950-1978) 21-41.

[12] George Bishop, data reported in Huba, S. 1999. "Biblical Version of Creation OK by Americans," *The Detroit News,* April 6.

[13] Michael Shermer, *How We Believe: The Search for God in the Age of Science* (New York: W. H. Freeman and Company, 2000), 72.

[14] Ibid 72-73.

[15] Michael Argyle in *Psychology and Religion: An Introduction* (London and New York: Routledge, 2001), 67, 224-225.

[16] Robert Sapolsky, *Biology and Human Behavior: The Neurological Origins of Individuality,* video lecture 8 by The Great Courses (The Teaching Company, Virginia, 2010).

[17] Lasagna et al., "A Study of the Placebo Response," *American Journal of Medicine* 16 (1954): 770-779.

[18] McCollough et al., "Religious Involvement and Mortality: A Meta-Analytic Review," *Health Psychology,* 19, no. 3 (2000): 211-227; Michael E. McCullough and Brian L. B. Willoughby, "Religion, Self-Regulation, and Self-Control: Associations, Explanations, and Implications," *Psychological Bulletin* 135, No. 1 (2009): 69–93;

Shanshan Li, Meir J. Stampfer, David R. Williams, et al., "Association of Religious Service Attendance With Mortality Among Women," JAMA Intern Med. 176 (2016): 777-785.

[19] Andrew Newberg, Eugene D'Aquili and Vince Rause, *Why God Won't Go Away: Brain Science and the Biology of Belief* (New York: Ballantine Books, 2001), 140.

[20] Sigmund Freud, The Future of an Illusion (London: Hogarth Press, 1927); Michael Argyle in *Psychology and Religion: An Introduction* (London and New York: Routledge, 2001), 101.

[21] Michael Argyle in *Psychology and Religion: An Introduction* (London and New York: Routledge, 2001), 16,17.

[22] Channa Ullman, "Cognitive and Emotional Antecedents of Religious Conversion," *Journal of Personality and Social psychology* 43, (1982): 183-192.

[23] Michael Shermer, *How We Believe: The Search for God in the Age of Science* (New York: W. H. Freeman and Company, 2000), 168-169

[24] Sandy and Y Rovner wrote about the problem in "Healthtalk; The '70s Feel-Good Pills and the '80s Addicts," in the *Washington Post* of November 30, 1979.

[25] Joanna Maselko, "Spirituality protects against depression better than church attendance," Temple University News Center, posted October 4, 2008, https://www.sciencedaily.com/releases/2008/10/081023120228.htm, accessed January 6, 2018; https://www.sciencedaily.com/releases/2008/10/081023120228.htm, accessed January 6, 2018; Sigmund Freud, *The Future of an Illusion* (Standard Edition), Translated by James Strachey (NY: W. W. Norton & Co., 1989), 56; *see also*: Shankar Vedantam, Hidden Brain Podcast, November 5, 2019: "Does Going to Church Improve Your Mental Health?" https://www.npr.org/2019/11/05/776270553/hidden-brain-does-going-to-church-improve-your-mental-health, accessed May 27, 2021.

[26] Andrew Newberg, Eugene D'Aquili and Vince Rause, *Why God Won't Go Away: Brain Science and the Biology of Belief* (New York: Ballantine Books, 2001), 129-130.

[27] Bronislaw Malinowski, *Magic, Science, and Religion* (New York: Doubleday, 1954), 17, 29, 139-140.

[28] Dimitra Papagiani and Michael A. Morse, *The Neanderthals Rediscovered: How Modern Science is Rewriting Their Story* (London: Thames and Hudson, 2013), 105.

[29] Andrew Newberg, Eugene D'Aquili and Vince Rause, *Why God Won't Go Away: Brain Science and the Biology of Belief* (New York: Ballantine Books, 2001), 132.

[30] Ibid, 139.

[31] Michael Shermer, *How We Believe: The Search for God in the Age of Science* (New York: W. H. Freeman and Company, 2000), 64.

[32] Helmut Hanisch, "Children's and Young People's Drawings of God," lecture given at the University of Gloucestershire in 2001, accessed February 21, 2016, http://www.uni-leipzig.de/~rp/vortraege/hanisch01.htm.1

[33] Michael Shermer, *How We Believe: The Search for God in the Age of Science* (New York: W. H. Freeman and Company, 2000), 168-169.

[34] Richard Dawkins, *The Selfish Gene* (Oxford: Oxford University Press, 1976), 1.

[35] Maggie Hyde and Michael McGuinness, *Introducing Jung* (Lanham, Maryland: Totem Books, 1997), 21.xxxx

[36] Ibid, 21, 33, 34. 39.

[37] Michael Argyle in *Psychology and Religion: An Introduction* (London and New York: Routledge, 2001), 105.

[38] Ibid, 107.

[39] Maggie Hyde and Michael McGuinness, *Introducing Jung* (Lanham, Maryland: Totem Books, 1997), 57.

[40] Charles S. Finch quotes Gerald Massey in "The Kamitic Genesis of Christianity," in *Nile Valley Civilizations: Proceedings of the Nile Valley Conference*, Atlanta, September 26-30, 1984, ed. by Ivan V. Sertima. Georgia: Morehouse College Edition, 1985, 179; W. M. Flinders Petrie, "The Gods of Ancient Egypt," in Hammerton's *Wonders of the Past* (New York, 1937), 667.

[41] Thomas Dawson in a six-part television and video series, *Dawn of Man: The Story of Human Evolution*, BBC Video (2000), vol. 3, pt. II, Human.

[42] Wayne Teasdale in Andrew Newberg, D'Aquili and Vince Rause, *Why God Won't Go Away: Probing the Biology of Religious Experience* (New York: Ballantine Books, 2001), 135-136, 171.

[43] Etzel Cardena, Thomas Dawson in a six-part television and video series, *Dawn of Man:* The Story of Human Evolution, BBC Video (2000), vol. 3, pt. II, Human.

[44] Michael Argyle in *Psychology and Religion: An Introduction* (London and New York: Routledge, 2001), 61.

[45] Ibid, 61.

[46] Edward O. Wilson, "Sociobiology," *Encyclopedia of Evolution* (New York: Facts on File, Inc., 2007), 382-384.

[47] Norman Doidge, *The Brain that Changes Itself: Stories of Personal Triumph from the Frontiers of Brain Science* (New York: Viking, 2007), 283-285.

[48] Jill Bolte Taylor, *My Stroke of Insight: A Brain Scientist's Personal Journey* (New York: Penguin Books, 2006).

[49] Newberg, D'Aquili and Vince Rause, *Why God Won't Go Away: Brain Science and the Biology of Belief* (New York: Ballantine Books, 2001), 7.

[50] Ibid, 137.

[51] Ibid, 113.

[52] Alethia Luna, "How to Induce a Trance State for Deep Psychospiritual Work." Accessed December 24th, 2022; https://lonerwolf.com/trance.

[53] Ibid, 135-136.

[54] Maggie Hyde and Michael McGuinness, *Introducing Jung* (Lanham, Maryland: Totem Books, 1997), 52-56.

[55] Philip Thody and Howard Read, "Nausea," *Introducing Sartre* (New York: Totem Books, 1998), 16-24.

[56] Michael Shermer, *How We Believe: The Search for God in the Age of Science* (New York: W. H. Freeman and Company, 2000), 66-67.

[57] Andrew Newberg, D'Aquili and Vince Rause, *Why God Won't Go Away: Brain Science and the Biology of Belief* (New York: Ballantine Books, 2001), 139.

[58] Ibid, 169.

CHAPTER 11 – FOREVER YOUNG

[1] Viktor E Frankl, *Life With Meaning*, (Belmont, C.A: Brookes/Cole, 1993), 21-22; Wolfgang Jilek and Loise Jilek-All, "Logotherapeutic Aspects of the North American Indian Guardian Spirit," in: Wawrytko, S.A. (Ed.), *Analecta Frankliana* (Logotherapy Press; Berkley 1982), 313.

[2] Marmot, M. G., Davey Smith, G., Stanfield, S., et al, "Health inequities among British civil servants: the Whitehall II study," *Lancet* 337: 1287-93.

[3] Matt Ridley on the relationship between rank, stress hormones, and heart disease in both monkeys and people in "Chromosome 10: Stress," *Genome* (New York: Perennial, 2000), 154-156.

[4] Jay Kaplan et al. and Michael McGuire and Raleigh in Ridley, "Chromosome 11: Personality," *Genome* (New York: Perennial, 2000), 170-171.

[5] Robert Sapolsky, *Why Zebras Don't Get Ulcers,* 3rd ed. (New York: Henry Holt/Owl Books, 2004), 359.

[6] Robert Sapolsky, "Sick of Poverty," *Scientific American,* December 2005, 92.

[7] S. Epel and co-workers showed at the University of California that stress accelerated cellular aging, shortening telomeres to a point the cell stopped dividing faster causing earlier death in "Accelerated Telomere Shortening in Response to Life Stress," *PNAS 101* (December 2004): 17312-17315.

[8] Cytokines comprise a range of proteins including Tumor Necrosis Factor alpha (TNF-α) and *interleukins* (IL), the name giving the chemicals' primary communication function: inter (among) leukins (leukocytes or white blood cells). Cytokines help fight injury short-term. But by patterning interleukins with short-term chemicals such as TNF-α, they may contribute to inflammation. Like IL-1 and IL-6, TNF-α may stimulate the hypothalamus, releasing cortisol long-term. Smoking, smoldering dental infections, and obesity (prolonged stress) may do the same, leading to sustained injury. Here, the chronically elevated hormone clobbers the immune system dampening it, preventing it from overshooting. It causes the body to attack itself, leading to autoimmune diseases mentioned in the text. Some of these conditions may respond to so called biologic agents Kineret or Enbrel that block IL-6 and TNF-α.

[9] Matt Ridley on the relationship between rank, stress hormones, and heart disease in both monkeys and people in "Chromosome 10: Stress," *Genome* (New York: Perennial, 2000), 154; Jill B. Becker, S. Marc Breedlove, and David Crews, *Behavioral Endocrinology* (Cambridge, MA: MIT Press, 1992).

[10] Sami Ouanes and Julius Popp, "High Cortisol and the Risk of Dementia and Alzheimer's Disease: A Review of the Literature," *Front Aging Neurosci.* 2019; 11:43. Published online 2019 Mar 1. doi: 10.3389/fnagi.2019.00043 Accessed October 20,2020. https://www.ncbi.nlm.nih.gov/pmc/articles/PMC6405479/; *see also,* Karen Weintraub, "'Stress Hormone' Cortisol Linked to Early Toll on Thinking Ability." Accessed October 25, 2020. https://www.scientificamerican.com/article/ldquo-stress-hormone-rdquo-cortisol-linked-to-early-toll-on-thinking-ability/

[11] Montefiore Medical Center, "Migraine attacks increase following stress 'let-down' *ScienceDaily* March 26, 2014. Accessed October 20/2020. https://www.sciencedaily.com/releases/2014/03/140326181915.htm#:~:text=The%20hormone%20cortisol%2C%20which%20rises,headache%20during%20periods%20of%20relaxation.%22

[12] Myrthala Moreno-Smith, Susan K Lutgendorf, and Anil K Sood, "Impact of stress on cancer metastasis," *Future Oncology.* 2010 Dec: 6(12): 1863-1881, doi: 10.2217/fon.10.142. Accessed October 20/2020. https://www.ncbi.nlm.nih.gov/pmc/articles/PMC3037818

[13] Robert Ader, "Conditioned Responses," in *Healing and the Mind,* ed. by Bill Moyer (New York: Broadway Books, 1993): 239-248.

[14] Michael Roizen, "Your RealAge," accessed August 26, 2019, https://www.sharecare.com/static/realage-oz

[15] Castillo-Richmond A, et al., "Effects of stress reduction on carotid atherosclerosis in hypertensive African Americans," *Stroke: Journal of the American Heart Association* 31 (2000): 568-573.

[16] M.J. Cooper, M.M. Aygen, "Transcendental Meditation in the management of hypercholesterolemia," *Journal of Human Stress* 5 (1979): 24-27; Zamarra, J. et al., "The usefulness of the Transcendental Meditation program in the treatment of patients with coronary artery disease," *American Journal of Cardiology* 77 (1996): 867-870.

[17] Katherine S. Pollard, "What Makes Us Human?" *Scientific American,* May 2009, 44. A gene E *(APOE),* that we share with the chimps, helps develop the workings of our immune system. When the body is invaded by microbes, the gene's human variant, APOEe4, vigorously turns on defense mechanisms, such as increased

temperature to kill viruses, to fend off the harmful invaders. Chimps lack this variant, so even before antibiotics, we survived longer and out-reproduced them.

[18] Heather Pringle, "Long Live the Humans," *Scientific American.* October 2013, 48.

[19] Erik Vance, "Why Nothing Works," *Discover,* July/Aug 2014, 42.

[20] Joshua Greene and Jonathan Haidt, "How (and where) does Moral Judgment Work?" *Trends in Cognitive Sciences,* 16 (2002): 517-523. Melinda W. Moyer, "Moral Animal," *Scientific American,* August 2010, 51; Green and Haidt, "How (and Where) does Moral Judgement Work?" 517-523.

[21] The Golden Rule, with some variations, states " Do to others as you would have them do to you," in Matthews 7:12, Luke 6:31, Leviticus 19:18, and by Confucius in *The Analects,* 1992, etc.

[22] Excellence Reporter, "Albert Einstein: On the Wisdom and the Meaning of Life," *Excellence Reporter*, June 5. 2019, accessed October 11, 2020, https://excellencereporter.com/2019/06/05/albert-einstein-on-the-wisdom-and-the-meaning-of-life/#:~:text=%E2%80%9CThe%20man%20who%20regards%20his,but%20almost%20disqualified%20for%20life.%E2%80%9D

[23] Ibid.

[24] Barry Bittman, et al., "Recreational music-making modulates the human stress response: a preliminary individualized gene expression strategy," *Med Sci Monitor*, 1, no, 2 (2005): BR31-40, archive, accessed February 08, 2015, http://www.MedSciMonit.com; Institute of Heartmath, "Music and the Immune System," accessed February 20, 2016, http://www.heartmath.org/research/research-abstracts/music and-immune.html; E. Glenn Schellenberg, University of Toronto at Mississauga, Ontario, "Music Lessons Enhance IQ," *Psychological Science* 15, no. 8 (2003): 511-514.

[25] Albert Einstein, "The World as I See It," *An Essay by Einstein*, accessed August 08, 2020, https://libquotes.com/albert-einstein/quote/lbg1o9c

Einstein also wrote in a 1950 letter to a grieving father that a human being was part of the whole, called by us "Universe," a part limited in time and space. He experienced himself, his thoughts, and feelings as something separate from the rest—a kind of optical delusion of his consciousness. The striving to free oneself from this delusion was the one issue of true religion. Not to nourish it but try to overcome it was the way to reach the attainable measure of peace of mind. Mike Zonk, "Albert Einstein's surprising Thoughts on the Meaning of Life (BIGTHINK.COM) *Bathtub*

Bulletin. Accessed October 28, 2020, https://bigthink.com/paul-ratner/albert-einsteins-surprising-thoughts-on-the-meaning-of-life

Einstein didn't have a personal God and was likely a deist. His God was like Baruch Spinoza's God. To Spinoza, nature was God. Einstein, it seems, believed in the universe. Like Spinoza, Einstein's beliefs also had mystical overtones, such as those of the mystics, the Kabbalists, the Dervishes, or the Buddhists. They all experienced union with the powers that be and would dissolve into the universe.

[26] Andrew Newberg, D'Aquili and Vince Rause, *Why God Won't Go Away: Brain Science and the Biology of Belief (*New York: Ballantine Books, 2001), 153-154.

[27] Albert Einstein quote. Accessed October 28, 2020, https://libquotes.com/albert-einstein/quote/lbg1o9c

[28] Fritjof Capra, *The Tao of Physics: An Exploration of the Parallels Between Modern Physics and Eastern Mysticism, 4th Ed.* (Boston: Shambala, 2000), 90-91

[29] Reinhold Niebuhr, "The Serenity Prayer," *Bartlett's Familiar Quotations,* 17th Ed. (New York: Little, Brown and Company, 2002), 735.

[30] Yohann Hari, *Chasing the Scream: The First and Last Days of the War on Drugs.* Kindle Ed. (New York: Bloomsbury, 2016), 293.

[31] Martin E. P. Seligman, *What You Can Change: The Complete Guide to Successful Self-Improvement* (New York: Ballantine Books, 1995), 5. Seligman addresses sexual identity and orientation and how much they are amenable to change or not in Chapter 11.

[32] Viktor E Frankl, *Life With Meaning,* (Belmont, C.A.: Brookes/Cole, 1993), 143-145.

[33] Ibid, 13.

CHAPTER 12 – WHAT, I'M NOT WHO I SAY?

[1] This quote is often attributed to William James, accessed August 10, 2020, https://philosiblog.com/2012/05/10/a-great-many-people-think-they-are-thinking-when-they-are-merely-rearranging-their-prejudices/. *See also:* Dean Hamer, *The God Gene: How Faith is Hardwired into our Genes* (New York: Doubleday, 2004), 161. Although the quote has also been credited to others, they are not as well-known as William James, accessed June 26, 2021, https://quoteinvestigator.com/2017/05/10/merely/.

[2] Na'im Akbar, "Nile Valley Origins of the Science of the Mind," in *Nile Valley Civilizations: Proceedings of the Nile Valley Conference,* September 26-30, 1984, ed. by Ivan V. Sertima (Atlanta, Georgia: Morehouse College Edition, 1985), 121.

[3] Faith Ministries Blog, An Online Study of the Bible, https://faithbibleministriesblog.com/2012/07/06/the-heart-and-the-mind-what-the-biblical-word-heart-means/ accessed September 14, 2018.

[4] David Suzuki in "Evolution and Perception," Discovery's video series, *The Brain: Our Universe Within; Matter Over Mind* (1996).

[5] Robert Sapolsky, "Sick of Poverty," *Scientific American,* December 2005, 92-99.

[6] James Olds and Peter Milner, "Positive reinforcement produced by electrical stimulation of septal area and other regions of rat brain." *J Comp Physiol Psychol.* December 1954, 419-427.

[7] Roy A. Wise, "Addictive Drugs and Brain Stimulation Reward," *Annual Review of Neuroscience*, vol. 340 (1996), 319-40; https://doi.org/10.1146/annurev.ne.19.030196.001535

[8] Matt Ridley, *Genome: The Autobiography of a Species in 23 Chapters* (New York: Perennial, 2000), 169-171; Thomas B. Czerner, *What Makes You Tick?* (New York: John Wiley & Sons, 2001), 200-201; Ronald Kotulak, *Inside the Brain* (Kansas City, MO: Andrews McMeel Publishing, 1996).

[9] Doblin, Rick, "Pahnke's 'Good Friday Experiment': a long-term follow-up and methodological critique," *Journal of Transpersonal Psychology,* 23 (1), 1991, 1–25.

[10] Dean Hamer, *The God Gene: How Faith is Hardwired into our Genes* (New York: Doubleday, 2004), 79-89; Doblin, Rick, "Pahnke's 'Good Friday Experiment': a long-term follow-up and methodological critique," *Journal of Transpersonal Psychology,* 23 (1), 1991, 1–25.

[11] Helen Fisher, *Why We Love: The Nature and Chemistry of Romantic Love* (New York: Holt and Company, 2004).

[12] Karen Armstrong, *A History of God: The 4000-Year Quest of Judaism, Christianity and Islam* (New York: Ballantine Books, 1993), 21.

[13] Following Einstein's death, Thomas Harvey examined his brain and found that Einstein had no parietal operculum in either hemisphere. A group at McMaster University, Ontario, Canada, studying Harvey's photographs found that in addition

to a vacant parietal operculum region in the inferior frontal gyrus in the frontal lobe, a part of a bordering region called the lateral sulcus, or Sylvian fissure, was absent. The vacancy may have enabled neurons in this part of the brain to communicate better. A part of the operculum, Broca's area, is important for language production and verb comprehension. It seems, to compensate for this absence in the speech area, the inferior parietal lobe was 15 percent wider than normal. This lobe is responsible for mathematical thought, visuospatial cognition and imagery of movement. Einstein himself claimed that he thought visually rather than verbally (Robert Lee Hotz, "Revealing thoughts on gender and brains," accessed October 15, 2020, https://www.seattletimes.com/nation-world/revealing-thoughts-on-gender-and-brains). A study of photographs of Einstein's brain found that Einstein's corpus callosum was thicker than that of controls, which showed that connectivity between his two hemispheres was also generally enhanced. (Weiwei Men, Dean Falk, Tao Sun, et. al., "The Corpus Callosum of Albert Einstein's brain: another clue to his high intelligence?" *Brain,* Vol. 137, April 2014, Page e268, https://doi.org/10.1093/brain/awt252

[14] Brain, W.R., "Visual disorientation with special reference to lesions of the right cerebral hemisphere," *Brain*, Volume 64, Issue 4, December 1941: 257, 264. https://doi.org/10.1093/brain/64.4.244

[15] Christian de Duve, *Life Evolving: Molecules, Life, and Meaning* (Oxford: Oxford University Press, 2002), 211.

[16] Johnjoe McFadden, *Quantum Evolution: The New Science of Life* (New York NY: W.W. Norton & Company, 2004), 292, 301-303.

[17] Antonio Damasio, *Looking for Spinoza: Joy, Sorrow and the Feeling Brain* (New York: Harcourt, 2003), 174; Spinoza: Richard H. Popkin, *Men are Deceived if They Think Themselves Free* (London: Oneworld, 2004), 100-101.
"Men are mistaken in thinking themselves free; their opinion is made up of consciousness of their own actions, and ignorance of the causes by which they are determined."

[18] Alexander Bain, *Mental Science* (New York: D. Appleton and Company, 1868), 414.

[19] Benjamin Libet, "Unconscious cerebral initiative and the role of conscious will in voluntary action," *Behavioral and Brain Sciences* (December 1985): 529-1985.

[20] Thomas B. Czerner, *What Makes You Tick?* (New York: John Wiley & Sons, 2001), 71.

[21] Sam Harris, *Free Will* (New York: Free Press, 2012), 5.

---

[22] In the study, Sperry briefly showed N.G. a picture of a cup to the right of a dot, and since the right optic nerve supplies the left brain, that side "saw" it and she said it was "a cup." Next, Sperry showed her a spoon to the left of the dot. The image likewise was seen by the right brain this time, and N.G. called it "nothing." Sperry then mixed the spoon with other objects under a screen and, when N. G. reached down with the left hand, she picked it out, but called it "a pencil," as her left-brain, without a clue what the left hand held, hazarded a guess.

[23] Roger Sperry's seminal work cited by Newberg, D'Aquili, and Vince Rause in, *Why God Won't Go Away: Brain Science and the Biology of Belief* (New York: Ballantine Books, 2001), 22-23.

[24] Hume, quoted in Bryan Magee, *The Story of Philosophy: The Essential Guide to the History of Western Philosophy,* 1st Ed. (New York NY: DK Publishing, 1998), 112-113; G.W. Leibniz "Monadologie (1714)," Nicholas Rechner, trans., *The Monadology: An Edition for Students* (Pittsburg: University of Pittsburg Press, 1991), 135.

CHAPTER 13 – THE EMPEROR'S NEW CLOTHES

[1] Stephen J. Gould, in *Wonderful Life: The Burgess Shale and the Nature of History* (NY: W.W. Norton and Company, 1990).

[2] In one of his manuscripts, Enoch goes up to the Tenth Heaven, https://paradiseandperdition.weebly.com

[3] James H. Austin, "The Septum and Pleasure," in *Zen and the Brain* (Cambridge, Massachusetts: MIT Press, 1998), 170-171.

[4] Dean Hamer, *The God Gene: How Faith is Hardwired into our Genes* (New York: Doubleday, 2004), 36.

[5] Remillard, Guy M. et al., "Sexual Ictal Manifestations Predominate in Women with Temporal Lobe Epilepsy: A Finding Suggesting Sexual Dimorphism in the Human Brain," *Neurology* (March 1983): 323-33.

[6] Karen Armstrong, *A History of God: The 4000-Year Quest of Judaism, Christianity, and Islam* ((New York: Ballantine Books, 1993), *xviii*.

[7] The Revelation, "Freemasonry proven to Worship Satan. Its symbols venerate the sex act," *Study of symbols* part 5 of 5, accessed February 8, 2015, http://www.theforbiddenknowledge.com/symbology/505.htm

[8] Manly P. Hall, *The Secret Teachings of all ages* (New York: Thatcher/Penguin, 2003), 290-294.

[9] Einstein, "The World As I See It," *An Essay by Einstein*, accessed October 28, 2020, https://history.aip.org/history/exhibits/einstein/essay.htm

[10] Lawrence Rifkin, "Is the Meaning of Life to Make Babies?" *Scientific American Blog* 175, accessed February 18, 2020, https://blogs.scientificamerican.com/guest-blog/is-the-meaning-of-your-life-to-make-babies/
See also: John Alcock, *The Triumph of Social Biology* (New York: Oxford University Press, 2001), 215.

[11] Andrew Newberg, D'Aquili and Vince Rause, *Why God Won't Go Away: Brain Science and the Biology of Belief (*New York: Ballantine Books, 2001), 7.

[12] Ibid, 164.

[13] Earlyne Chaney and William L. Messinick, *Kundalini and the Third Eye* (City of Commerce, CA: Stockton Trade Press, 1980). Chaney and her associate Messinick are the founders of Astara, a fraternity based on the ancient Egyptian teaching of Thoth.

[14] William James, *Varieties of Religious Experience* (Charleston, S.C.: Bibliobazaar, 2003), 64.

[15] Austin, "Lateralizable Functions and the EEG," *Zen and the Brain* (Cambridge, Massachusetts: MIT Press, 1998)*,* 364.

[16] Ian Cotton, "Dr Persinger's God Machine," *The Independent,* Sunday 2 July 1995, accessed February 08, 2015, www.independent.co.uk > Arts + Ents.

[17] Gospel by Mary Magdalene, 4:34, accessed April 16, 2020, http://online.fliphtml5.com/snac/xsqy/#p=1

[18] "In Seventh Heaven or 'What Enoch Did Next,'" accessed April 28, 2019, https://paradiseandperdition.weebly.com; Joseph B. Lumpkin, *The Books of Enoch: The Angels, The Watchers and The Nephilim,* 2nd Ed. (Blountsville, AL: Fifth Estate, 2011), 177-186.

[19] Hans Christian Andersen, "The Emperor's New Clothes," (1837) in Eventyr, fortalte for Børn. - Tredie Hefte (Fairy-tales told to the children. - Third booklet), (C.A. Reitzel Publishers, 1837).

[20] Dan Brown, *The Da Vinci Code,* special illustrated ed.,1st ed. (New York: Doubleday, 2003), 316-319.

[21] Andrew Newberg, D'Aquili and Vince Rause, *Why God Won't Go Away: Brain Science and the Biology of Belief* (New York: Ballantine Books, 2001), 9, 125, 126.

[22] Timothy Freke, Peter Gandy, *The Jesus Mysteries: Was the "Original Jesus" a Pagan God?* (New York: The Three Rivers Press, 1999), 67.

[23] Ibid, 63.

[24] F. F. Powell describes in "Saint Paul's Homage to Plato" that Paul, the first Christian missionary and writer of a significant part of the New Testament, fused the philosophy of Plato with the story of Jesus. Accessed February 11, 2015, http://www.worldandi.com/newhome/public/2004/April/mtpub2.asp

[25] Timothy Freke, Peter Gandy, *The Jesus Mysteries: Was the "Original Jesus" a Pagan God?* (New York: The Three Rivers Press, 1999), 63.

[26] Normative Systems (LEP Library of Exact Philosophy), Carlos E. Atchourron, Eugenio Bulygia.

[27] Joseph Atwill, *Caesar's Messiah: The Roman Conspiracy to invent Jesus* (Charleston, SC: CreateSpace, 2011), 25-27.

[28] Timothy Freke, Peter Gandy, *The Jesus Mysteries: Was the "Original Jesus" a Pagan God?* (New York: The Three Rivers Press, 1999), 71.

[29] Ibid.

[30] Bryan Magee, *The Story of Philosophy: The Essential Guide to the History of Western Philosophy,* 1st Ed. (New York NY: DK Publishing, 1998), 40-41.

[31] Gerald Massey's Published Lectures (1), "The Historical Jesus and Mythical Christ,"
paragraph 92, accessed June 29, 2021,
https://minorvictorianwriters.org.uk/massey/dpr_01_historical_jesus.htm. *See also*: Gerald Massey, *The Historical Jesus* (2000), 150-185.

[32] Paul C. Boyd, *The African Origin of Christianity: A Biblical and Historical Account,* vol. 1 (London: Karia Press, 1991), 37.

[33] Sir Flinders Petrie, "The Gods of Ancient Egypt," in John. A. Hammerton, *Wonders of the Past* (New York: Wise & Co. 1937), 667, 678.

[34] Paul C. Boyd, *The African Origin of Christianity: A Biblical and Historical Account,* vol. 1 (London: Karia Press, 1991), 37.

[35] Bart D. Ehrman, *Lost Gospels: Books that did not Make it Into the New Testament* (New York: Oxford University Press, 2003), 20, verse 7.

[36] Paul C. Boyd, *The African Origin of Christianity: A Biblical and Historical Account* (London: Karia Press, 1991), 95. Ibid, 28, verse 108.

[37] Earlyne Chaney and William L. Messinick, *Kundalini and the Third Eye* (City of Commerce, CA: Stockton Trade Press, 1980). Foreword, 8.

[38] "Jesus Myth hypothesis," *Wikipedia: The Free Encyclopedia*, accessed February 11, 2015, http://en.wikipedia.org/wiki/jesus_myth

[39] *Acts of Peter,* 3.

[40] Peter Joseph, "The Greatest Story Ever Told" in *Zeitgeist, The Movie* (2007) part I, accessed February 02, 2020, http://zeitgeistmovie.com; David W. Deley, "Solar Mythology and the Jesus Story." Accessed October 25, 2020, http://www.solarmythology.com/lessons/christ2002.htm

[41] Gerald Massey, *The Historical Jesus and the Mythical Christ: Separating Fact From Fiction* (Escondido, California: The Book Tree, 2000), 7.

[42] Ernest L Martin, "A 'Peter' was in Rome Two Thousand Years B.C!" Simon Magus Series, accessed October 17, 2020, https://www.hwalibrary.com/cgi-bin/mobile/m_hwa.cgi?action=getmagazine&InfoID=1389878915; The origin of the Catholic Church and the name Peter, which was the nickname for Simon the Apostle, is revealed, "A 'Peter' Was in Rome Two Thousand Years BC! accessed February 11, 2015. http://historicist.info/articles2/peterrome.htm

[43] Ibid.

[44] Ibid.

[45] "Simoni deo Sancto," *Dictionary of Christian Biography,* vol. 4, 682.

[46] Simon Magus may have been a cypher for Paul of Tarsus. Both were reviled at the time for their Gnostic leanings and may have been interchanged often. https://www.newworldencyclopedia.org/entry/Simon_Magus, accessed November 4th, 2022.

[47] Timothy Freke, Peter Gandy, *The Jesus Mysteries: Was the "Original Jesus" a Pagan God?* (New York: The Three Rivers Press, 1999), 10.

[48] "Talent" as used in this context, is mystic parlance for the ability or proclivity to induce spiritual experiences.

[49] Andrew Newberg, D'Aquili and Vince Rause, *Why God Won't Go Away: Brain Science and the Biology of Belief* (New York: Ballantine Books, 2001), 172.

[50] Paul Broom presents evidence indicating we have an inborn moral sense in *Just Babies: The Origins of Good and Evil* (New York: Crown, 2013).

[51] Bryan Magee, *The Story of Philosophy: The Essential Guide to the History of Western Philosophy,* 1st Ed. (New York NY: DK Publishing, 1998), 18; The Sophists, *Stanford Encyclopedia of Philosophy,* Stanford University, 2020. Accessed 10/24/2020, https://plato.stanford.edu/entries/sophists/#pagetopright.

[52] Hume, cited in Bryan Magee, *The Story of Philosophy: The Essential Guide to the History of Western Philosophy,* 1st Ed. (New York NY: DK Publishing, 1998), 112-113.

[53] Joel Osteen Lakewood Church, accessed February 11, 2015, http://www.joelosteen.lakewood.cc. Osteen has several best-selling books under his belt.

[54] Daniel G. Amen, *Change Your Brain, Change Your Life: The Breakthrough Program for Conquering Anxiety, Depression, Obsessiveness, Lack of Focus, Anger, and Memory Problems,* (New York: Harmony Books, 2015), 159-160. Amen describes a 12-minute meditation called Kirtan Kriya (KK). It involves chanting simple sounds while doing repetitive finger movements with both hands: Touch thumbs to index fingers while chanting "saa." Touch the thumbs to middle fingers while chanting "taa." Touch thumbs to ring fingers while chanting "naa." Touch thumbs to pinkie fingers while chanting "maa." Repeat the sounds for 2 minutes aloud and for 2 minutes in a whisper. Repeat the sounds for 4 minutes silently. Repeat the sounds for 2 minutes in a whisper. Repeat the sounds for 2 minutes aloud. When you finish, sit quietly for 1 to 2 minutes. Try to hold on to your calmed mind and body throughout the day.

[55] Daniel J. Fairbanks, *Relics of Eden: The Powerful Evidence of Evolution in Human DNA,* (Amherst, New York: Prometheus Books, 2007).

[56] Fritjof Capra, *The Tao of Physics: An Exploration of the Parallels Between Modern Physics and Eastern Mysticism,* 4th Ed. (Boston: Shambala, 2000), 90.

[57] Joseph Atwill, *Caesar's Messiah: The Roman Conspiracy to Invent Jesus* (Charleston, SC: CreateSpace, 2011), 28-29.

[58] Gerald Massey, *The Historical Jesus and the Mythical Christ: Separating Fact from Fiction* (Escondido, California: The Book Tree, 2000), 3.

[59] Penn State, Department of Energy and Mineral Engineers, "Are We Running out of Oil?" Accessed October 27, 2020. https://www.e-education.psu.edu/eme801/node/486.

[60] Edward N. Lorenz in a paper he presented at the American Association for the Advancement of Science, 139th Meeting in Washington on December 29, 1972.

[61] Daily, G.C., Ehrlich, A.H. & Ehrlich, P.R. Optimum human population size. *Popup Environ* **15,** 469–475 (1994). Accessed January 30, 2020, https://doi.org/10.1007/BF02211719).

[62] Joseph B. Lumpkin, *The Books of Enoch: The Angels, The Watchers and The Nephilim,* 2nd Ed., Book 1 Enoch (Blountsville, AL: Fifth Estate, 2011), 186 (22: 1-3).